MULTIMEDIA SECURITY:
WATERMARKING, STEGANOGRAPHY, AND FORENSICS

EDITED BY

FRANK Y. SHIH

CRC Press
Taylor & Francis Group
Boca Raton   London   New York

CRC Press is an imprint of the
Taylor & Francis Group, an **informa** business

CRC Press
Taylor & Francis Group
6000 Broken Sound Parkway NW, Suite 300
Boca Raton, FL 33487-2742

© 2013 by Taylor & Francis Group, LLC
CRC Press is an imprint of Taylor & Francis Group, an Informa business

First published in paperback 2017

No claim to original U.S. Government works
Version Date: 20120531

ISBN 13: 978-1-138-07254-1 (pbk)
ISBN 13: 978-1-4398-7331-1 (hbk)

Library of Congress Cataloging-in-Publication Data

Multimedia security : watermarking, steganography, and forensics / editor, Frank Y. Shih.
   p. cm.
   Summary: "This book provides technical information on multimedia security and steganography as well as the fundamental theoretical framework for developing the extensive advanced techniques. By comprehensively considering the essential principles of the multimedia security and steganographic systems, readers can not only obtain the novel ideas in implementing the advanced algorithms, but also discover the new problems. The book consists of many technical tutorials from various professionals. The text is illustrated with plentiful graphs and examples in order to simplify the problems, so readers can easily understand even complicated theories"-- Provided by publisher.
   Includes bibliographical references and index.
   ISBN 978-1-4398-7331-1 (hardback)
   1. Multimedia communicatios--Security measures. I. Shih, Frank Y.

TK5105.15.M84 2012
006.7--dc23
                                                         2012021069

Visit the Taylor & Francis Web site at
http://www.taylorandfrancis.com

and the CRC Press Web site at
http://www.crcpress.com

To my loving wife and children, to my parents who encouraged me through the years, and to those who helped in the process of writing this book

Contents

PART I   Multimedia Mining and Classification

PART II   Watermarking

PART III   *Steganography*

PART IV   *Forensics*

Preface

Digital watermarking, steganography, and forensics are important topics because digital multimedia are widely used and the Internet is rapidly growing. This book intends to provide a comprehensive overview of the different aspects of mechanisms and techniques for information security. It has been written for students, researchers, and professionals who take related courses, want to improve their knowledge, and want to gain experience pertaining to the role of digital watermarking, steganography, and forensics.

Digital watermarking technology can be used to guarantee authenticity and can be applied as proof that the content has not been altered since insertion.

Steganographic messages are often first encrypted by some traditional means, and then a covert text is modified in some way to contain the encrypted message.

Digital forensics utilizes computational techniques to process multimedia content and then applies the results to help crime investigations. The need for information security exists everywhere every day.

This book aims to provide students, researchers, and professionals with technical information regarding digital watermarking, steganography, and forensics, as well as instruct them on the fundamental theoretical framework in developing the extensive advanced techniques. By comprehensively considering the essential principles of digital watermarking, steganography, and forensics systems, one can not only obtain novel ideas in implementing the advanced algorithms, but can also discover the new problems. The principles of digital watermarking, steganography, and forensics in this book are illustrated with plentiful graphs and examples in order to simplify the problems, so readers can easily understand even complicated theories. Several robust algorithms that are presented in this book to illustrate the framework provide assistance and tools in understanding and implementing the fundamental principles.

OVERVIEW OF THE BOOK

The book is divided into four parts: Multimedia Mining and Classification (Chapters 1 through 3), Watermarking (Chapters 4 through 10), Steganography (Chapters 11 through 14), and Forensics (Chapters 15 through 20). In Chapter 1, the feasibility, techniques, and demonstrations of discovering hidden knowledge by applying multimedia duplicate mining methods to the massive multimedia content are introduced. Three promising knowledge-discovery applications are demonstrated to show the benefits of duplicate mining. In Chapter 2, a new framework is proposed for video concept classification with the help of discriminative learning and multiple correspondence analysis. In Chapter 3, an improved bag of feature and feature vocabulary-based technique of replacing the Harris-affine detection method by a random sampling procedure together with an increased number of sample points is presented. It is shown that the method improves categorization accuracy on a five-category problem using the Caltech-4 dataset.

Chapters 4 through 10 address digital watermarking techniques. In Chapter 4, the problems of detecting visible watermarks from images and removing the watermarks to generate watermark-free images are described. A new Fourier-based image alignment method with iterative refinement for the watermark detection is presented. In Chapter 5, a digital watermarking technique based on a chaotic map and a reference register is presented. A block-based chaotic map, which outperforms the traditional one by breaking local spatial similarity, is used to increase the amount of significant coefficients in the transformed image. A reference register is then employed to locate specific coefficients of a container efficiently for watermark embedding and extraction. In Chapter 6, a pseudo-random pixel rearrangement algorithm to improve the security of most image watermarking techniques is introduced. It rearranges image pixels based on the properties of Gaussian integers and results in a more random-looking image transformation that, in turn, significantly improves the security of the embedded watermark. In Chapter 7, an overview of existing fragile reversible data-hiding schemes is provided. Basic schemes are introduced and examples are illustrated to explain some complicated schemes. In Chapter 8, a novel semi-fragile spatial watermarking method based on local binary pattern (LBP) operators by using local pixel contrast for the embedding and extraction of watermarks is presented. A general framework for multi-level image watermarking is also extended. In Chapter 9, an efficient authentication method for JPEG images based on genetic algorithms (GA) is presented. A two-level detection strategy is also introduced to reduce the false acceptance ratio of invalid blocks. In Chapter 10, an efficient block-based fragile watermarking system for tamper localization and recovery of images is proposed. The cyclic redundancy checksum (CRC) is used to authenticate the feature of a block stored in a pair of mapping blocks. The proposed scheme localizes the tampered area irrespective of the gray value, and the detection strategy is designed in such a way to reduce false alarms.

Chapters 11 through 14 cover digital steganography techniques. Chapter 11 surveys image steganography and steganalysis. It introduces the key concepts behind image steganography and steganalysis, the history and origin of steganography, and the steganography and steganalysis tools currently available. In Chapter 12, different types of image steganographic schemes based on vector quantization (VQ) are presented. The concept of VQ is introduced first, and the main components, such as the input image to be compressed, the VQ codebook, the VQ indices of input image, and the reconstructed image, are applied to digital steganography. In Chapter 13, a different evolutionary approach, named differential evolution (DE), is used to increase the performance of the steganographic system. The key element that distinguishes DE from other population-based approaches is differential mutation, which aims to find the global optimum of a multidimensional, multimodal function. In Chapter 14, a robust steganographic system is presented by artificially counterfeiting statistic features instead of the traditional strategy by avoiding the change of statistic features. A GA-based methodology by adjusting gray values of a cover-image while creating the desired statistic features to generate the stego-images that can break the inspection of steganalytic systems is described.

Chapters 15 through 20 discuss digital forensic techniques. In Chapter 15, a robust image inpainting algorithm to determine a new filling order based on the structure priority value is introduced. The algorithm can fill the missing area with more accurate structure information. In addition, the dynamic searching range is used to improve the efficiency of finding the best patches in the source region. In Chapter 16, the techniques of copy–cover image forgery are discussed and four detection methods for copy–cover forgery detection, which are based on PCA, DCT, spatial domain, and statistical domain, are compared. Their effectiveness and sensitivity under the influences of Gaussian blurring and lossy JPEG compressions are investigated. In Chapter 17, an algorithm for both modification detection and localization, whose structure can support parallel-processing mode, is proposed. The mechanism of both changeable-parameter and self-synchronization is used to achieve all the performance requirements of hash function. In Chapter 18, video forensics is discussed, in which a major proportion of related research work is to perform mining for criminal evidence in videos recorded by a heterogeneous collection of surveillance camcorders. This is a new interdisciplinary field, and people working in the field need video-processing skills as well as an in-depth knowledge of forensic science; hence, the barrier for entering the field is high. In Chapter 19, the problems of vulnerability in the original multi-chaotic systems–based image encryption scheme are analyzed. A self-synchronizing method is proposed as an enhancement measure to solve the problems and defeat cryptanalysis. In Chapter 20, recent advances on human motion behavior modeling, including abilities to detect changes in human motion behavior, and to classify the types of small group motion behavior, are presented. Experimental results on several public human action datasets are provided to demonstrate the effectiveness of the presented methods in addressing these human behavior modeling challenges.

FEATURES OF THE BOOK

- New state-of-the-art techniques for digital watermarking, steganography, and forensics
- Numerous practical examples
- A more intuitive development to the complex technology
- Updated bibliography
- Extensive discussion on watermarking, steganography, and forensics
- Inclusion of watermarking, steganalysis, and forensics techniques and their counter-examples

FEEDBACK ON THE BOOK

It is my hope that an opportunity is given to correct any error in this book; therefore, please provide a clear description of any error that you may find.

Your suggestions on how to improve the textbook are always welcome. For this, use either email (shih@njit.edu) or regular mail to the author:

Dr. Frank Y. Shih
College of Computing Sciences
New Jersey Institute of Technology
University Heights
Newark, NJ 07102-1982

Acknowledgments

My most heartfelt thanks and all of my love go to my wife and two children for their great encouragement, patience, and sacrifice in allowing me to work at all hours from home. I would like to thank my parents for educating me from early childhood to motivate my writing and realize my own potential.

I am very lucky to have world-renown professors and researchers contribute their experiences in the book chapters. They ended up writing almost half of this book and improving the book's overall quality by several folds. I also want to thank the editors who helped with the book production and advised on sections that needed clarification.

Portions of the book appeared in earlier forms as conference papers, journal papers, or theses by my students at the New Jersey Institute of Technology. Additionally, portions of the book are contributed by various experts in the fields.

I gratefully acknowledge the following publishers for giving me permission to reuse texts and figures that appeared in some of these earlier publications: the Institute of Electrical and Electronic Engineers (IEEE), Bentham Open, Elsevier, Springer, and Wiley Publishers.

Editor

Frank Y. Shih received his BS degree from the National Cheng Kung University, Tainan, Taiwan, in 1980, an MS degree from the State University of New York, Stony Brook, USA, in 1983, and a PhD degree from Purdue University, West Lafayette, Indiana, USA, in 1987. He is currently a professor jointly appointed in the Department of Computer Science, the Department of Electrical and Computer Engineering, and the Department of Biomedical Engineering at New Jersey Institute of Technology, Newark, New Jersey. He is also director of the Computer Vision Laboratory, New Jersey Institute of Technology, Newark, New Jersey. Dr. Shih has held visiting professor positions at Princeton University; Columbia University; National Taiwan University; National Institute of Informatics, Tokyo; Conservatoire National Des Arts Et Metiers, Paris; and Central South University, Changsha, China. He is an internationally renowned scholar and serves as the editor-in-chief for *International Journal of Multimedia Intelligence and Security (IJMIS)*. Dr. Shih is currently on the editorial board of *Pattern Recognition, Pattern Recognition Letters, International Journal of Pattern Recognition and Artificial Intelligence, Journal of Information Hiding and Multimedia Signal Processing, Recent Patents on Engineering, Recent Patents on Computer Science, The Open Nanoscience Journal, International Journal of Internet Protocol Technology, Journal of Internet Technology, ISRN Signal Processing*, and *ISRN Machine Vision*. He has served as a steering member, committee member, and session chair for numerous professional conferences and workshops. Dr. Shih has received numerous grants from the National Science Foundation, Navy and Air Force, and industry. He was the recipient of the Research Initiation Award of the NSF in 1991 and the Board of Overseers Excellence in Research Award of NJIT in 2009.

Dr. Shih has made significant contributions to information hiding, focusing on the security and robustness of digital watermarking and steganography. He has authored three books: *Digital Watermarking and Steganography, Image Processing and Mathematical Morphology*, and *Image Processing and Pattern Recognition*. He has published 115 journal papers, 95 conference papers, and 10 book chapters. His current research interests include image processing, computer vision, watermarking and steganography, digital forensics, sensor networks, pattern recognition, bioinformatics, information security, robotics, fuzzy logic, and neural networks.

Contributors

Mayra Bachrach received her MS in computer science from New Jersey Institute of Technology in 2011. She is a teacher at Bergen County Academies, Hackensack, New Jersey. Her research interests include image processing and databases.

Chin-Chen Chang received his BS in applied mathematics in 1977 and MS in computer and decision sciences in 1979, both from the National Tsing Hua University, Hsinchu, Taiwan. He received his PhD in computer engineering in 1982 from the National Chiao Tung University, Hsinchu, Taiwan. During the academic years of 1980–1983, he was on the faculty of the Department of Computer Engineering at the National Chiao Tung University. From 1983 to 1989, he was on the faculty of the Institute of Applied Mathematics, National Chung Hsing University, Taichung, Taiwan. From August 1989 to July 1992, he was the head of, and a professor in, the Institute of Computer Science and Information Engineering at the National Chung Cheng University, Chiayi, Taiwan. From August 1992 to July 1995, he was the dean of the College of Engineering at the same university. From August 1995 to October 1997, he was the provost at the National Chung Cheng University. From September 1996 to October 1997, Dr. Chang was the acting president at the National Chung Cheng University. From July 1998 to June 2000, he was the director of the Advisory Office of the Ministry of Education of the People's Republic of China. From 2002 to 2005, he was a chair professor of National Chung Cheng University. Since February 2005, he has been a chair professor of Feng Chia University. In addition, he has served as a consultant to several research institutes and government departments. His current research interests include database design, computer cryptography, image compression, and data structures.

I-Cheng Chang received a BS in nuclear engineering in 1987, and an MS and PhD in electrical engineering in 1991 and 1999, respectively, all from National Tsing Hua University, Hsinchu, Taiwan. In 1999, he joined Opto-Electronics & Systems Laboratories, Industrial Technology Research Institute, Hsinchu, Taiwan, as an engineer and project leader. He is currently an associate professor in the Department of Computer Science and Information Engineering, National Dong Hwa University, Hualien, Taiwan. His research interests include video and image processing, surveillance network system, computer vision, and multimedia system design. Dr. Chang received the Annual Best Paper Award from the *Journal of Information Science and Engineering* in 2002, and research awards from the Industrial Technology Research Institute in 2002 and 2003. He is a member of the IEEE, and the IPPR of Taiwan, People's Republic of China.

Shu-Ching Chen has been a full professor in the School of Computing and Information Sciences (SCIS), Florida International University (FIU), Miami since August 2009. Prior to that, he was an assistant/associate professor in SCIS at FIU from 1999. He received master's degrees in computer science, electrical engineering, and civil engineering in 1992, 1995, and 1996, respectively, and a PhD in electrical and computer engineering in 1998, all from Purdue University, West Lafayette, Indiana, USA. His main research interests include content-based image/video retrieval, distributed multimedia database management systems, multimedia data mining, multimedia systems, and disaster information management. Dr. Chen has authored or coauthored more than 240 research papers in journals, refereed conference/symposium/workshop proceedings, book chapters, and one book. Dr. Chen received the best paper award from the 2006 IEEE International Symposium on Multimedia. He was awarded the IEEE Systems, Man, and Cybernetics (SMC) Society's Outstanding Contribution Award in 2005 and was the co-recipient of the IEEE Most Active SMC Technical Committee Award in 2006. He was also awarded the Inaugural Excellence in Graduate Mentorship Award from FIU in 2006, the University Outstanding Faculty Research Award from FIU in 2004, the Excellence in Mentorship Award from SCIS in 2010, the Outstanding Faculty Service Award from SCIS in 2004, and the Outstanding Faculty Research Award from SCIS in 2002. He is the chair of IEEE Computer Society Technical Committee on Multimedia Computing and co-chair of IEEE Systems, Man, and Cybernetics Society's Technical Committee on Knowledge Acquisition in Intelligent Systems. He is a fellow of Society for Information Reuse and Integration (SIRI).

Ya-Yun Cheng received a BS from the University of Kaohsiung, Taiwan, in 2010. She is currently a master's student in the Institute of Information Systems and Applications at National Tsing Hua University. Her research interest includes computer vision, machine learning, medical image analysis, and image segmentation.

Venkata Gopal Edupuganti received Bachelor of Technology and Master of Technology degrees in Information Technology from the University of Hyderabad, Hyderabad, India, in 2004 and 2007, respectively. He is currently working toward his PhD in the Department of Computer Science, New Jersey Institute of Technology, Newark, New Jersey. His research interests include image processing, image retrieval, watermarking, and pattern recognition.

Chia-We Hsu received her BS in computer science and information engineering from Da Yeh University, Changhua, Taiwan, in 2004, and her MS in computer science and information engineering from Dong Hwa University, Hualien, Taiwan, in 2007. Her research interests include image/video processing and pattern recognition.

Aleksey Koval received his PhD in computer science from the New Jersey Institute of Technology (NJIT) in 2011. His primary field of study is cryptography and security. Dr. Koval has vast work experience in the IT industry working for AT&T and IBM. He currently works for IBM.

Shang-Hong Lai received his BS and MS in electrical engineering from the National Tsing Hua University, Hsinchu, Taiwan, and a PhD in electrical and computer engineering from the University of Florida, Gainesville, in 1986, 1988, and 1995, respectively. He joined Siemens Corporate Research in Princeton, New Jersey, as a member of the technical staff in 1995. In 1999, he became a faculty member in the Department of Computer Science, National Tsing Hua University, Taiwan. He is currently a professor in the same department and the director of the computer and communication center in the university. In 2004, he was a visiting scholar at Princeton University. Dr. Lai's research interests include computer vision, visual computing, pattern recognition, medical imaging, and multimedia signal processing. He has authored more than 180 papers published in related international journals and conferences. He holds many patents for inventions related to computer vision and medical image analysis. He has been a program committee member of several international conferences, including CVPR, ICCV, ECCV, ACCV, ICPR, and ICME.

Hong-Yuan Mark Liao received his BS in physics from National Tsing-Hua University, Hsin-Chu, Taiwan, in 1981, and an MS and PhD in electrical engineering from Northwestern University in 1985 and 1990, respectively. In July 1991, he joined the Institute of Information Science, Academia Sinica, Taiwan. Currently he is a research fellow. In 2008, he became the division chair of the Computer Science and Information Engineering Division II, National Science Council of Taiwan. He is jointly appointed as a professor in the Computer Science and Information Engineering Department of National Chiao-Tung University and the Department of Electrical Engineering of National Cheng Kung University. Since February 2009, he has been jointly appointed as the multimedia information chair professor of National Chung Hsing University. In August 2010, he was appointed as an adjunct chair professor of Chung Yuan Christian University. His current research interests include multimedia signal processing, video-based surveillance systems, video forensics, and multimedia protection.

Dr. Liao was a recipient of the Young Investigators' award from Academia Sinica in 1998. He received the Distinguished Research Award from the National Science Council of Taiwan in 2003 and 2010, and the National Invention Award of Taiwan in 2004. In 2008, he received a Distinguished Scholar Research Project Award from National Science Council of Taiwan. In 2010, he received the Academia Sinica Investigator Award. In June 2004, he served as the conference co-chair of the 5th International Conference on Multimedia and Exposition (ICME) and technical co-chair of the 8th ICME held in Beijing. From 2006 to 2008, Dr. Liao was the president of the Image Processing and Pattern Recognition Society of Taiwan.

Dr. Liao is on the editorial boards of the *IEEE Signal Processing Magazine*, *IEEE Transactions on Image Processing*, and *IEEE Transactions on Information Forensics and Security*. He was an associate editor of *IEEE Transactions on Multimedia* during 1998–2001.

Xiaofeng Liao received his PhD from the University of Electronic Science and Technology of China in 1997. He is a professor in the Department of Computer Science and Engineering, Chongqing University, Chongqing, China. His research interests include information security, image processing, neural networks, and artificial intelligence.

Shinfeng D. Lin received his BS in automatic control engineering from Feng Chia University, Taichung, Taiwan, in 1980, and an ME and PhD in electrical engineering from Mississippi State University in 1985 and 1991, respectively. He is currently a professor, chairman, and deputy dean in the Department of Computer Science and Information Engineering, College of Science and Engineering, National Dong Hwa

University, Taiwan. Dr. Lin was the director of the Bureau of Education, Hualien County, Taiwan, from January 2002 to September 2003. His research interests include signal/image processing, image/video compression, and information security. He won the Gold Medal Award at the 2005 International Trade Fair "Ideas–Inventions–New Products" (IENA), Nuremberg, Germany. Dr. Lin is a member of Tau Beta Pi and Eta Kappa Nu.

Hong Man received his PhD in electrical engineering from Georgia Institute of Technology in 1999. He joined the Stevens Institute of Technology in 2000. He is currently an associate professor in the Electrical and Computer Engineering Department, the director of the Visual Information Environment Laboratory, and the director of the undergraduate computer engineering program at Stevens. He served on the Organizing Committees of IEEE ICME (2007 and 2009), IEEE MMSP (2002 and 2005), and the Technical Program Committees of various IEEE conferences. He is a member of the IEEE Technical Committees on Multimedia (MMTC) and Communications and Information Security (CISTC). His research interests include signal and image processing, pattern recognition and data mining, on which he has published more than 70 technical journal and conference papers. He is a senior member of IEEE and a member of ACM, ASEE.

Sebastien Poullot obtained his PhD in computer science in 2009 after earning two master's degrees in bio-informatics (2004) and computer sciences (2005). He was invited by Professor Satoh's team at the National Institute of Informatics, and later worked as a project researcher (2009–2011). Since 2011, he has been employed at INRIA Rocquencourt in the IMEDIA team. His research is mainly focused on video and image retrieval, especially the scalability of applications.

Shin'ichi Satoh received his BE in electronics engineering in 1987, and his ME and PhD in information engineering in 1989 and 1992 at the University of Tokyo. He joined the National Center for Science Information Systems (NACSIS), Tokyo, in 1992. He has been a full professor at the National Institute of Informatics (NII), Tokyo, since 2004. He was a visiting scientist at the Robotics Institute, Carnegie Mellon University, from 1995 to 1997. His research interests include image processing, video content analysis, and multimedia database. Currently, he is leading the video-processing project at NII.

Alexander Sheppard received a BS in computer science from the New Jersey Institute of Technology. He is currently working toward a PhD in computer science at the University of North Carolina, Chapel Hill. His research interests include image processing and pattern recognition.

Shih-Chieh Shie received his BS from the Department of Computer Science and Engineering, Tatung Institute of Technology, Taipei, Taiwan, People's Republic of China, in 1996, and an MS and PhD from the Department of Computer Science and Information Engineering, National Dong Hwa University, Hualien, Taiwan, People's Republic of China, in 2000 and 2005, respectively. Dr. Shie is currently an assistant professor in the Department of Computer Science and Information Engineering, National Formosa University, Taiwan, People's Republic of China. His research interests include information hiding, image processing, image compression, and image watermarking.

Mei-Ling Shyu has been an associate professor in the Department of Electrical and Computer Engineering (ECE), University of Miami (UM), since June 2005. Prior to that, she was an assistant professor in ECE at UM dating from January 2000. She received her PhD from the School of Electrical and Computer Engineering, Purdue University, West Lafayette, Indiana, USA in 1999 and three master's degrees in computer science, electrical engineering, and restaurant, hotel, institutional, and tourism management from Purdue University in 1992, 1995, and 1997. Her research interests include multimedia data mining, multimedia information management and retrieval, and network security. She has coauthored more than 200 technical papers published in prestigious journals, book chapters, and refereed conference/workshop/symposium proceedings. She is the vice chair of the IEEE Computer Society Technical Committee on Multimedia Computing. She received the Best Student Paper Award with her student at the Third IEEE International Conference on Semantic Computing in September 2009, the Best Published Journal Article in *International Journal of Multimedia Data Engineering and Management (IJMDEM)* for 2010 Award from IGI Global in April 2011, and the Johnson A. Edosomwan Scholarly Productivity Award from the College of Engineering at UM in 2007. She is a fellow of SIRI.

Hong-Ren Su received his BS in biomedical imaging and radiological science from the National Yang Ming University, Taiwan, in 2000 and an MS in nuclear science from the National Tsing Hua University, Taiwan, in 2002. He is currently working toward obtaining a PhD from the Institute of Information Systems and Applications at National Tsing Hua University. He is a student member of the IEEE. His research interests includes computer vision, machine learning, and medical image analysis. His current address is Institute of Systems and Applications, National Tsing Hua University, Hsinchu, Taiwan.

Ming-Ting Sun received his BS from the National Taiwan University in 1976, an MS from the University of Texas at Arlington in 1981, and a PhD from the University of California, Los Angeles in 1985, all in electrical engineering. Dr. Sun joined the faculty of the University of Washington in September 1996. Before that, he was the director of the Video Signal Processing Group at Bellcore (now Telcordia). He has been a chair professor at Tsing Hwa University, and a visiting professor at Tokyo University, National Taiwan University, and National Chung-Cheng University. His research interests include machine learning and video processing. Dr. Sun has been awarded 11 patents and has published more than 200 technical publications, including 15 book chapters in the area of video technology. He was actively involved in the development of H.261, MPEG-1, and MPEG-2 video-coding standards. He has also coedited a book titled *Compressed Video Over Networks*. He was the editor-in-chief of *IEEE Transactions on Multimedia* during 2000–2001. He received an IEEE CASS Golden Jubilee Medal in 2000, and he was a co-chair of the SPIE VCIP

(Visual Communication and Image Processing) 2000 Conference. He was the editor-in-chief of *IEEE Transactions on Circuits and Systems for Video Technology (T-CSVT)* during 1995–1997 and the *Express Letter* editor of T-CSVT during 1993–1994. He was a corecipient of the T-CSVT Best Paper Award in 1993. From 1988 to 1991, he served as the chairman of the IEEE CAS Standards Committee and established an IEEE Inverse Discrete Cosine Transform Standard. He received an Award of Excellence from Bellcore in 1987 for his work on the digital subscriber line. Dr. Sun was elected as a Fellow of the IEEE in 1996.

Hsiao-Rong Tyan received her BS in electronic engineering from Chung Yuan Christian University, Chungli, Taiwan, in 1984, and an MS and PhD in computer science from Northwestern University, Evanston, Illinois, in 1987 and 1992, respectively. She is an associate professor in the Department of Information and Computer Engineering, Chung Yuan Christian University, Chungli, Taiwan, where she currently conducts research in the areas of computer security, computer networks, and real-time intelligent systems.

Boris S. Verkhovsky is a professor in the Computer Science Department at the New Jersey Institute of Technology (NJIT). He received his PhD in computer science jointly from Latvia State University and from the Academy of Sciences of the USSR (Central Institute of Mathematics and Economics, Moscow). From his prior affiliations at the Scientific Research Institute of Computers (Moscow), Princeton University, IBM Thomas J. Watson Research Center, Bell Laboratories and since 1986 at the NJIT, he acquired research interests and experience in cryptography, communication security, large-scale systems optimal design and control, optimization, and algorithm design.

For the last 20 years, Dr. Verkhovsky's research activity has been centered on the design and analysis of cryptographic systems and information assurance algorithms. Prior to NJIT he worked at the Academy of Sciences of the USSR, IBM Thomas J. Watson Research Center, was associate professor at Princeton University, a member of Technical Staff at Bell Labs, and was Endowed Chair Professor, University of Colorado. Professor Verkhovsky is listed in *Marquis Who's Who in America*. Professor Verkhovsky is editor-in-chief of the *International Journal of Communications, Network and System Sciences*.

Zhi-Hui Wang received a BS in software engineering in 2004 from the North Eastern University, Shenyang, China. She received an MS in software engineering in 2007 and a PhD in computer software and theory in 2010, both from the Dalian University of Technology, Dalian, China. Since November 2011, she has been a visiting scholar of the University of Washington. Her current research interests include information hiding and image compression.

Xiaomeng Wu is a project researcher at the National Institute of Informatics, Japan. He graduated from the University of Shanghai for Science and Technology with a baccalaureate degree in thermal energy engineering in 2001. He graduated from the University of Tokyo with a doctoral degree in information and communication engineering in 2007. His research interests include multimedia indexing, multimedia retrieval, and near-duplicate detection. He is a member of the IEEE Computer Society and ACM.

Yi-Ta Wu received his BS in physics from Tamkang University, Taipei, Taiwan, in 1995, and an MS in computer science from National Dong-Hwa University, Hualien, Taiwan, in 1997. Dr. Wu received his PhD from the Department of Computer Science, New Jersey Institute of Technology, in May 2005. He was a research fellow at the University of Michigan, Ann Arbor, and is currently affiliated with Industrial Technology Research Institute, Taiwan. His research interests include image/video processing, mathematical morphology, image watermarking, steganography, surveillance system, robot vision, pattern recognition, shortest path planning, and artificial intelligence.

Di Xiao received a BS from Sichuan University, Chengdu, China in 1997. He obtained both an MS and PhD from Chongqing University, Chongqing, China, in 2002 and 2005, respectively. He is currently a professor in the College of Computer Science and Engineering, Chongqing University, Chongqing, China. His research interests include chaotic dynamical systems-based cryptography, information authentication, digital watermark, and digital image forensics.

Yafeng Yin is currently a PhD candidate in the Electrical and Computer Engineering Department at Stevens Institute of Technology, and a research intern at Bell Laboratories, Alcatel-Lucent. He received his bachelor's from the Department of Automatic Control at Beijing Institute of Technology in 2005. His research interests include image processing, pattern recognition and machine learning, with applications to video-based object recognition, tracking, and human group behavior analysis.

Yuan Yuan obtained both an BS and MS from the Department of Computer Science and Technology, Zhejiang University of Technology, Hangzhou, People's Republic of China, in 2004 and 2006, respectively. He secured a PhD in computer science from the New Jersey Institute of Technology, Newark, New Jersey, in 2011. He is currently affiliated with Amazon Corporation, Seattle, Washington, DC.

Wenyin Zhang is an assistant professor in the School of Informatics, Linyi University, Linyi, Shandong, People's Republic of China. He obtained his master's in 2002 from Shandong University of Science and Technologies and was awarded a doctorate by the Chinese Academy of Sciences in 2005. His research interest includes image processing, information hiding, digital watermarking.

Qiusha Zhu is currently a PhD candidate at the Department of Electrical and Computer Engineering (ECE), University of Miami. She received her MS in electronic engineering from School of Electronic, Information and Electrical Engineering, Shanghai Jiao Tong University in March 2009 and a BS in electronic engineering from Shanghai University as an outstanding graduate in June 2006. She worked as a database apprentice at DOW Chemical, Shanghai from November 2007 to June 2008. Her research interests include multimedia information retrieval and data mining, especially how to effectively retrieve online images and videos, and how to fuse information available from different sources to boost this process.

Part I

Multimedia Mining and Classification

1 Multimedia Duplicate Mining toward Knowledge Discovery

Xiaomeng Wu, Sebastien Poullot,
and Shin'ichi Satoh

CONTENTS

1.1   INTRODUCTION

The spread of digital multimedia content and services in the field of broadcasting and on the Internet has made multimedia data mining an important technology for transforming these sources into business intelligence for content owners, publishers, and distributors. A recent research domain known as multimedia duplicate mining (MDM) has emerged largely in response to this technological trend. The "multimedia duplicate mining" domain is based on detecting image, video, or audio copies from a test collection of multimedia resources. One very rich area of application is digital rights management, where the unauthorized or prohibited use of digital media on file-sharing networks can be detected to avoid copyright violations. The primary thesis of MDM in this application is "the media itself is the watermark," that is, the media (image, video, or audio) contains enough unique information to be used to detect copies (Hampapur et al., 2002). The key advantage of MDM over other technologies, for example, the watermarking, is the fact that it can be introduced after copies are made and can be applied to content that is already in circulation.

Monitoring commercial films (CFs) is an important and valuable task for competitive marketing analysis, for advertising planning, and as a barometer of the advertising industry's health in the field of market research (Li et al., 2005; Gauch and Shivadas, 2006; Herley, 2006; Berrani et al., 2008; Dohring and Lienhart, 2009; Putpuek et al., 2010; Wu and Satoh, 2011). In the field of broadcast media research, duplicate videos shared by multiple news programs imply that there are latent semantic relations between news stories. This information can be used to define the similarities between news stories; thus, it is useful for news story tracking, threading, and ranking (Duygulu et al., 2004; Zhai and Shah, 2005; Wu et al., 2008a; Wu et al., 2010). From another viewpoint, duplicate videos play a critical role in assessing the novelty and redundancy among news stories, and can help in identifying any fresh development among a huge volume of information in a limited amount of time (Wu et al., 2007a; Wu et al., 2008b). Additionally, MDM can be used to detect filler materials, for example, opening CG shots, anchor person shots, and weather charts in television, or background music in radio broadcasting (Satoh, 2002).

This chapter discusses the feasibility, techniques, and demonstrations of discovering hidden knowledge by applying MDM methods to the massive amount of multimedia content. We start by discussing the requirements and selection criteria for the duplicate mining methods in terms of the accuracy and scalability. These claims involve the sampling and description of videos, the indexing structure, and the retrieval process, which depend on the application purposes. We introduce three promising knowledge-discovery applications to show the benefits of duplicate mining. The first application (Wu and Satoh, 2011) is dedicated to fully unsupervised TV commercial mining for sociological analysis. It uses a dual-stage temporal recurrence hashing algorithm for ultra-fast detection of identical video sequences. The second application (Wu et al., 2010) focuses on news story retrieval and threading: it uses a one-to-one symmetric algorithm with a local interest point index structure to accurately detect identical news events. The third application (Poullot et al., 2008, 2009) is for large-scale cross-domain video mining. It exploits any weak geometric consistencies between near-duplicate images and addresses the scalability issue of a near-duplicate search. Finally, a discussion on these techniques and applications is given.

1.2 SELECTION CRITERION OF DUPLICATE MINING METHODS

Choosing a duplicate mining method can be difficult because of the variety of methods currently available. The application provider must decide which method is best suited to their individual needs, and, of course, the type of duplicate that they want to use as the target. In this sense, the definition of a duplicate is generally subjective and, to a certain extent, does depend on the type of application being taken into consideration. The application ranges from exact duplicate mining, where no changes are allowed, to a more general definition that requires the resources to be of the same scene, but with possibly strong photometric or geometric transformations.

1.2.1 EXACT DUPLICATE MINING

One direction for MDM is to mine duplicate videos or audios derived from the original resource without any or with very few transformations. This type of duplicate is known as an exact duplicate or exact copy. Typical cases include TV CFs and file footages used in news. Most existing studies on exact duplicate mining introduce the concept of fingerprint or "hash" functions. This is a signal that has been extracted from each sample of a video. It uniquely identifies the sample as genuine and cannot be decoded to get the original one, because the fingerprint is made by some of the features of the sample.

The fingerprints are computed on the sampled frames given the video dataset in order to construct a hash table, into which the frames are inserted by regarding the corresponding fingerprint as an inverted index. An important assumption made here is that identical or approximately identical frames of all the video duplicates have exactly the same fingerprint, so a hash collision occurs in the corresponding hash bucket. Based on this assumption, approximately identical frames of all the exact duplicates can be efficiently retrieved by using a collision attack process.

In most cases, the sampling rate, which defines the number of sampled frames per unit of time (usually seconds) taken from a continuous video to make a discrete fingerprint, is as dense as the allowed maximal frame frequency of the video, for example, 29.97 (NTSC) or 25 (PAL) frames per second for an MPEG-2 video. This high-density sampling can accommodate the fingerprinting errors, that is, different fingerprints derived from identical frames, and allows the detection to be precise in order to locate the boundaries of the duplicates. The description of the video, based on which the fingerprint is extracted, can be color histogram (Li et al., 2005), ordinal signature (Li et al., 2005), color moments (Gauch and Shivadas, 2006), gradient histogram (Dohring and Lienhart, 2009), luminance moments (Wu and Satoh, 2011), and acoustic signatures (Haitsma and Kalker, 2002; Cano et al., 2005). All these global descriptors have been effectively demonstrated, thanks to the limited number of transformations between the original resource and the duplicate. Therefore, most researchers attach more importance to the efficiency and scalability than the effectiveness of the exact duplicate mining methods.

1.2.2 NEAR-DUPLICATE MINING

Near duplicates are considered the slight alterations of videos or images issued from the same original source. In other literature, near duplicates are also known as copies. The alterations are obtained from an original using various photometric or geometric transformations, for example, cam cording, picture in picture, insertions of patterns, strong reencoding, gamma changes, decreases in quality, postproduction transformations, and so on. There is significant diversity in the nature and amplitude of the encountered transformations. Near duplicates are widespread among the current worldwide video pool. A typical example is the digital copies of copyrighted materials transferred on file-sharing networks or video-hosting services, for example, Flickr, YouTube, and so on. Two examples are shown in Figure 1.1. The transformation of the example on the left is cam-cording, and those of the right one include a change of contrast, blurring, noising, and scaling.

Near-duplicate mining is considered a more difficult case than exact duplicate mining because the videos or images can be strongly transformed. The assumption of identical or approximately identical images from among all the duplicates with exactly the same fingerprint (Section 1.2.1) becomes too weak to make when detecting near duplicates. Most existing studies on near-duplicate mining use an image distance measure instead of the fingerprinting-based hash collision attack process for finding approximately identical images. An image distance measure compares the similarities between two images in various dimensions, for example, the color, texture, and shape. Most of these studies are designed for images, but can be extended to the video (an image stream) domain by embedding the temporal consistency constraints between duplicate images in the mining process.

A general representation of the images must be chosen to compute the similarities between two images. A picture may be represented by a single global feature (e.g., color histogram, edge orientation histogram, Fourier, or Hough) or by many small local ones (e.g., SIFT, Harris, or dipoles: Gouet and Boujemaa, 2001; Lowe, 2004; Mikolajczyk et al., 2005; Joly et al., 2007). The main difference consists in the precision of the description and in the scalability of the approach, two inversely proportional parameters. Methods with global features usually have a higher efficiency than those with local ones. However, local descriptions have recently become more popular because of their superiority over the global ones in terms of their discriminative power, especially in the most difficult cases where the duplicate videos are short and strongly transformed (Law-To et al., 2007).

Given a video dataset, the local descriptors are computed on the sampled frames in order to construct a descriptor database. Most methods choose a sampling rate

FIGURE 1.1 Two examples of near duplicates. (From NISV, Sound and Vision Video Collection 2008. With permission.)

(e.g., one frame per second) that is much lower than the allowed maximal frame frequency of the video to ease the potentially high computation burden of the image distance measure. The descriptor database can be queried with a video or a picture, where its local features (Gouet and Boujemaa, 2001; Ke et al., 2004; Jegou et al., 2008) are used as queries, or be crossed over itself by using each of its features as a query. A set of candidates is returned for each feature query. A vote is then conducted to choose the better candidates. Some of these solutions have shown great scalability concerning the descriptor matching, but the last decision step has a prohibitive cost when performing mining. In the early 2000s, the bag of words (BoW) representation (Philbin et al., 2007; Wu et al., 2007b; Yuan et al., 2007) raised up. It was inspired from the inversed lists used in text search engines. Here, a global representation of an image is computed by using the local descriptions. A visual vocabulary is first computed on the sampled local features, by using a K-means algorithm. Then, for a given image, all its local features are quantized on this vocabulary; the resulting description is a histogram with a dimensionality equal to the vocabulary size. There are two major advantages to this approach, the compactness of the feature, and the suppression of the last decision step. It usually leads to a more scalable solution while offering good-quality results. One of the notorious disadvantages of BoW is that it ignores the spatial relationships among the patches, which is very important in image representation. Therefore, a decision process based on the spatial configuration (e.g., with RANSAC) is usually performed as well on the top candidates (Fischler and Bolles, 1981; Matas and Chum, 2005; Joly et al., 2007).

1.3 TV COMMERCIAL MINING FOR SOCIOLOGICAL ANALYSIS

1.3.1 BACKGROUND

TV commercials, also known as CFs, are a form of communication intended to persuade an audience to consume or purchase a particular product or service. It is generally considered the most effective mass-market advertising format of all the different types of advertising. CF mining has recently attracted the attention of researchers. This task aims at developing techniques for detecting, localizing, and identifying CFs broadcast on different channels and in different time slots. CF removal based on CF detection and localization is an important preprocessing technique in broadcast media analysis, where CFs are usually considered to be an annoyance. In market research, monitoring CFs is an important and valuable task for competitive marketing analysis, for advertising planning, and as a barometer of the advertising industry's health. Another promising application of CF mining is to furnish consumers with a TV CF management system, which categorizes, filters out, and recommends CFs on the basis of personalized consumer preference.

CF mining techniques can be categorized into knowledge- and repetition-based techniques. The knowledge-based ones (Albiol et al., 2004; Hua et al., 2005; Duan et al., 2006; Li et al., 2007) use the intrinsic characteristics of CF (e.g., the presence of monochrome, silence, or product frames) as prior knowledge for detecting them. Most knowledge-based techniques are efficient but not generic enough because of the data-dependent prior knowledge they use. For example, most Japanese TV stations do not use monochrome or silence frames to separate two consecutive CF sequences. Other characteristics (e.g., product frame, shot information, and motion information) normally vary with the nations and regions. Meanwhile, CF can be regarded as repetition in the stream. Some recent studies (Li et al., 2005; Gauch and Shivadas, 2006; Herley, 2006; Berrani et al., 2008; Dohring and Lienhart, 2009; Putpuek et al., 2010) have used this property to detect CFs. Conversely, repetition-based ones are unsupervised and more generic, but are a large computational burden. In some studies, the required system task is given a test collection of videos and a reference set of CFs to determine for each CF the location at which the CF occurs in the test collection. These studies are not fully unsupervised considering the construction of the reference set. In most cases, the boundaries of the CFs in the reference set are known. The increase in processing time is linear to that of the size of the test collection. A more challenging task is, when given a test collection of videos, how to determine the location, if any, at which some CF occurs. This is also the main problem that the technique introduced in this section aims to solve. Neither the reference database nor the CF boundary is provided beforehand. The system should be fully unsupervised. The algorithm needs to be much more scalable because the processing time will increase as the square of the test collection size increases.

1.3.2 Temporal Recurrence Hashing Algorithm

The target is to output a set of CF clusters when given a broadcast video stream as the input. A CF cluster is a set of multiple identical CF sequences and corresponds to a specific commercial, for example, an Apple Macintosh commercial. A CF sequence, which is composed of one or more consecutive fragments, is a long and complete sequence corresponding to an instance of a commercial. A fragment, which is similar to but much shorter than a shot or subshot, is a short segment, composed of one or more consecutive and identical frames. In this study, the visual and audio features are extracted from all the frames of a broadcast video stream. Consecutive frames with identical low-level features are assembled into a fragment. An example of the fragments derived from two recurring CF sequences is shown in Figure 1.2, from where we can observe a strong temporal consistency among the pairs of recurring fragments. For instance, the temporal positions of the fragments themselves are consecutive, and the temporal intervals between each two fragments are identical to each other. This temporal consistency is exceptionally useful for distinguishing the recurring sequences from the nonduplicate ones. The CF mining task can thus be formulated into searching for recurring fragment pairs with high-temporal consistency from all the pairs of recurring fragments derived from the entire stream.

In this study, two temporal recurrence fingerprints are proposed for capturing the temporal information of each recurring fragment pair. The first fingerprint corresponds to the temporal position of the former fragment in the recurring fragment pair, the resolution of which is set to a given set of minutes so that the consecutive fragment pairs derived from two recurring CF sequences can be mapped to the same fingerprint or neighboring ones. The second fingerprint corresponds to the temporal interval between the two fragments, the resolution of which is set to a given set of

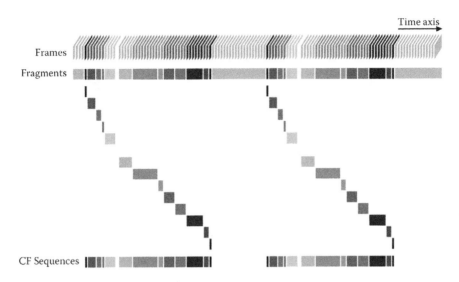

FIGURE 1.2   Recurring fragment pairs derived from two CF sequences of the same CF cluster.

seconds. As shown in Figure 1.3, all the recurring fragment pairs derived from the stream are inserted into a 2-dimensional hash table based on the two fingerprints. By doing so, the recurring fragment pairs with high-temporal consistency can be automatically assembled into the same bin. These fragments, the assembly of which corresponds to a pair of long and complete CF sequences, can thus be easily detected by searching for the hash collisions from the hash table. More details about this algorithm can be found in Wu and Satoh (2011).

This temporal recurrence hashing algorithm was tested by using a 10-h broadcast video stream with a manually annotated ground truth for the accuracy evaluation. The sequence-level accuracy was 98.1%, which is enough to evaluate how precisely the algorithm can detect and identify recurring CF sequences. The frame-level accuracy was 97.4%, which evaluates how precisely the algorithm can localize the start and end positions of the CF sequences in the stream. Three state-of-the-art studies were implemented for comparison, including two video-based techniques (Berrani et al., 2008; Dohring and Lienhart, 2009) and one audio-based one (Haitsma and Kalker, 2002). The temporal recurrence hashing algorithm outperformed the related studies in terms of both the sequence- and the frame-level accuracy. The algorithm mined the CFs from the 10-h stream in <4 s, and was more than 10 times faster than those in the related studies.

1.3.3   KNOWLEDGE DISCOVERY BASED ON CF MINING

CF mining is an important and valuable task for competitive marketing analysis and is used as a barometer for advertising planning. We discuss a promising application of CF mining in this section that visualizes the chronological distribution of a massive amount of CFs and enables the discovery of hidden knowledge in the advertising market. This application is illustrated in Figure 1.4. It visualizes nine

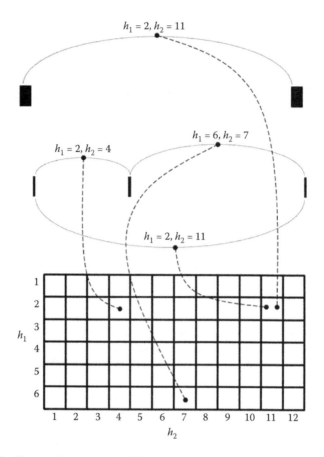

FIGURE 1.3   Temporal recurrence hashing.

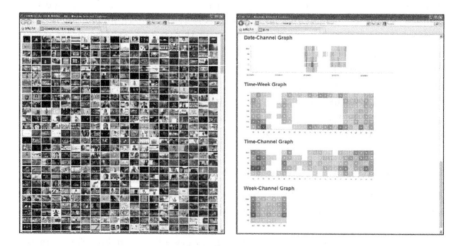

FIGURE 1.4   Application of knowledge discovery based on CF mining.

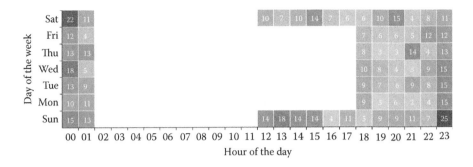

FIGURE 1.5   Asahi Breweries CF chronological distribution.

types of chronological distributions involving five types of information including broadcast frequency, date, hour of the day, day of the week, and channel. The CFs were mined from a 4-month and 5-channel broadcast video archive, the total length of which was 20 months. The algorithm mined CFs from this 20-month stream in around 5 h.

Figure 1.5 visualizes the chronological distribution of a beer CF. The horizontal axis indicates the hour of the day, and the vertical one indicates the day of the week. The darker-shaded blocks indicate the time zones with a higher broadcast frequency. A well-regulated distribution can be observed from this graph. The distributions with the same figuration were also found in the graphs of other alcohol-related CFs. This is because there is a voluntary restriction in Japan, which prohibits the broadcast of alcohol CFs from 5 a.m. to 5 p.m. on weekdays and from 5 a.m. to 11 a.m. on weekends.

Figure 1.6 shows an example of a car CF. Three time zones with higher broadcast frequencies can be seen in Figure 1.6a. The first time zone is from 6 a.m. to 8 a.m., which is when most salaried employees are having breakfast. The second one is around 12 a.m., which is lunch time. The third one is after 6 p.m., which is when most salaried employees have finished their work and are watching TV. This observation can also be found in Figure 1.6b. One explanation of this observation could be that this car targets toward male rather than female drivers. Another explanation could be that in most Japanese families, the fathers rather than the mothers manage the money, so the car dealer prefers the most popular TV slots for male viewers rather than those for female viewers.

By analyzing these statistics derived from CF mining, the actual conditions of the business management, which could not be efficiently grasped by using conventional technology, can be determined in a business-like atmosphere. Furthermore, the influences that the information sent from producers, distributors, and advertisers had on the audiences and consumers can be circumstantially analyzed by analyzing the relationship between the statistics from the TV commercials and those from the Web sources, including blogs, and Twitter. We believe that this kind of knowledge can be transformed into business intelligence for producers, distributors, and advertisers.

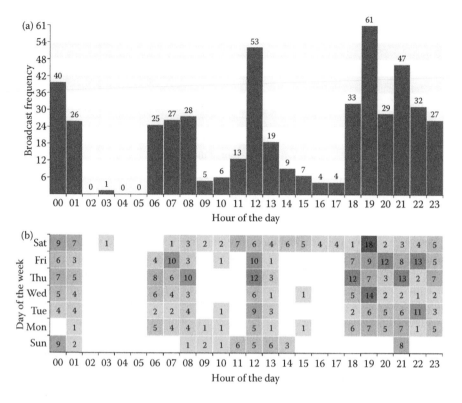

FIGURE 1.6 Daihatsu Tanto Exe CF chronological distribution. (a) Hour–frequency histogram. (b) Hour–day frequency diagram.

1.4 NEWS STORY RETRIEVAL AND THREADING

1.4.1 Background

The spread of digital multimedia content and services in the field of broadcasting and on the Internet has made data retrieval and its reuse an important methodology for helping a user to more easily grasp the availability of these sources. A recent research domain known as news story retrieval has emerged largely in response to this methodological trend, and it is a fundamental step toward semantic-based video summarization and browsing. It tracks the development of evolving and historical news stories originating from a specific news topic, mines their dependencies from different sources, and represents their temporal flow in a semantic way.

News story retrieval is normally studied as a Query-By-Example (QBE) topic with textual information as the underlying cues. A QBE parser parses the query, for example, the text of a full document obtained from the given Web pages or the closed captions of a news story, and looks for keywords on the basis of which similar documents are searched for (Chieu and Lee, 2004; Ide et al., 2004; Kumar et al., 2004; Lee et al., 2010). The main limitation of this research domain is that the textual information is normally not discriminating enough to distinguish between the documents of similar but irrelevant topics. For example, a news story on the "Trial of Saddam Hussein" was broadcast on November 6, 2006 during the Japanese news program "FNN SPEAK." The keywords parsed from this story included "sentence" (8), "this go-round" (4), "former president Hussein" (3), "President Bush" (3), "United States" (3), and "November 5th" (3), with the parenthetic numbers being the frequencies of the keywords. By using these keywords as the query and Ide et al.'s work (Ide et al., 2004) as the search engine, the news stories on irrelevant topics were output as the results, including the "Sentence of Homicidal Criminal Yasunori Suzuki," "United States Midterm Election," "Sentence of Criminal Miyoko Kawahara," and "Sentence of Homicidal Criminal Shizue Tamura." The reason for this is because the keywords parsed from the query were limited and not informative enough to represent the characteristics of the corresponding news topic. In addition to the textual transcripts, news videos provide a rich source of visual information. In broadcast videos, there are also a number of visual near duplicates, which appear at different times and dates and across various broadcast sources. These near duplicates basically form pairwise equivalent constraints that are useful for bridging evolving news stories across time and sources.

The aim of this research is news story retrieval and reranking using visual near-duplicate constraints. A news story is formally defined as a semantic segment within a news program that contains a report depicting a specific news topic. A news topic is a significant event or incident that is occurring in the contemporary time frame. A news story can be described in the form of multiple video segments, for example, shots, and closed captions. Each of the segments and sentences in a news story must be semantically identical with respect to its news topic. For instance, a news program can typically be regarded as an assembly of individual news stories that are presented by one or more anchors and collectively cover the entire duration of the program. Adjacent stories within a news program must be significantly different in topic. Figure 1.7 shows an example of two news stories reported by different TV stations and on different days.

The first task of the proposed news story retrieval and reranking approach involves a news story retrieval process based on the textual information. Given a news story $\vec{s}_q$ as the query, the objective is to return a set of candidate news stories $\mathbf{S} = \{\vec{s}_i \mid i \in \{1, 2, ..., n_s\}\}$, satisfying the following criterion:

$$\mathrm{Sim}(\vec{s}_q, \vec{s}_i) > \sigma_q \quad \forall i \in \{1, 2, ..., n_s\} \tag{1.1}$$

$\mathrm{Sim}(\vec{s}_i, \vec{s}_j)$ denotes the textual similarity between $\vec{s}_i$ and $\vec{s}_j$, σ_q a threshold, and n_s the number of candidate stories. The second task involves a news story reranking process based on the pseudo-relevance feedback, which is the main emphasis of this approach. The objective of this task is to find a subset $\tilde{\mathbf{S}} \subset \mathbf{S}$, then to assume that all $\vec{s} \in \tilde{\mathbf{S}}$ are semantically identical to $\vec{s}_q$ in the topic, and finally, to do a query expansion under this assumption to perform a second-round retrieval. The problems that need to be solved here include: (1) how to automatically find a proper subset $\tilde{\mathbf{S}}$ out of $\mathbf{S}$ satisfying the assumption described above, and (2) how to expand the query so that the relevant stories low-ranked in the initial round can then be retrieved at a higher rank to improve the overall performance.

... Lewinsky is claiming that during that visit the president told her that she could testify in Mrs Jones' lawsuit that her visits to him at the White House were to see his secretary, Betty....

1998/02/24 ABC

... Annan said the "qualitative difference" of this agreement from previous ones is that this one has been negotiated with Saddam Hussein himself, which means that "the leadership has ....

FIGURE 1.7 Two news stories originating from "Lewinsky Scandal" and "UNSCOM" news topics. (From LDC, TRECVID 2004 Development Data. With permission.)

1.4.2 STORYRANK

We propose a novel approach called StoryRank in this study, which solves the first problem by exploiting the visual near-duplicate constraints. After retrieving news stories based on the textual information, near duplicates are mined from the set of candidate news stories. A one-to-one symmetric algorithm with a local interest point index structure (Ngo et al., 2006) is used because of its robustness to any variations in translation and scaling due to video editing and different camerawork. Figure 1.8 shows some examples of visual near duplicates within the set of candidate news stories depicting the "Lewinsky Scandal." Note that some of these examples are derived from news stories reported by different TV stations or on different days.

Visual near-duplicate constraints are integrated with textual constraints so that the news story reranking approach can guarantee a high-relevance quality for

FIGURE 1.8 Visual near duplicates existing across multiple news stories for same topic. (From LDC, TRECVID 2004 Development Data. With permission.)

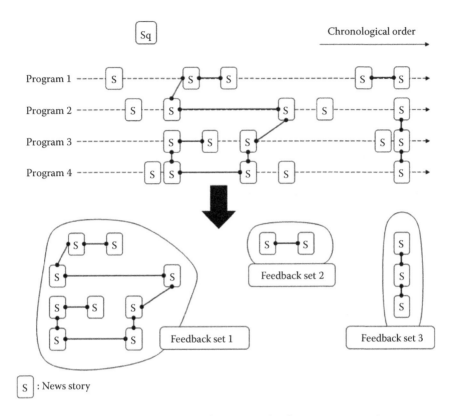

FIGURE 1.9 Exploit must-link near-duplicate constraints between news stories.

pseudo-relevance feedback. As shown in Figure 1.9, two news stories are linked together if they share at least one pair of near duplicates. This pair of stories can be regarded as having a must-link constraint, which indicates that they discuss the same topic. News stories are then clustered into groups called feedback sets by applying transitive closure to these links. One of the characteristics of these feedback sets is that the visual near-duplicate constraints within each feedback set usually form a strongly connected network so that the news stories that are identical to the query in topic comprise a majority of all the stories in this network. Another characteristic is that the dominant feedback set is usually identical in topic to the query. Therefore, the dominant feedback set is chosen as the pseudo-relevance feedback and used for performing the second-round news story retrieval.

Within graph theory and network analysis, there are various measures for the centrality of a vertex within a graph that determines the relative importance of a vertex within the graph, for example, how important a person is within a social network, or in the theory of space syntax, how important a room is within a building or how well used a road is within an urban network. One of these measures is called the Eigenvector centrality. It assigns relative scores to all the nodes in a network based on the principle as follows: connections to high-scoring nodes contribute more, to the score of the node in question, than equal connections to low-scoring nodes.

PageRank (Brin and Page, 1998) is a variant of the Eigenvector centrality measure, the most famous application of which is the Google Internet search engine. In this study, PageRank is used as a pseudo-relevance feedback algorithm and integrated with visual near-duplicate constraints. Each news story is regarded as a node and the whole candidate story set a node-connected network. The news stories in the dominant feedback set are regarded as high-quality nodes, which are different from the classic PageRank, and for each node in the entire network, only the connections between the node and the high-quality nodes are used as votes of support to iteratively define the overall relevancy of the story.

1.4.3 EXPERIMENTATION EVALUATION

The approach was tested by using a very-large-scale broadcast video database comprised of videos broadcast from April 1, 2005 to March 31, 2009. These videos were broadcast from eight news programs produced by four Japanese TV stations. The database was comprised of 100,573 news stories, and 20 queries were selected for experimentation, as listed in Table 1.1, including 11 Japanese and nine foreign news stories. The design of these queries is based on the most important topics from both domestic and international news stories during these 4 years. The duration within which the search was conducted varied from 1 to 15 months. The average

TABLE 1.1
Twenty Queries Selected for Experimentation

ID	Topic	Start Date	Duration	
T01	7 July 2005 London Bombings	July 7, 2005	1	Foreign
T02	Trial of Saddam Hussein	October 19, 2005	15	Foreign
T03	Architectural Forgery in Japan	November 17, 2005	2	Domestic
T04	Murder of Airi Kinoshita	December 1, 2005	1	Domestic
T05	Murder of Yuki Yoshida	December 2, 2005	1	Domestic
T06	Fraud Allegations of Livedoor	January 16, 2006	2	Domestic
T07	Murder of Goken Yaneyama	May 18, 2006	1	Domestic
T08	2006 North Korean Missile Test	July 6, 2006	1	Foreign
T09	2006 North Korean Nuclear Test	October 4, 2006	1	Foreign
T10	Trial of Saddam Hussein	November 5, 2006	2	Foreign
T11	2007 Chuetsu Offshore Earthquake	July 16, 2007	1	Domestic
T12	2007 Burmese Anti-Government Protests	September 23, 2007	1	Foreign
T13	Tainted Chinese Dumplings	January 30, 2008	2	Domestic
T14	Tokitsukaze Stable Hazing Scandal	February 7, 2008	1	Domestic
T15	2008 Sichuan Earthquake	May 20, 2008	1	Foreign
T16	Akihabara Massacre	June 8, 2008	1	Domestic
T17	Wakanoho Toshinori	August 18, 2008	1	Domestic
T18	2008 Chinese Milk Scandal	September 22, 2008	1	Foreign
T19	Murder of Former MHLW Minister	November 18, 2008	1	Domestic
T20	2008 Mumbai Attacks	November 27, 2008	1	Foreign

precision (Turpin and Scholer, 2006) was computed to evaluate the performance of the news story reranking.

The Mean Average Precision of the news story reranking approach was 96.9%, which was higher than all the related studies based on textual information. An intuitive reason for this is that the reranking approach based on PageRank improves the informativeness and representativeness of the original query, and the visual near-duplicate constraints guarantee a higher relevance quality than the textual constraints used in pseudo-relevance feedback algorithms. Additionally, evolving news stories usually repeatedly use the same representative videos even if the airdates and consequently the subtopics of these stories are distant from each other. Therefore, the feedback set formulated from visual near duplicates can guarantee a larger coverage of subtopics than those derived from textual constraints, so the news reranking approach is more robust against the potential variation of the subtopics of the evolving news stories.

1.5   INDEXING FOR SCALING UP VIDEO MINING

1.5.1   BACKGROUND

A major and common issue of works focused on videos and images is to deal with some very large amounts of data. Millions of hours of video and billions of images are now available on the Internet and in the archives. Various research domains share the necessity of finding some similarities in datasets in order to highlight the links for categorization, classification, and so on. Performing a rough similarity comparison is a basic quadratic operation. Consequently, applying such a brute force approach on a very large pool may become hopeless. Some subsets can and must be predefined, but, even in very specialized domains, the accumulation of data and knowledge pushes the time consumption of algorithms beyond an acceptable limit. On the other hand, parallelism can also be performed, but this usually induces a massive expenditure of money and energy. Moreover, it is a linear solution, and some other bottlenecks may appear. One solution more than the others has been found to quickly perform a similarity search. The core of the proposition is to perform an approximative similarity search, which, for a low loss of quality in the results, may offer a tremendous reduction of time consumption, which is better than for an exact similarity search. A large part of these approximative approaches use an index structure (Datar et al., 2004; Joly et al., 2007; Lv et al., 2007). This type of structure is used to filter out the database, and thus shrinks the pool of candidates.

Different approaches must be investigated depending on resources and requirements in order to organize a video database. The similarities between videos can be computed between their annotations, their audio component, or their image component.

A textual search engine can be set up if some annotations are available. The main advantage of such an approach is the powerful scalability. It can also deal with the semantic aspect provided by the users. The drawbacks are the sparsity and low resolution of the annotations. The audio part of a video can also be processed (Haitsma and Kalker, 2002); lots of speech-to-text applications are gaining popularity, and a more accurate extraction of the meaning and knowledge is being attained. Videos sharing the same sound line, or even subject, but different visual contents can be gathered. However, when the sound line is changed between a copy and an original, it becomes harder to gather similar ones. In this case, the image part is processed (Ke et al., 2004; Joly et al., 2007; Law-To et al., 2007; Poullot et al., 2008). The advantages and drawbacks of audio and image approaches are somehow opposite to the textual ones: a potential tight resolution, but a lack of scalability. The following part of this section is dedicated to one of the possible image-based solutions.

In this proposition, the goal is to find videos with a very strong similarity: copies, that is, near duplicates. Digging out copies can be useful in many automatic processes for organizing and preserving the contents: merge annotations, clustering, cleaning (by deleting some duplicates), and classifying (using nodes and links attributes). Moreover, once links are found, some mining works can be done to characterize them and the videos. Now, given the massive amount of video available contents, a scalable approach is mandatory for finding the links between videos.

A cross-dimensional index structure (Poullot et al., 2008, 2009) based on the local description and image plane is introduced here. It is designed for images but can be used for video (an image stream) by extension. The structure can easily be generalized to many applications; here, it is applied to a content-based video copy detection framework.

1.5.2 GLOCAL DESCRIPTION

The Glocal description relies on a binary quantization of the local description space. This space is considered a cell, and each step of the quantization splits the cell(s) of the previous step along one dimension. The number of cell is $N_c = 2^h$, where h is the quantization step. Figure 1.10 shows three possible quantizations of a description space (symbolized in 2D here, and should be 128D for a SIFT case) containing six local descriptors. The resulting Glocal vector is binary, the presence or absence of descriptors in the cells is quoted. It is not a histogram, so it is not influenced by the textures and repeated patterns, which is the inverse to standard BoW features.

The Dice coefficient is used to measure the similarity between Glocal signatures, $S_{\text{Dice}}(g_1, g_2) = 2 \times (G_1 \cap G_2)/(|G_1| + |G_2|)$, where G_i is a set of bits set to one in the signature g_i and $|G_i|$ denotes the set cardinality.

There must be a balance between the number of local features used and the quantization in order to obtain discriminant vectors. Therefore, sparsity is needed, for example, in Figure 1.10, $N_c = 16$ at minimum is needed. For the case of a copy

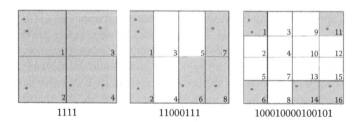

FIGURE 1.10   Three different quantizations of six local descriptors at $N_c = 4, 8, 16$.

detection, using $N_a \in [100, 200]$ local descriptors per keyframe and $N_c = 1024$ has shown good results. However, for an object search application, a lot more local descriptors must be picked up in the frames, hundreds (or more) of them, and thus N_c must be increased to more than a thousand in order to preserve the sparsity and discriminating power of the vectors.

Two types of descriptors were used in our experiments, a local jet spatio-temporal descriptor, which is a 20-dimensional vector (see Gouet and Boujemaa, 2001 for details) and SIFT. The first one is computed at the Point of Interest (PoI) detected by the Harris Corner Detector. SIFT is computed at PoI detected by using the Difference of Gaussian. However, using raw SIFTs for building the Glocal descriptors is not a good option, because of the sparsity of the SIFT description space itself. So, the SIFTs are first randomly projected in a subspace of 32 dimensions that maintains the intrinsic dimensionality of the SIFTs.

Then, for the rough local jet and projected SIFT descriptors, the components are ordered by decreasing uniformity (computed on a set of random descriptors). Indeed, as only the 10th or so first dimensions are used for the quantization, it is better to work on the ones that offer the best balance and separability.

Only the keyframes are exploited to create the descriptors to improve the scalability of the system. They are automatically detected in the videos using the global luminance variation, and the appearance rate is 3000 per hour (a little <1 keyframe per second, very much less than the usual 25-fps rate).

1.5.3 Cross-Dimensional Indexing

Once the Glocal descriptors have been computed for the entire video set, the indexing is performed. The mining task has a quadratic complexity ($\Theta(n^2)$) where n is the number of items). Only the keyframes and a Glocal description are used to reduce n and the complexity of the task. Indexing aims at further reducing the final complexity. The idea is to construct buckets based on the N-grams of a bit set to one in the binary descriptor. This scheme was already tested before by picking the bits set to one depending on their position in the binary vector itself. For instance, take the positions of the three following bits set to one to build a triplet. This triplet is associated to the bucket where the Glocal descriptor is inserted. In that way, the triplets 1-6-11, 6-11-14, and 11-14-16 from Figure 1.10 can be obtained, and four buckets in which the Glocal descriptor of the keyframe is inserted.

In the case of strong occlusions (half picture for example), this generally leads to a loss of detection. That is not surprising as the local descriptors are not associated considering their position in the keyframe but in the description space for indexing. A dual space indexing which uses both the position in the keyframe and in the description space is then more likely to be relevant, that is, stronger to many postproduction transformations observed on copies, as occlusion, crop, subtitles, inlays, and so on. The recall gets better, and, crossing two types of information for indexing precision does as well. Note also that in the case of an object search, this solution will also be much more relevant, and will be able to focus on only a small part of the image.

The local descriptors in a keyframe may describe the background, a moving object, a standing character, and so on. Linking the local descriptors to objects is very difficult, and thus, the community for this issue has not yet been determined. It would require perfect segmentation and good tracking. Moreover, these tasks are likely to have higher computational costs, and so, are not suitable for the scalability issue. However, using these relations between the local descriptors would be very powerful. In order to exploit this relation, a simple assumption can be made: *if the positions of the PoI are close to each other the local descriptors computed around are more likely to describe the same object in the keyframe.* Consequently, descriptors that are close must be associated, and we propose using these associations in the indexing scheme. The quantizations of the local features (the bits set to one in the binary vector) are thus associated depending on the positions of their PoI in the keyframe for building the N-grams. The indexing relies on a 2-fold association that crosses the description and keyframe spaces.

Let say $N = 3$, the 3-grams are built this way: take a PoI, find its nearest neighbor in the keyframe (using L2 distance in the image plane), then associate it with 1-NN and 2-NN, and with 3-NN and 4-NN, and so on. The resulting 3-grams are the quantizations of these local descriptors associations. Of course N can be increased, but the higher it is, the stronger the constraints are: N PoI neighbors must be preserved and the descriptors computed at these positions must be quantized in the same way.

As PoI or SIFT matching can be lost between two versions of a video, creating only one N-gram per local descriptor is still quite risky and unstable for retrieval purposes. Some more redundancy can be injected in the index for improving the recall. For each descriptor, an unlimited number of N-grams can be set up using more or less neighbors. The more N-grams that are chosen, the better the recall is and the lower the scalability is. In Figure 1.11, on the left is a set of PoI, and three N-grams from two of them, and on the right the N-grams resulting from the deletion

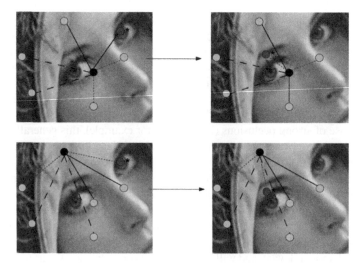

FIGURE 1.11 Set of points with three triplet associations, centered on two different PoI. The original associations are on the left, and the deletions and insertions of one point are on the right.

of one PoI and the insertion of another one (quoted with *). It can be seen that at the top one out of three combinations is kept, and at the bottom, two out of three are kept.

It was also quite interesting to find that depending on the PoI, the N-grams are not redundant between each other, offering a widespread of the keyframes in the index.

The sparseness of the database can be evaluated along with the cost of a query in such a system. Let us assume that 150 PoI are extracted per keyframe, and 3 grams are built for each, then the Glocal descriptor is then inserted in a maximum of 450 buckets. For $N_c = 2048$, there are $N_b = (2048!/3!) \approx 1.42 \times 10^9$ possible buckets. For a 15,000-h video database, about 4.5×10^7 keyframes are extracted, so about 2×10^9 Glocal descriptors are stored in the buckets. If equiprobability of the quantization is respected, each bucket on an average contains about 15 Glocal descriptors. Now, for a picture query, under the same conditions, 450 buckets, and 6750 comparisons will be performed to find similar keyframes. If we compare this to the number of keyframes in the database, and so to the number of necessary comparisons in the case of a sequential search, we have a computation ratio of $r_c = 1.5 \times 10^{-4}$ (0.015% of the computations are performed compared to the sequential scan).

Some statistics were computed on a real 15,000-h database in order to get the true distribution of the bucket population. It contains 5.2×10^7 KF and 1.2×10^{10} Glocal descriptors. Gathering the population of the 450 hundred most populated buckets (the worst scenario) leads to a set of 4×10^7 Glocal descriptors. In the worst case, 77% of the exhaustive comparisons are performed for an image query. In particular, when performing a set of 600 queries, we have an average of 73,000 comparisons per picture, which is far below the extreme value: 0.14% of the comparisons that are performed for each picture. However, this value is still 10 times larger than the expected theoretical one due to the imbalance in the number of buckets.

However, we can apply a cut in the database by removing the bigger buckets. They can actually be seen as "stop words," very noisy, and not informative. If 99.9% of the information (population of the database) is kept, the 450 most populated buckets contain 1.6×10^6 Glocal descriptors. Then, in the worst case, only 3.5% of the sequential comparisons are computed.

1.5.4 SHAPE EMBEDDING

A shape descriptor is embedded during the indexing to more precisely improve our results. The shape descriptor aims at characterizing the spatial conformation in the keyframe of the associated local descriptors used for defining an N-gram. A fuzzy comparison of two shape descriptors can then be used to discard some false-positive matching, and thus, improving the precision. Moreover, this test can be done before the similarity computation, as a filtering step, and, so, reduces the processing time.

We propose using $r_{ij} = (sizeOfLongerSide/sizeOfSmallerSide)$, where *sizeOfSmallerSide* is the smaller length between the PoI used for building the N-gram and *sizeOfLongerSide* is the longer one. Currently, 3 grams are mainly used, so this corresponds to the ratio between the smaller on the longer edges of the triangle. It must be noted that this shape descriptor is robust to scale changes, rotations, and small affine transformations. This descriptor is then a metadata associated with a Glocal

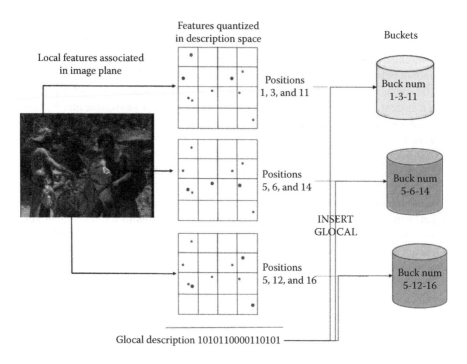

FIGURE 1.12 Indexing of the Glocal description of a keyframe in three different buckets along with three different shape codes. (From NISV, Sound and Vision Video Collection 2008. With permission.)

descriptor for a given N-gram. For indexing, the Glocal descriptor is inserted in the bucket corresponding to this N-gram along with this shape descriptor. Figure 1.12 sums up the indexing process.

In the mining stage, or query stage, when comparing two Glocal descriptors lying in the same bucket, a fuzzy comparison between their shape descriptors is performed first. Basically, the test is to see if $|(s_1 - s_2)| \le \theta$ returns true, where s_1 is the shape descriptor of Glocal descriptor 1 and s_2 the shape descriptor of Glocal descriptor 2. If the test is successful (if the shape descriptors are more or less equal), then the similarity between the two Glocal descriptors is computed. The fuzzy test adds some robustness to the shape descriptor. It also can be adjusted for more or less tolerance. Typically $\theta = 0.5$ is used for our test on the CBCD TRECVID task. For object search, it should be lower than for video copy detection.

In the 15,000-h database, the number of avoided distance computations between the Glocal descriptors was counted in the real case of 600 query pictures. Using all the buckets, 75% of the computations were skipped, and 47% were skipped when removing the "stop words" bucket. These statistics show that the matchings due to "stop words" can also be discarded by using the shape description.

Finally, if a similarity between two Glocal descriptors is computed, in order to limit the number of results, a threshold Th is used, and the pair of frame (F_1, F_2) is kept if $S_{\text{Dice}}(G_{F_1}, G_{F_2}) \ge$ Th where G_{F_i} is the Glocal descriptor of frame F_i.

1.5.5 Temporal Consistency

The connections identified between individual keyframes are used to delimit and link together the video sequences that are transformed versions of the same content. Starting from the two connected keyframes, two adjoined sequences are built by using the stepwise addition of the other connected keyframes (with increasing time codes) that verify the temporal consistency conditions. These conditions make the detection more robust to the absence of a few connected keyframes and to the presence of some false-positive detections.

The first requirement is that the temporal gap between the last keyframe in a sequence (with time code $Tc_{x,l}$) and a candidate keyframe to be added to the same sequence (with time code $Tc_{x,c}$) should be lower than a given threshold g: $Tc_{x,c} - Tc_{x,l} < g$. The gaps are due to the absence of several connected keyframes, as a result of the postprocessing operations (addition or removal of several frames), of any instabilities of the keyframe detector, or of any false negatives in the detection of connected keyframes.

The second requirement bounds the variation of the temporal offset (jitter) between the connected keyframes of two different sequences of ID_x and ID_y. Jitter is caused by postprocessing operations or by instabilities in the keyframe detector. If $Tc_{x,l}$ is the time code of the last keyframe in ID_x, $Tc_{y,l}$ is the time code of the last keyframe in ID_y, $Tc_{x,c}$ and $Tc_{y,c}$ are the time codes of the candidate keyframes, then the condition for upper bounding the jitter by using τ_j is: $|(Tc_{x,c} - Tc_{x,l}) - (Tc_{y,c} - Tc_{y,l})| < \tau_j$.

The candidate keyframes are added at the end of the current sequences only if both the gap and the jitter conditions are satisfied. And, finally, the third condition is that the sequences should be longer than a minimal value τ_j to be considered valid; this removes the very short detections, which are typically false positives.

1.5.6 Experiments and Results

Since the scalability is our main focus in this work, the tests were mainly driven on a laptop PC, a Samsung Q310, processor dual core Q8600 at 2.6 GHz with 4 Gb of RAM. No parallel version of the methods for using both of the cores was used.

1.5.6.1 Quality

We used the TRECVID 2008 CBCD benchmark to experiment with the proposed description and indexing approach. This benchmark is composed of a 207-h video reference database and 2010 artificially generated query video segments (or 43 h), 1340 of which contain copies. One query can only correspond to one segment of one video from the reference database. A query not only can contain a copied part of a reference video, but also the video segments that are not related to the reference database. The queries are divided into 10 groups of 134 queries each, each group corresponding to one type of transformation ("attack"): camcording (group 1), picture-in-picture type 1 (group 2), insertion of pattern (group 3), strong reencoding (group 4), change of gamma (group 5), three combined basic transformations (group 6), five combined basic transformations (group 7), three combined postproduction transformations (group 8), five combined postproduction transformations (group 9),

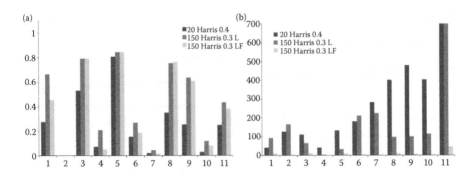

FIGURE 1.13 Comparison of recall on TV2008 CBCD benchmark (a); comparison of absolute number of false alarms (b). The 10 first sets of columns correspond to the attack groups, and the 11th one is the average (a) and absolute sum (b).

and a combination of all the transformations (group 10). The "basic" transformations are: blur, gamma, frame dropping, contrast, compression, scaling, and noise. The postproduction transformations are: crop, shift, contrast, caption, flip, insertion of pattern, and picture in picture (insertion of a shrunk version of an original video in a larger one). Since many queries include a left-right flip to which the local features used here are not robust, the features were detected and described on both the initial queries and a flipped version of the queries (so there are 4020 queries and a total of 293 h of video).

Three different methods were used for testing our proposals. The first one is the reference method described by Poullot et al. (2008) with the following settings: $N_a = 20$, $N_c = 256$, Th = 0.4. The other two methods use the following settings: $N_a = 150$, $N_c = 1024$, Th = 0.3, and cross-space indexing. However, one uses shape embedding (denoted as "locality and shape") while the other does not (denoted as "locality"). These settings were chosen because they offered the best compromises between precision and recall (a study was first driven and $N_a \in [20, 200]$, $N_c \in [256, 512, 1024]$, Th $\in [0.2, 0.5]$ were tested). The feature used was the local jet spatio-temporal descriptor computed at Harris Corner positions.

Figure 1.13 shows the recall and number of false alarms for each method. It can be seen that the recall is improved by using more local features and the locality-based bucket selection (L); the addition of local geometric information (distance ratios in LF) has a weak impact on the recall, lowering it for some specific transformations (especially camcording and strong reencoding). The most important contribution of the geometric information was the significant reduction in the number of false positives, despite a lower decision threshold.

The examples in Figure 1.14 were not detected if cross-dimensional hashing was not used; we should note that large inlays were inserted in these examples, and thus we found that using the locality information is indeed very relevant.

A comparison between two features (Section 1.5.2) is shown in Figure 1.15, for the method using both the locality and distance ratios (LF). We set $N_a = 150$ and, since the precision was high, the decision threshold was reduced to Th = 0.2. With these parameter values, the SIFT features provided on average a slightly lower overall recall and equivalent precision.

FIGURE 1.14   Three examples of detected copies (original image is on right). (From NISV, Sound and Vision Video Collection 2008. With permission.)

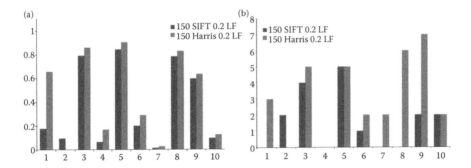

FIGURE 1.15   (a) comparison of recall using SIFT and Local Jets. (b) comparison of absolute number of false alarms. The 10 groups correspond to the attack groups.

1.5.6.2   Scalability

We used two larger datasets, which were some continuous recordings of five different Japanese TV channels over several days for experimenting scalability of the approach. Overall, the databases, respectively, contained 1000 and 3000 h of video. In this set, we found a lot of redundancy (jingles, advertising, credits, weather forecast, reporting, and so on), but most of the transformations were light and corresponded to the insertion of logos or design changes and also to time-line editing. For performing a real large-scale test, a self-join of the database was performed: use all the videos inserted in the database as a query. Once the database is built, the Glocal descriptors in each bucket, according to the shape descriptor condition, are compared to the lying ones. Then, the time consistency step (Section 1.5.5) is applied.

Concerning some main memory issues, the 3000-h database was built on a computer with an X7460 single core CPU at 2.66 GHz and 8 Gb of RAM. For the time comparison with the building step, the self-join was also performed on the same computer. Table 1.2 shows the computation time for these operations. The time required to build the database depends only on the number of features selected on the frames and the number of frames, and this is a linear dependency. The self-join similarity is supposed to have a quadratic complexity, but when using an index structure, the real complexity is far below this limit. Using the locality for indexing does not change this complexity; it just improves the recall, as seen in the previous quality part. On the other hand, using the shape description, the complexity is drastically reduced, and thus, so is the required time. In this case, many of the distance computations between Glocal Descriptors are cut off (Section 1.5.4).

TABLE 1.2
Building and Self-Join Operation on Large
Video Databases

Method	Operation	Database Size	
		1000 h	3000 h
20_Harris_0.4	Build	15 min	21 min
	Self-join	2 h 01 min	12 h 00 min
100_Harris_0.3	Build	1 h 57 min	3 h 04 min
Locality	Self-join	7 h 30 min	40 h 00 min
Locality and shape	Self-join	1 h 17 min	0 h 54 min
150_Harris_0.3	Build	3 h 11 min	5 h 01 min
Locality	Self-join	8 h 48 min	41 h 10 min
Locality and shape	Self-join	2 h 21 min	1 h 34 min

1.5.7 CONCLUSIONS

Content-based video copy detection can provide useful information for browsing and structuring video databases, for large institutional archives as well as for video-sharing websites. The challenge is to be both fast and reliable even when the transformations between original videos and copies are strong.

On the basis of the local descriptions of the video frames and a BoF approach, information regarding the geometric configuration of these features in the image plane is also exploited in order to obtain a sufficient level of reliability and scalability, resulting in a cross-dimensional hashing process. Since many transformations like strong cropping and video inlays alter the longer-range structure of the frames but maintain part of the short-range structure, we suggest taking into account the locality in the image plane (nearest neighbors of a feature) when indexing the video frames. Moreover, some simple local geometric data can be used to further discriminate the frames hashed in the same bucket and accelerate the matching computations. These data are selected to be as robust as possible to the most common types of image transformations.

An experimental evaluation of the detection quality of our proposal was conducted on the TRECVID 2008 copy-detection benchmark, using two different types of features, and we found that there was a significant improvement over the previous method. The scalability was then assessed on larger databases of up to 3000 h of video and the fact that the computation time was greatly reduced by the filtering operation exploiting the local geometric information was highlighted.

1.6 CONCLUSIONS AND FUTURE ISSUES

The main elements of an MDM system include the sampling and description of videos, the indexing structure, and the retrieval process, which depend on the application purposes. On the basis of these claims, this chapter discusses the requirements and selection criteria of MDM approaches in terms of the robustness, discriminative power, scalability, and so on. It also provided a review of three promising MDM systems and their applications to knowledge discovery. The experimental evaluations and the application cases of these systems demonstrated the high applicability and potential of this research domain. Future research is expected to contribute toward achieving a good tradeoff between the robustness and discriminative power, reducing the storage space and memory usage required for video sampling and indexing, and getting a good trade-off between the effectiveness and efficiency. It will be also interesting to see what will happen if we combine the watermarking and MDM strategies in the rich area of digital rights management and copyright violation detection. Although MDM is still in its early stages, the urgent need for practical applications in this area as well as in market research and broadcast media research makes it a hot topic, and numerous innovations are expected in the near future. We hope this chapter will provide some useful information to readers interested in this research field.

ACKNOWLEDGMENT

Sound and Vision video is copyrighted. The Sound and Vision video used in this work is provided solely for research purposes through the TREC Video Information Retrieval Evaluation Project Collection.

TERMINOLOGY INDEXES

Commercial film
Cross dimensional indexing
Data mining
Duplicate detection
Fast quantization
Fingerprinting
Hashing algorithm
News story reranking
News story tracking
PageRank
Shape embedding

REFERENCES

Albiol, A., Ch, M. J., Albiol, F. A., and Torres, L., Detection of TV commercials, *IEEE International Conference on Acoustics, Speech, and Signal Processing*, Montreal, Qeubec, Canada, Vol. 3, pp. 541–544, May 2004.

Berrani, S.-A., Manson, G., and Lechat, P., A non-supervised approach for repeated sequence detection in TV broadcast streams, *Image Commun.*, 23(7), 525–537, 2008.

Brin, S. and Page, L., The anatomy of a large-scale hypertextual Web search engine, *Comput. Netw. ISDN Syst.*, 30(1–7), 107–117, 1998.

Cano, P., Batlle, E., Kalker, T., and Haitsma, J., A review of audio fingerprinting, *J. VLSI Signal Process. Syst.*, 41(3), 271–284, 2005.

Chieu, H. L. and Lee, Y. K., Query based event extraction along a timeline, *Proceedings of the 27th Annual International ACM SIGIR Conference on Research and Development in Information Retrieval*, Sheffield, UK, pp. 425–432, July 2004.

Datar, M., Immorlica, N., Indyk, P., and Mirrokni, V. S., Locality-sensitive hashing scheme based on p-stable distributions, *Proceedings of the Twentieth Annual Symposium on Computational Geometry*, Brooklyn, NY, USA, pp. 253–262, June 2004.

Dohring, I. and Lienhart, R., Mining TV broadcasts for recurring video sequences, *Proceedings of the ACM International Conference on Image and Video Retrieval*, Santorini, Fira, Greece, pp. 1–8, July 2009.

Duan, L.-Y., Wang, J., Zheng, Y., Jin, J. S., Lu, H., and Xu, C., Segmentation, categorization, and identification of commercial clips from TV streams using multimodal analysis, *Proceedings of the 14th Annual ACM International Conference on Multimedia*, Santa Barbara, CA, USA, pp. 201–210, October 2006.

Duygulu, P., Pan, J.-Y., and Forsyth, D. A., Towards auto-documentary: Tracking the evolution of news stories, *Proceedings of the 12th Annual ACM International Conference on Multimedia*, New York, NY, USA, pp. 820–827, October 2004.

Fischler, M. A. and Bolles, R. C., Random sample consensus: A paradigm for model fitting with applications to image analysis and automated cartography, *Commun. ACM*, 24(6), 381–395, 1981.

Gauch, J. M. and Shivadas, A., Finding and identifying unknown commercials using repeated video sequence detection, *Comput. Vis. Image Underst.*, 103(1), 80–88, July 2006.

Gouet, V. and Boujemaa, N., Object-based queries using color points of interest, *Proceedings of the IEEE Workshop on Content-based Access of Image and Video Libraries*, Kauai, Hawaii, pp. 30–36, December 2001.

Haitsma, J. and Kalker, T., A highly Robust audio fingerprinting system, *Third International Conference on Music Information Retrieval*, Paris, France, October 2002.

Hampapur, A., Hyun, K., and Bolle, R. M., Comparison of sequence matching techniques for video copy detection, *Storage and Retrieval for Media Databases*, San Jose, CA, USA, pp. 194–201, January 2002.

Herley, C., ARGOS: Automatically extracting repeating objects from multimedia streams, *IEEE Trans. Multimed.*, 8(1), 115–129, 2006.

Hua, X.-S., Lu, L., and Zhang, H.-J., Robust learning-based TV commercial detection, *IEEE International Conference on Multimedia and Expo*, Amsterdam, The Netherlands, pp. 149–152, July 2005.

Ide, I., Mo, H., Katayama, N., and Satoh, S., Topic threading for structuring a large-scale news video archive, *Image and Video Retrieval, Lecture Notes in Computer Science*, Springer, Berlin, Vol. 3115, pp. 123–131, 2004.

Jegou, H., Douze, M., and Schmid, C., Hamming embedding and weak geometric consistency for large scale image search, *Proceedings of the 10th European Conference on Computer Vision: Part I*, Marseille, France, pp. 304–317, October 2008.

Joly, A., Buisson, O., and Frelicot, C., Content-based copy retrieval using distortion-based probabilistic similarity search, *IEEE Trans. Multimed.*, 9(2), 293–306, February 2007.

Ke, Y., Sukthankar, R., and Huston, L., An efficient parts-based near-duplicate and sub-image retrieval system, *Proceedings of the 12th Annual ACM International Conference on Multimedia*, New York, NY, USA, pp. 869–876, October 2004.

Kumar, R., Mahadevan, U., and Sivakumar, D., A graph-theoretic approach to extract storylines from search results, *Proceedings of the Tenth ACM SIGKDD International Conference on Knowledge Discovery and Data Mining*, Seattle, WA, USA, pp. 216–225, August 2004.

Law-To, J., Chen, L., Joly, A., Laptev, I., Buisson, O., Gouet-Brunet, V., Boujemaa, N., and Stentiford, F., Video copy detection: A comparative study, *Proceedings of the 6th ACM*

International Conference on Image and Video Retrieval, Amsterdam, The Netherlands, pp. 371–378, July 2007.

Lee, Y., Jung, H., Song, W., and Lee, J.-H., Mining the blogosphere for top news stories identification, *Proceedings of the 33rd International ACM SIGIR Conference on Research and Development in Information Retrieval*, Geneva, Switzerland, pp. 395–402, July 2010.

Li, Y., Jin, J. S., and Zhou, X., Matching commercial clips from TV streams using a unique, robust and compact signature, *Proceedings of the Digital Image Computing on Techniques and Applications*, Cairns, Australia, pp. 355–362, December 2005.

Li, Y., Zhang, D., Zhou, X., and Jin, J. S., A confidence based recognition system for TV commercial extraction, *Proceedings of the Nineteenth Conference on Australasian Database—Volume 75*, Gold Coast, Australia, pp. 57–64, January 2007.

Lowe, D. G., Distinctive image features from scale-invariant keypoints, *Int. J. Comput. Vis.*, 60(2), pp. 91–110, November 2004.

Lv, Q., Josephson, W., Wang, Z., Charikar, M., and Li, K., Multi-probe LSH: Efficient indexing for high-dimensional similarity search, *Proceedings of the 33rd International Conference on Very Large Data Bases*, Vienna, Austria, pp. 950–961, September 2007.

Matas, J. and Chum, O., Randomized RANSAC with sequential probability ratio test, *Proceedings of the Tenth IEEE International Conference on Computer Vision—Volume 2*, Rio de Janeiro, Brazil, pp. 1727–1732, October 2005.

Mikolajczyk, K., Tuytelaars, T., Schmid, C., Zisserman, A., Matas, J., Schaffalitzky, F., Kadir, T., and Gool, L. V., A comparison of affine region detectors, *Int. J. Comput. Vis.*, 65(1–2), pp. 43–72, November 2005.

Ngo, C.-W., Zhao, W.-L., and Jiang, Y.-G., Fast tracking of near-duplicate keyframes in broadcast domain with transitivity propagation, *Proceedings of the 14th Annual ACM International Conference on Multimedia*, Santa Barbara, CA, USA, pp. 845–854, October 2006.

Philbin, J., Chum, O., Isard, M., Sivic, J., and Zisserman, A., Object retrieval with large vocabularies and fast spatial matching, *IEEE Conference on Computer Vision and Pattern Recognition*, Minneapolis, Minnesota, USA, pp. 1–8, June 2007.

Poullot, S., Crucianu, M., and Buisson, O., Scalable mining of large video databases using copy detection, *Proceedings of the 16th ACM International Conference on Multimedia*, Vancouver, British Columbia, Canada, pp. 61–70, October 2008.

Poullot, S., Crucianu, M., and Satoh, S., Indexing local configurations of features for scalable content-based video copy detection, *Proceedings of the First ACM Workshop on Large-Scale Multimedia Retrieval and Mining*, Beijing, China, pp. 43–50, October 2009.

Putpuek, N., Cooharojananone, N., Lursinsap, C., and Satoh, S., Unified approach to detection and identification of commercial films by temporal occurrence pattern, *Proceedings of the 2010 20th International Conference on Pattern Recognition*, Istanbul, Turkey, pp. 3288–3291, August 2010.

Satoh, S., News video analysis based on identical shot detection, *2002 IEEE International Conference on Multimedia and Expo*, Lausanne, Switzerland, Vol. 1, pp. 69–72, August 2002.

Turpin, A. and Scholer, F., User performance versus precision measures for simple search tasks, *Proceedings of the 29th Annual International ACM SIGIR Conference on Research and Development in Information Retrieval*, Seattle, Washington, USA, pp. 11–18, August 2006.

Wu, X., Hauptmann, A. G., and Ngo, C.-W., Novelty detection for cross-lingual news stories with visual duplicates and speech transcripts, *Proceedings of the 15th International Conference on Multimedia*, Augsburg, Germany, pp. 168–177, September 2007a.

Wu, X., Hauptmann, A. G., and Ngo, C.-W., Practical elimination of near-duplicates from web video search, *Proceedings of the 15th International Conference on Multimedia*, Augsburg, Germany, pp. 218–227, September 2007b.

Wu, X., Ngo C.-W., and Hauptmann, A.G., Multimodal news story clustering with pairwise visual near-duplicate constraint, *IEEE Trans. Multimed.*, 10(2), pp. 188–199, February 2008a.

Wu, X., Hauptmann, A. G., and Ngo, C.-W., Measuring novelty and redundancy with multiple modalities in cross-lingual broadcast news, *Comput. Vis. Image Underst.*, 110(3), pp. 418–431, June 2008b.

Wu, X., Ide, I., and Satoh, S., PageRank with text similarity and video near-duplicate constraints for news story re-ranking, *Advances in Multimedia Modeling, Lecture Notes in Computer Science*, Springer, Berlin, Vol. 5916, pp. 533–544, 2010.

Wu, X. and Satoh, S., Temporal recurrence hashing algorithm for mining commercials from multimedia streams, *2011 IEEE International Conference on Acoustics Speech and Signal Processing*, Prague, Czech Republic, pp. 2324–2327, May 2011.

Yuan, J., Wu, Y., and Yang M., Discovery of collocation patterns: From visual words to visual phrases, *IEEE Conference on Computer Vision and Pattern Recognition*, Minneapolis, MN, USA, pp. 1–8, June 2007.

Zhai, Y. and Shah, M., Tracking news stories across different sources, *Proceedings of the 13th Annual ACM International Conference on Multimedia*, Hilton, Singapore, pp. 2–10, November 2005.

2 Discriminative Learning-Assisted Video Semantic Concept Classification

Qiusha Zhu, Mei-Ling Shyu, and Shu-Ching Chen

CONTENTS

2.1 INTRODUCTION

With the fast development and wide usage of digital devices, more and more people like to record their daily lives as videos and share them on websites such as YouTube, Yahoo Video, and Youku, to name a few. To get a better understanding of such video data and discover useful knowledge from them, effective and automatic multimedia semantic analysis becomes crucial. However, a manual analysis of multimedia data can be very expensive or simply not feasible when the time is limited or when the amount of data is enormous. Hence, the multimedia research community faces a major challenge: how to effectively and efficiently organize these videos. Data mining techniques have been successfully utilized to provide solutions to such a challenge. An example is video concept classification, which has been an attractive research focus in the past 10 years (Chen et al., 2006a; Shyu et al., 2008).

When extracting useful knowledge from the content of a video, it further brings up several issues such as semantic gap, imbalanced data, high-dimensional feature space, and varied qualities of the media (Chen et al., 2006b). Semantic gap refers to the gap between low-level visual features like color, texture, shape, etc., and high-level semantic concepts like boats, outdoors, streets, etc. (Hauptmann et al., 2007), which is produced by "the lack of coincidence between the information that one can extract from the visual data and the interpretation that the same data have for a user in a given situation" (Smeulders et al., 2000, p. 1353). A lot of effort has been put into bridging this semantic gap, but it is still difficult to conquer (Naphade et al., 2006). Next is the data imbalance issue that one usually encounters in video concept classification. It is very often that the number of videos of a particular concept we are interested in (called target/positive concept) is much less than the number of videos of the nontarget/negative concepts. The training process using a very small number of positive training data instances (i.e., data instances belonging to the target concept) and an extremely large number of negative (and usually noisy) training data instances (i.e., data instances belonging to nontarget concepts) typically results in poor or unsatisfactory classification performance. In the meantime, several new features are being developed to try to capture the semantic meaning of a video. For example, scale-invariant feature transform (SIFT) is one of the most well-known feature descriptors (Lowe, 2004), which uses the bag-of-words (BOW) model that usually represents an image in thousands of visual words. This could easily result in a high-dimensional feature space and cause overfitting due to "curse of dimensionality."

To overcome the aforementioned issues, one approach is to select a good representative subset of features from the original feature space. Such a process can reduce the feature space by removing irrelevant, redundant, or noisy features to improve classification performance and train a model to be more cost-effective. Feature selection can also preserve the semantics of the features compared with other dimensionality reduction techniques. Instead of altering the original representation of features like those based on projection (e.g., principal component analysis (PCA)) and compression (e.g., information theory) (Saeys et al., 2007), feature selection eliminates those features with little predictive information, keeps those with a better representation of the underlying data structure, and thus enhances the discriminative ability and semantic interpretability of the model. An effective representative subset of features should not contain (i) noisy features that decrease the classification accuracy, or (ii) irrelevant features that increase the computation time. Thus, a good feature selection method can intrinsically help video concept classification overcome these challenges. Feature selection techniques are also adopted in other areas as a pre-process step in order to improve model performance. An extensive comparison of 12 feature selection metrics for the high-dimensional domain of text classification was presented in (Forman, 2003), including chi-square measure (CHI), information gain (IG), document frequency, bi-normal separation (BNS), etc. These methods were evaluated on a benchmark of 229 text classification problem instances and the results showed that BNS was the top choice for evaluation criteria like accuracy, F1_score, and recall. Hua et al. (2009) compared several famous feature selection methods in the area of bioinformatics, including IG, gini index, t-test, sequential forward selection (SFS), and so on. Feature selection in this area is inevitable but quite challenging because biological technologies usually produce data with thousands of features but a relatively small sample size. Experiments on both synthetic and real data showed that none of these methods performed best across all scenarios, and revealed some trends relative to the sample size and relations among the features.

Another approach to overcome the challenges in multimedia semantic analysis is to improve inductive learning techniques and construct more powerful classifiers. Traditional classifiers like Decision Tree (DT), Native Bayes (NB), k-Nearest Neighbor (k-NN), and so on are considered as weak classifiers due to their low complexity, or unstable or poor performance. Multiple approaches have been proposed to build more powerful classifiers by combining weak classifiers. Boosting and bagging are two of the most popular combining strategies. The idea of bagging (Bressan, 1996) is to generate random bootstrap replicates from the training set, construct a classifier on each subset, and then combine them using the majority vote. The original boosting was proposed by Freund and Schapire (1997) which incrementally builds the final classifier by adding a new weak classifier at each step. When adding the weak classifiers, they are typically weighted in some way that is usually related to the weak classifiers' accuracy. The data are reweighed too: samples that are misclassified gain weights and samples that are classified correctly lose weights, so future weak classifiers could focus more on the "difficult" samples. After repeating this step a certain number of times, the final decision rule is constructed by weighting the weak classifiers at each step.

In this chapter, a new framework is proposed for video concept classification with the help of discriminative learning and multiple correspondence analysis (MCA) (Greenacre and Blaslus, 2006). Since features extracted from raw videos are numeric (e.g., feature "shot length" has many values: 2.5, 24.12, 6.0, etc.), as a straightforward approach, PCA has been used as a feature selection method in many literatures (Lu et al., 2007). However, it alters the original representation of the features by projecting all of them into a low-dimensional space and combining them linearly. MCA, on the other hand, is designed for nominal data (e.g., feature "outlook" has three items: "sunny," "overcast," and "rainy"), but has been proved to be able to effectively capture the correlation between a feature and a class (Lin et al., 2008a,b; Zhu et al., 2010). Thus, MCA could be considered as a potentially better approach since, by choosing a subset from the original feature space, the semantic meaning of the feature is retained. The cosine value of the angle between a feature item and a class generated by MCA can indicate the strength of the relation between these two variables. In statistics, another property of a relation is reliability, which is called p-value or statistical significance. This motivates us to integrate correlation and reliability information represented by the cosine angle values and p-values to measure the relation between a feature and a class. Please note that the focus is on the problem of supervised learning, which means that the classification is carried out with the labels of the training data set known beforehand. Based on the metric that integrates the cosine angle values and p-values generated from MCA, a score is calculated for each feature with respect to each class, and therefore two feature lists are generated according to the ranking scores, one for each class. After employing MCA to select two discriminative sets of features from the original extracted features of the training data set, the features for the target concept are used to train the MCA-based positive model; while the features for the nontarget concepts are used to train the MCA-based negative model. To reach a better performance, a strategy is introduced that utilizes the dual-model for concept classification in the testing data set.

To evaluate the proposed discriminative learning framework, four widely used feature selection methods for supervised learning, namely the IG, CHI, correlation-based feature selection (CFS), and relief filter (REF) methods, available in WEKA (Witten and Frank, 2005) are used as a pre-processing step for five well-known classifiers, respectively: DT, Rule-based JRip (JRip), NB, Adaptive Boosting (Adaboost), and k-NN classifiers. The data used in the experiments is from TRECVID 2009 high-level feature extraction task, where the high-level features are actually the concepts to be detected in video shots. The overall experimental results demonstrate that the proposed framework which integrates the angle values and p-values generated from MCA together with the utilization of a dual model performs better than the other four feature selection methods in combination with the five classifiers in terms of the F1_scores.

This chapter is organized as follows. Related work is introduced in Section 2.2. The proposed framework is presented in Section 2.3, followed by an analysis of the experimental results in Section 2.4. Finally, Section 2.5 concludes the chapter.

2.2 RELATED WORK

Since feature selection and classification are two main components of the proposed framework, this section introduces related work in these two areas. It first describes different categories of supervised feature selection methods and then focuses on four commonly used ones: the IG, CHI, CFS, and REF. After feature selection, the training data set with selected features is used to train the classifiers which would classify the testing data set. Five well-known basic classifiers, namely DT, JRip, NB, Adaboost, and k-NN, are introduced, which are used in comparison with the proposed framework in Section 2.4.

2.2.1 FEATURE SELECTION

Depending on how it is combined with the construction of the classification model, supervised feature selection can be further divided into three categories: wrapper methods, embedded methods, and filter methods. Wrappers choose feature subsets with high prediction performance estimated by a specified learning algorithm which acts as a black box, and thus wrappers are often criticized for their massive amounts of computation which are not necessary. Similar to wrappers, embedded methods incorporate feature selection into the process of training a given learning algorithm, and thus they have the advantages of interacting with the classification model and meanwhile being less computationally intensive than wrappers. These two categories usually yield better classification results than the filter methods, since they are tailored to a specific classifier, but the improvements of the performance are not always significant because of "curse of dimensionality" and the fact that the specific-tuned classifiers may overfit the data. In contrast, filter methods are independent of the classifiers and can be scaled to high-dimensional datasets while remaining computationally efficient. In addition, filtering can be used as a pre-processing step to reduce space dimensionality and overcome the overfitting problem. Therefore, filter methods only need to be executed once, and then different classifiers can be evaluated based on the selected feature subsets.

Filter methods can be further divided two main subcategories. The first one is univariate methods which consider each feature with the class separately and ignore the interdependence between the features. Representative methods in this category include IG and CHI, both of which are widely used to measure the dependence of two random variables. IG evaluates the importance of features by calculating their IG with the class, but this method is biased to features with more values. In (Lee and Lee, 2006), a new feature selection method was proposed which selected features according to a combined criterion of IG and novelty of information. This criterion strives to reduce the redundancy between features, while maintaining IG in selecting appropriate features. In contrast, CHI calculates the chi-square statistics between each feature and the class, and a large value indicates a strong correlation between them. Although this method does not adhere strictly to the statistics theory because the probability of errors increases when a statistical test is used multiple times, it is applicable as long as it only ranks features with respect to their usefulness (Manning et al., 2008). Jiang et al. (2010) used the bag-of-visual-words features to represent keypoints in images for semantic concept detection. Feature selection applied the CHI to calculate the chi-square statistics between a specific visual word and a binary label of an image class, and eliminated those virtual words with chi-square statistics below a threshold. Extensive experiments on the TRECVID data indicated that the bag-of-visual-words features with an appropriate feature selection choice could produce highly competitive results.

The second subcategory is the multivariate methods which take features' interdependence into account. However, they are slower and less-scalable compared with the univariate methods. CFS is one of the most popular methods. It searches among the features according to the degree of redundancy between them in order to find a subset of features that are highly correlated with the class, yet uncorrelated with each other (Hall, 2000). Experiments on natural datasets showed that CFS typically eliminated over half of the features, and the classification accuracy using the reduced feature set was usually equal to or better than the accuracy using the complete feature set. The disadvantage is that CFS degrades the performance of classifiers in cases where some eliminated features are highly predictive of very small areas of the instance space. This kind of cases could be frequently encountered when dealing with imbalanced data. Relief (Robnik-Sikonja and Kononenko, 1997) is another commonly used method whose idea is to choose the features that can be most distinguishable between classes. It evaluates the worth of a feature by repeatedly sampling an instance and considering the value of the given feature for the nearest instance of the same and different classes. However, relief lacks a mechanism to deal with the outlier instances, and according to Hua et al. (2009), it has worse performance than the univariate filter methods in most cases. Sun (2007) proposed an iterative relief (I-Relief) method by exploring the framework of the Expectation-Maximization algorithm. Large-scale experiments conducted on nine UCI datasets and six microarray datasets demonstrated that I-Relief performed better than relief without introducing a large increase in computational complexity.

According to the form of the outputs, the four aforementioned feature selection methods can also be categorized into ranker and nonranker. A nonranker method provides a subset of features automatically without giving an order of the selected features such as CFS. On the other hand, a ranker method provides a ranking list by scoring the features based on a certain metric, to which IG, CHI, and relief belong. Then different stopping criteria can be applied in order to get a subset from it. Most commonly used criteria include forward selection, backward elimination, bidirectional search, setting a threshold, genetic search, and so on.

2.2.2 CLASSIFIER LEARNING

There are several categories of inductive learning algorithms, such as Bayesian classifiers, trees, rules, and lazy classifiers. A popular data mining software WEKA has the implementation of several classification algorithms for each category. The five classifiers used in this chapter are listed as follows:

- *DT*: J48 implements C4.5 (Quinlan, 1993) which is a tree structure where leaf nodes represent class labels and branches represent conjunctions of features that lead to those class labels. At each node of the tree, C4.5 chooses one attribute of the data that most effectively splits its set of samples into subsets enriched in one class or the other. Its criterion is the normalized IG (difference in entropy) that results from choosing an attribute for splitting the data. The attribute with the highest normalized IG is chosen to make the decision. The C4.5 algorithm then recurs on the smaller subtrees.
- *JRip*: implements a propositional rule learner, repeated incremental pruning to produce error reduction (RIPPER), which was proposed by Cohen (1995). The algorithm contains a rule building stage and a rule optimization stage. The building stage repeats the rule grow and prune phase until the description length (DL) of the rule set and the examples is 64 bits greater than the smallest DL met so far, or there are no positive examples, or the error rate $\geq 50\%$.
- *Naïve Bayes*: John and Langley (1995) implemented the probabilistic Naive Bayes classifier and uses kernel density estimators to improve the performance if the normality assumption is grossly incorrect. In spite of its naive design and apparently over-simplified assumptions, naive Bayes classifier only requires a small amount of training data to estimate the parameters (means and variances of the variables) necessary for classification. Because independent variables are assumed, only the variances of the variables for each class need to be determined and not the entire covariance matrix.
- *Adaptive boost (Adaboost)*: Being considered as the best method for boosting (Freund and Schapire, 1997) which is used to combine weak classifiers, it assigns weights to the training data instances and iteratively combines the output of individual classifiers. AdaBoost is adaptive in the sense that subsequent classifiers built are tweaked in favor of those instances misclassified by the previous classifiers. It can only handle nominal class problems and is sensitive to noisy data and outliers, but it often dramatically improves performance.

- *k-NN*: is a type of instance-based learning or lazy learning which stores the training data instances and does no real work until the classification time. IBK (Aha and Kibler, 1991) is a *k*-nearest-neighbor classifier that finds the *k* training data instances closest in the Euclidean distance to the testing data instance. A testing data instance is classified by a majority vote of its neighbors, with the instance being assigned to the class most common amongst its *k* nearest training neighbors. Predictions from *k* neighbors can be weighted according to their distances to the testing data instance.

2.3   THE DISCRIMINATIVE LEARNING FRAMEWORK

The proposed discriminative learning framework is shown in Figure 2.1. It mainly contains two components: MCA-based feature selection and MCA-based dual-model classification. First, visual features are extracted from raw videos. In order to evaluate the framework, three-fold cross validation is adopted to split the data into training data set and testing data set. So the whole data set of each concept is randomly split into three sets with an approximately equal number of instances and equal positive to negative ratio. Each fold uses two of three sets as the training data set and the remaining one as the testing data set. Information entropy maximization (IEM) (Fayyad and Irani, 1993), a widely used supervised discretization based on minimum description length (MDL), also available in WEKA, is applied to the training set to discretize numeric features into nominal ones, and the same partitions are applied on the testing set.

Next, MCA-based feature selection is performed on the training set, which is enclosed in the dashed rectangular boxes labeled as (1). The correlation and reliability information generated from MCA are utilized to select two discriminative sets of features, one set for the positive class and the other one for the negative class. For the testing set, two sets of features same as the ones got from the training set are selected. The component of MCA-based dual-model classification is enclosed in the dashed rectangular boxes labeled as (2). It contains two MCA-based classifiers, a positive model and negative model, which are trained by the two sets of features from training set respectively, and a strategy is introduced to fuse these two models into a more powerful classifier to predict the class labels of the testing data instances. The following paragraphs are the detailed explanation about MCA theory used in the framework and the two components: MCA-based feature selection and MCA-based dual-model classification.

2.3.1   MULTIPLE CORRESPONDENCE ANALYSIS

Standard correspondence analysis (CA) is a descriptive/exploratory technique designed to analyze simple two-way contingency tables containing some measure of correspondence between the rows and columns. MCA can be considered as an extension of the standard CA to more than two variables (Greenacre and Blaslus, 2006). Meanwhile, it also appears to be the counterpart of PCA for nominal (categorical) data.

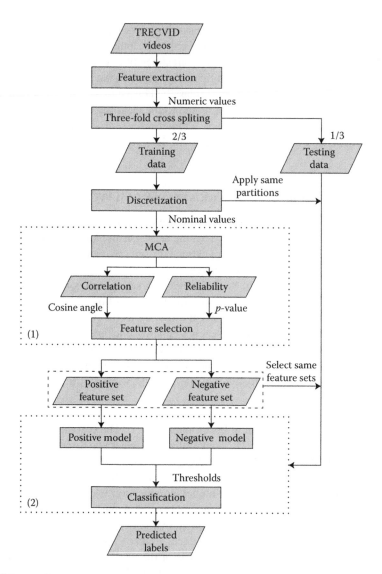

FIGURE 2.1 The proposed framework.

MCA constructs an indicator matrix with instances as rows and categories of valuables as columns. In order to apply MCA, each feature needs to be first discretized into several intervals/items or nominal values (called feature-value pairs in our study), and then each feature is combined with the class to form an indicator matrix. Assuming there are totally K features, and the kth feature has l_k feature-value pairs and the number of classes is m, then the indicator matrix is denoted by Z with size $n \times (l_k + m)$, where n is the number of data instances. Instead of performing on the indicator matrix which is often vary large, MCA analyzes the inner product of this indicator matrix, that is, $Z^T Z$, called the Burt Table which is symmetric with size

$(l_k + m) \times (l_k + m)$. The grand total of the Burt Table is the number of instances which is n, then $P = Z^T Z/n$ is called the correspondence matrix with each element denoted as P_{ij}. Let r_i and c_j be the row and column masses of P, that is, $r_i = \sum_j p_{ij}$ and $c_j = \sum_i p_{ij}$. The centering involves calculating the differences $(P_{ij} - r_i c_j)$ between the observed and expected relative frequencies, and normalization involves dividing these differences by $\sqrt{r_i c_j}$, leading to a matrix of standardized residuals $s_{ij} = (p_{ij} - r_i c_j)/\sqrt{r_i c_j}$. The matrix notation of this equation is presented in Equation 2.1.

$$S = D_r^{-1/2}(P - rc^T)D_c^{1/2} \tag{2.1}$$

where r and c are vectors of row and column masses, and D_r and D_c are diagonal matrices with these masses on the respective diagonals. Through singular value decomposition (SVD), $S = U\Sigma V^T$ where Σ is the diagonal matrix with singular values, the columns of U is called left singular vectors, and those of V are called right singular vectors. The connection of the eigenvalue decomposition and SVD can be seen through the transformation in Equation 2.2:

$$SS^T = U\Sigma V^T V\Sigma U^T = U\Sigma^2 U^T = U\Lambda U^T \tag{2.2}$$

Here, $\Lambda = \Sigma^2$ is the diagonal matrix of the eigenvalues, which is also called principal inertia. Thus, the summation of the principal inertia is the total inertia which is also the amount that quantifies the total variance of S. The geometrical way to interpret the total inertia is that it is the weighted sum of squares of principal coordinates in the full S-dimensional space, which is equal to the weighted sum of squared distances of the column or row profiles to the average profile. This motivates us to explore the distance between feature-value pairs and classes represented by the rows of principal coordinates in the full space. Since in most cases, over 95% of the total variance can be captured by the first two principal coordinates, the chi-square distance between a feature-value pair and a class can be well represented by the Euclidean distance between them in the first two dimensions of their principal coordinates. Thus, a geometrical representation, called the symmetric map, can visualize a feature-value pair and a class as points in the two-dimensional map.

As shown in Figure 2.2, a nominal feature F_k with three feature-value pairs corresponds to three points in the map, namely F_k^1, F_k^2, and F_k^3 respectively. Considering

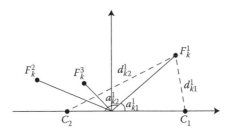

FIGURE 2.2   The geometrical representation of MCA.

a binary class, it is represented by two points lying in the x-axis, where C_1 is the positive class and C_2 is the negative class. Take F_k^1 as an example. The angle between F_k^1 and C_1 is a_{k1}^1, and the distance between them is d_{k1}^1, the angle and distance of F_k^1 with C_2 is a_{k2}^1 and d_{k2}^1. Similar to standard CA, the meaning of a_{k1}^1 and d_{k1}^1 in MCA can be interpreted as follows:

- *Correlation*: The cosine value of the angle between a feature-value pair and a class in the symmetric map indicates the strength of their relation. It represents the percentage of the variance that the feature-value pair point is explained by the class point. A larger cosine value which is equal to a smaller angle indicates a larger correlation between the feature-value pair and the class.
- *Reliability*: As stated before, chi-square distance could be used to measure the dependence between a feature-value pair point and a class point. Here, a derived value from chi-square distance called the p-value is used because it is a standard measure of the reliability of a relation, and a smaller p-value indicates a higher level of reliability.

More technically, p-value represents a decreasing index of the reliability of a result. The higher the p-value, the less we can believe that the observed relation between variables in the sample is a reliable indicator of the relation between the respective variables in the population. Assume that the null hypothesis H_0 is true, which here represents there is no relation between a feature-value pair and a class. Thus, p-value tells us the probability of error involved in rejecting H_0. Generally, one rejects H_0 if the p-value is smaller than or equal to the significance level, which means the smaller the p-value, the higher possibility of the correlation between a feature-value pair and a class is true. p-Value can be calculated through the chi-square cumulative distribution function (CDF) and the degree of freedom is (*number of feature-value pairs* $-$ 1) $\times$ (*number of classes* $-$ 1). For example, the chi-square distance between F_k^1 and C_1 is d_{k1}^1 and their degree of freedom is $(3-1) \times (2-1)$, and then their p-value is $1 - cdf(d_{k1}^1, 2)$. Therefore, correlation and reliability are from different points of view, and can be integrated together to represent the relation between a feature and a class.

2.3.2 MCA-Based Feature Selection Metric

For a feature F_k, the cosine angle values and p-values of each feature-value pair of this feature to the positive and negative classes are calculated, corresponding to correlation and reliability, respectively. If the angle of the jth feature-value pair F_k^j with the positive class is <90 degrees, it indicates this feature-value pair is correlated with the positive class, and negatively correlated with the negative class to the same degree, so Equation 2.3 stands.

$$\cos(a_{k1}^j) = -\cos(a_{k2}^j) \tag{2.3}$$

For p-value, since a smaller p-value indicates a higher reliability, p_{k1}^j can be used as the error of accepting the correlation between F_k^j and C_1, and p_{k2}^j is the error of accepting the correlation between F_k^j and C_2, except for the situation of Jeffrey–Lindley paradox (Lindley, 1957). This paradox describes a situation when the p-value is very close to zero but the probability of the correlation being true is very close to zero as well. Such scenario could happen when the prior distribution is the sum of a sharp peak at H_0 with probability p and a broad distribution with the rest of the probability $1 - p$. In our experiments (to be discussed in Section 2.4), it occurs when the count of the cross-table constructed by the feature-value pairs and the classes is <1% of the count of the corresponding class, which also makes sense since a rare occurrence can be considered as a "fluke."

After getting the correlation and reliability information of each feature-value pair with the class from the geometrical representation from MCA, a score can be calculated for each feature. Assume there are totally n data instances with K features. For the kth feature with l_k feature-value pairs, the angles and p-values for the jth feature-value pair are a_{k1}^j and p_{k1}^j for the positive class C_1, and a_{k2}^j and p_{k2}^j for the negative class, respectively. Considering Equation 2.3, $|\cos(a_{k1}^j)|$ or $|\cos(a_{k2}^j)|$ can be used to indicate the correlation between F_k^j with either C_1 or C_2, but p_{k1}^j is not equal to p_{k2}^j. Thus, there are two scores for each feature, one is for the positive class pos_score_k and one is for the negative class neg_score_k. $|\cos(a_{k1}^j)|$ is added to the feature score if its p-value is smaller than a threshold, which is defined in Equation 2.4a as $TH(p_1)$ for C_1, and in Equation 2.4b as $TH(p_2)$ for C_2, where $\underset{\Sigma l_k}{\mathrm{median}}(p_{k1})$ and $\underset{\Sigma l_k}{\mathrm{max}}(p_{k1})$ is the median and maximum of all the p-values related to C_1 (Σl_k, k is from 1 to K), and $TH(p_1)$ is set to the median of these two values based on the empirical study.

$$TH(p_1) = \mathrm{median}(\underset{\Sigma l_k}{\mathrm{median}}(p_{k1}), \underset{\Sigma l_k}{\mathrm{max}}(p_{k1})) \qquad (2.4a)$$

$$TH(p_2) = \mathrm{median}(\underset{\Sigma l_k}{\mathrm{median}}(p_{k2}), \underset{\Sigma l_k}{\mathrm{max}}(p_{k2})) \qquad (2.4b)$$

The following pseudo-code is developed to integrate the angle value and p-value as a feature-scoring metric (considering Jeffrey–Lindley paradox). According to the scores of the features, two ranking lists can be generated in descending order of the scores, one list of features is specially discriminative to positive class while the other list of features is specially discriminative to negative class. Forward selection is adopted to select the best feature subset for each model. It starts with an empty feature subset, and then at each time, it takes out the feature at top of the ranking list and adds it to the feature subset which is evaluated by the model. Repeat this until all the features in the ranking list are added into the feature subset. The subset that produces the best F1_score is selected as the feature subset for that model.

```
1    for k=1 to K
2      pos_score_k=0;
3      neg_score_k=0;
4      for j=1 to l_k
5        if p_{k1}^j < TH(p_1) and count_{k1}^j > 0.01 × n
6          pos_score_k += |cos (a_{k1}^j)|;
7        end
8        if p_{k2}^j < TH(p_2) and count_{k2}^j > 0.01 × n
9          neg_score_k += |cos (a_{k2}^j)|;
10       end
11     end
12     pos_score_k=pos_score_k/l_k;
13     neg_score_k=neg_score_k/l_k;
14   end
```

2.3.3 MCA-BASED DUAL-MODEL CLASSIFICATION

A data instance that consists of a certain feature-value pair from each feature can be viewed as a transaction in the database. For the training set, a transaction is an instance without the class label, shown in Table 2.1, each row is a transaction. Cosine value of the angles can be used as weights of feature-value pairs to indicate their correlation with the classes. If the value is close to 1, then this feature-value pair is highly correlated to the positive class. On the contrary, if the value is close to −1, then the feature-value pair is highly correlated to the negative class. If it is close to 0, it means this feature-value pair does not have much discriminative ability. Based on this, for instance, if the sum of all weights of its feature-value pairs is a relatively large positive value, the label of this instance is probably positive, vice versa. Thus, for the training set with K features, a transaction weight for each of its data instances TW_i can be calculated, denoted in Equation 2.5, where j is a value between 1 to l_k, and a threshold can be set that optimizes the classifier. For a testing data instance, its transaction weight is first calculated in the same way, and then it is compared with the threshold. If it is smaller than the threshold, then it is predicted as a negative instance, otherwise predicted as a positive instance.

$$TW_i = \sum_{k=1}^{K} \cos(a_k^j)$$

(2.5)

TABLE 2.1

Transactions of the Training Data Set

Feature 1	Feature 2	...	Feature K
F_1^3	F_2^1	...	F_K^4
F_1^1	F_2^1	...	F_K^1
...	...	...	...

After the feature selection phase, two subsets of features are selected to train two classifiers, which are based on MCA transaction weights. In the above Equation 2.5, K is adjusted to the number of features in the selected feature set for each classifier. In addition, only positive correlated feature-value pairs are considered, which means for the positive model, $\cos(a_k^j)$ is added to TW_i only if $\cos(a_k^j) > 0$. On the opposite, for the negative model, $\cos(a_k^j)$ is added to TW_i only if $\cos(a_k^j) < 0$. In this way, the weights of the feature-value pairs of a transaction will not cancel each other out. A threshold for the positive model pos_TH and a threshold for the negative model neg_TH need to be tuned, respectively, which involve finding two parameters: t_1 and t_2, shown in Equations 2.6a and 2.6b. Based on the dataset and empirical studies, t_1 is set to 1 while t_2 is set to 0.06 in the experiment in Section 2.4.

$$pos\_TH = pos\_mean + t_1 \times pos\_std \qquad (2.6a)$$

$$neg\_TH = neg\_mean - t_2 \times neg\_std \qquad (2.6b)$$

However, since the number of features selected for each classifier is different, normalization is needed to be applied to both models to ensure the transaction weights lie in the same range and suitable for later processes. Z-score normalization is adopted here. Thus, after normalization, pos_TH is equal to t_1, while neg_TH is equal to t_2. A strategy is introduced to fuse these two classifiers into a more powerful one. When a testing data instance i comes, two transaction weights are calculated, one by each model: pos_TW_i and neg_TW_i.

```
1   for each testing instance i
2     if pos_TW_i ≥ pos_TH and neg_TW_i ≥ neg_TH
3       instance_i ← positive
4     else if pos_TW_i ≤ pos_TH and neg_TW_i ≤ neg_TH
5       instance_i ← negative
6     else
7       if |pos_TW_i - pos_TH| < |neg_TW_i - neg_TH|
8         instance_i ← positive
9       else
10        instance_i ← negative
11      end
12    end
13  end
```

The above pseudo-code presents the rules to classify this testing data instance. Note that $|pos\_TW_i - pos\_TH|$ is the distance to the threshold of the positive model, and $|neg\_TW_i - neg\_TH|$ is the distance to the negative model, they are able to compare with each other after the normalization applied before.

2.4   EXPERIMENTS AND RESULTS

To evaluate the proposed framework, experiments are conducted using 10 concepts from TRECVID 2009. Each concept data set has more than 12,000 instances and 48

TABLE 2.2

Concepts to Be Evaluated

No.	Concept Name	P/N Ratio
1	Chair	0.07
2	Traffic-intersection	0.01
3	Person-playing-musical-instrument	0.04
4	Person-playing-soccer	0.01
5	Hand	0.25
6	People-dancing	0.02
7	Night-time	0.06
8	Boat-ship	0.06
9	Female-human-face	0.10
10	Singing	0.07

low-level visual features. These 10 concepts range from slightly imbalanced (e.g., hand) to highly imbalanced (e.g., traffic-intersection) with positive to negative (*P/N*) ratio as shown in Table 2.2. Five classifiers, namely DT, JRip, NB, Adaboost which uses DT as the basic classifier, and *k*-NN where *k* is set to 3, are used as comparisons, and the input of these five classifiers are the features selected by four feature selection methods: IG, CHI, CFS, and REF. So there are 20 combinations used as comparisons of the proposed framework, each with one feature selection method combined with one classifier. Precision (pre), recall (rec), and F1_score (F1), which is the harmonic mean of precision and recall, are adopted as the evaluation metrics for classification, their calculations are defined in Equations 2.7a, 2.7b, and 2.7c, where *TP*, *FP*, and *FN* represent true positive, false positive, and false negative, respectively. The final result of a classifier is the average classification results of three folds.

TABLE 2.3

Classification Results of Adaboost

No.	CFS pre	CFS rec	CFS F1	IG pre	IG rec	IG F1	CHI pre	CHI rec	CHI F1	REF pre	REF rec	REF F1
1	0.46	0.31	0.37	0.58	0.31	0.40	0.61	0.31	0.41	0.55	0.33	0.41
2	0.93	0.17	0.28	0.67	0.25	0.36	0.79	0.25	0.38	0.58	0.26	0.36
3	0.72	0.57	0.63	0.76	0.58	0.66	0.77	0.58	0.66	0.76	0.58	0.66
4	0.63	0.44	0.52	0.71	0.54	0.61	0.67	0.50	0.57	0.69	0.49	0.57
5	0.36	0.18	0.24	0.33	0.27	0.30	0.33	0.24	0.30	0.34	0.28	0.31
6	0.48	0.18	0.27	0.55	0.26	0.35	0.54	0.26	0.35	0.50	0.29	0.36
7	0.38	0.22	0.28	0.44	0.23	0.30	0.40	0.23	0.29	0.36	0.23	0.28
8	0.38	0.19	0.25	0.55	0.20	0.29	0.55	0.21	0.30	0.50	0.20	0.29
9	0.52	0.21	0.29	0.49	0.34	0.41	0.49	0.34	0.40	0.49	0.33	0.40
10	0.38	0.20	0.26	0.42	0.24	0.30	0.46	0.23	0.30	0.47	0.22	0.30

TABLE 2.4

Classification Results of DT

No.	CFS			IG			CHI			REF		
	pre	rec	F1	pre	rec	F1	pre	rec	F1	pre	rec	F1
1	0.70	0.23	0.34	0.69	0.23	0.34	0.68	0.24	0.35	0.69	0.25	0.37
2	0.93	0.17	0.28	0.93	0.22	0.36	0.93	0.22	0.36	0.86	0.24	0.37
3	0.76	0.51	0.61	0.79	0.51	0.62	0.79	0.51	0.62	0.80	0.51	0.62
4	0.65	0.20	0.30	0.78	0.26	0.39	0.77	0.25	0.36	0.75	0.30	0.40
5	0.50	0.09	0.15	0.45	0.15	0.22	0.46	0.15	0.22	0.44	0.16	0.23
6	0.60	0.11	0.19	0.60	0.17	0.27	0.58	0.18	0.27	0.68	0.19	0.29
7	0.54	0.15	0.23	0.44	0.19	0.26	0.45	0.19	0.26	0.46	0.19	0.27
8	0.57	0.15	0.23	0.52	0.20	0.29	0.52	0.21	0.30	0.55	0.20	0.29
9	0.50	0.18	0.27	0.56	0.28	0.37	0.56	0.28	0.37	0.55	0.26	0.36
10	0.64	0.15	0.25	0.59	0.15	0.24	0.56	0.16	0.24	0.54	0.15	0.24

$$\text{precision} = \frac{TP}{TP + FP} \qquad (2.7a)$$

$$\text{recall} = \frac{TP}{TP + FN} \qquad (2.7b)$$

$$\text{F1\_score} = 2 \times \frac{\text{precision} \times \text{recall}}{\text{precision} + \text{recall}} \qquad (2.7c)$$

Tables 2.3 through 2.7 show the precision, recall, and F1_score of five classifiers using four different feature selection methods. Since all the selection metrics only

TABLE 2.5

Classification Results of JRip

No.	CFS			IG			CHI			REF		
	pre	rec	F1	pre	rec	F1	pre	rec	F1	pre	rec	F1
1	0.69	0.26	0.37	0.64	0.29	0.40	0.66	0.29	0.40	0.63	0.29	0.40
2	0.92	0.18	0.30	0.89	0.22	0.35	0.82	0.22	0.35	0.79	0.23	0.36
3	0.73	0.57	0.64	0.76	0.58	0.65	0.78	0.54	0.64	0.74	0.60	0.66
4	0.62	0.35	0.44	0.55	0.56	0.55	0.53	0.56	0.55	0.62	0.55	0.58
5	0.47	0.10	0.16	0.43	0.14	0.21	0.47	0.15	0.23	0.46	0.13	0.20
6	0.52	0.15	0.24	0.55	0.21	0.30	0.50	0.22	0.30	0.48	0.23	0.31
7	0.51	0.21	0.30	0.45	0.20	0.33	0.48	0.25	0.33	0.44	0.27	0.33
8	0.59	0.20	0.30	0.55	0.24	0.33	0.53	0.24	0.33	0.59	0.23	0.32
9	0.50	0.23	0.32	0.53	0.37	0.43	0.49	0.38	0.43	0.53	0.38	0.44
10	0.51	0.13	0.21	0.54	0.18	0.27	0.52	0.20	0.29	0.51	0.20	0.28

TABLE 2.6

Classification Results of *k*-NN

No.	CFS			IG			CHI			REF		
	pre	rec	F1	pre	rec	F1	pre	rec	F1	pre	rec	F1
1	0.71	0.27	0.39	0.68	0.31	0.43	0.69	0.31	0.43	0.69	0.30	0.42
2	0.95	0.20	0.33	0.91	0.23	0.36	0.91	0.23	0.36	0.83	0.23	0.35
3	0.80	0.53	0.64	0.85	0.58	0.69	0.85	0.56	0.68	0.85	0.57	0.68
4	0.85	0.39	0.52	0.88	0.49	0.62	0.89	0.50	0.64	0.87	0.47	0.61
5	0.43	0.11	0.18	0.42	0.16	0.24	0.42	0.16	0.24	0.43	0.16	0.24
6	0.61	0.14	0.23	0.65	0.16	0.26	0.70	0.16	0.27	0.66	0.15	0.25
7	0.46	0.16	0.23	0.47	0.19	0.27	0.46	0.19	0.27	0.47	0.19	0.27
8	0.59	0.16	0.25	0.62	0.18	0.28	0.59	0.18	0.27	0.60	0.18	0.28
9	0.52	0.23	0.32	0.52	0.31	0.39	0.52	0.31	0.39	0.56	0.30	0.39
10	0.60	0.16	0.25	0.63	0.20	0.30	0.61	0.19	0.29	0.61	0.20	0.30

generate the ranking list of the features except CFS which directly provides the selected features, forward selection mentioned in Section 2.3.2 is used to find the feature subset that gives the best performance of a classifier, evaluated by F1_score. For each feature selection method, the best performance produced by a feature subset is recorded in the tables.

Table 2.8 shows the performance of the proposed framework (denoted as MCA). The same forward selection is used in the two models to select the feature sets. As can be seen from Figure 2.3 which shows the F1_scores of MCA and the best F1_scores from the five tables (cells shaded gray) of each classifier for all 10 concepts, MCA performs the best in all 10 concepts regarding to F1_score which is the most important metric taking both precision and recall into account, followed by

TABLE 2.7

Classification Results of NB

No.	CFS			IG			CHI			REF		
	pre	rec	F1	pre	rec	F1	pre	rec	F1	pre	rec	F1
1	0.50	0.32	0.39	0.48	0.33	0.39	0.47	0.34	0.39	0.46	0.35	0.40
2	0.43	0.25	0.31	0.45	0.19	0.26	0.52	0.20	0.28	0.27	0.23	0.24
3	0.60	0.60	0.59	0.59	0.62	0.60	0.61	0.60	0.60	0.59	0.64	0.61
4	0.33	0.74	0.46	0.30	0.81	0.44	0.30	0.80	0.44	0.31	0.80	0.44
5	0.34	0.29	0.31	0.33	0.37	0.35	0.33	0.37	0.35	0.33	0.37	0.35
6	0.30	0.35	0.32	0.31	0.31	0.31	0.31	0.31	0.31	0.14	0.53	0.22
7	0.27	0.60	0.37	0.31	0.43	0.36	0.32	0.42	0.36	0.27	0.52	0.36
8	0.30	0.40	0.34	0.37	0.35	0.36	0.39	0.35	0.37	0.39	0.34	0.37
9	0.48	0.33	0.39	0.45	0.48	0.46	0.45	0.48	0.46	0.34	0.63	0.45
10	0.43	0.30	0.35	0.38	0.34	0.36	0.38	0.35	0.36	0.38	0.33	0.35

TABLE 2.8

Classification Results of the Proposed Framework

No.	MCA		
	pre	**rec**	**F1**
1	0.58	0.41	0.48
2	0.62	0.35	0.45
3	0.77	0.69	0.73
4	0.82	0.63	0.71
5	0.58	0.27	0.37
6	0.41	0.34	0.37
7	0.49	0.35	0.41
8	0.55	0.29	0.38
9	0.40	0.68	0.50
10	0.53	0.36	0.43

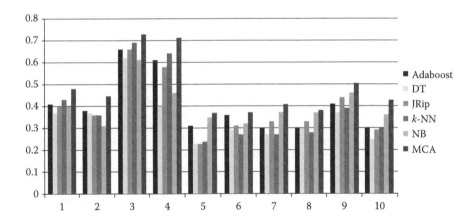

FIGURE 2.3 Comparison of best F1_scores among six classifiers.

NB, k-NN, and Adaboost. DT performs the worst. MCA outperforms the second best result in an average of 4% in F1_scores, and outperforms the worst one in an average of 15%.

In terms of classification performance on the features selected by the four feature selection methods, CFS usually has the worst results except when using NB, four out of 10 concepts achieve the best F1_scores. The performance of IG and CHI are quite similar, and REF produces a slightly better result in DT compared with IG and CHI. In general, for each classifier, the margin of the performance resulted by these three methods is quite small.

2.5   CONCLUSIONS

In this chapter, a new dual-model discriminative learning framework for video semantic classification is proposed to address the challenges such as semantic gap, imbalanced data, and high-dimensional feature space in automatic multimedia semantic analysis. Our proposed framework utilizes MCA to calculate the relation between a feature-value pair and a class. The correlation and reliability information, represented by the cosine angles values and p-values, respectively, are integrated into a single metric to score the features. In addition, the correlation information is reutilized to build two models based on the transaction weights. A strategy is also proposed to fuse these two models into a more powerful classifier. The results of the experiments on 10 concepts from TRECVID 2009 demonstrate that the proposed framework shows promising results compared with 20 combinations of five well-known classifiers and four widely used feature selection methods.

REFERENCES

Aha, D. and Kibler, D., Instance-based learning algorithms, *Machine Learning*, 6, 37–66, 1991.

Bressan, M., Bagging predictors, *Machine Learning*, 24(3), 123–140, 1996.

Chen, S.-C., Chen, M., Shyu, M.-L., Zhang, C., and Chen, M., A multimodal data mining framework for soccer goal detection based on decision tree logic, *International Journal of Computer Applications in Technology, Special Issue on Data Mining Applications*, 27(4), 312–323, 2006a.

Chen, S.-C., Chen, M., Shyu, M.-L., and Wickramaratna, K., Semantic event detection via temporal analysis and multimodal data mining, *IEEE Signal Processing Magazine, Special Issue on Semantic Retrieval of Multimedia*, 23, 38–46, 2006b.

Cohen, W. W., Fast Effective Rule Induction, in *International Conference on Machine Learning*, Tahoe City, California, USA, pp. 115–123, 1995.

Fayyad, U. M. and Irani, K. B., Multi-interval discretization of continuous-valued attributes for classification learning, in *Proceedings of International Joint Conference on Artificial Intelligence*, Chambéry, France, pp. 1022–1027, 1993.

Forman, G., An extensive empirical study of feature selection metrics for text classification, *The Journal of Machine Learning Research*, 3, 1289–1305, 2003.

Freund, Y. and Schapire, R. E., A decision-theoretic generalization of on-line learning and an application to boosting, *Journal of Computer and System Sciences*, 55(1), 119–139, 1997.

Greenacre, M. J. and Blaslus, J., *Multiple Correspondence Analysis and Related Methods*. Chapman and Hall/CRC, London, UK, 2006.

Hall, M. A., Correlation-based feature selection for discrete and numeric class machine learning, in *Proceedings of International Conference on Machine Learning*, Stanford, California, USA, pp. 359–366, 2000.

Hauptmann, A. G., Yan, R., Lin, W., Christel, M. G., and Wactlar, H. D., Can high-level concepts fill the semantic gap in video retrieval? A case study with broadcast news, *IEEE Transactions on Multimedia*, 9(5), 958–966, 2007.

Hua, J., Tembe, W. D., and Dougherty, E. R., Performance of feature selection methods in the classification of high-dimension data, *Pattern Recognition*, 42(3), 409–424, 2009.

Jiang, Y. G., Yang, J., Ngo, C. W., and Hauptmann, A. G., Representations of keypoint-based semantic concept detection: A comprehensive study, *IEEE Transactions on Multimedia*, 12(1), 42–53, 2010.

John, G. H. and Langley, P., Estimating continuous distributions in Bayesian classifiers, in *Conference on Uncertainty in Artificial Intelligence*, Montreal, Quebec, Canada, pp. 338–345, 1995.

Lee, C. and Lee, G. G., Information gain and divergence-based feature selection for machine learning-based text categorization, *Information Processing and Management*, 42(1), 155–165, 2006.

Lin, L., Ravitz, G., Shyu, M.-L., and Chen, S.-C., Correlation-based video semantic concept detection using multiple correspondence analysis, in *Proceedings of IEEE International Symposium on Multimedia*, Berkeley, California, USA, pp. 316–321, 2008a.

Lin, L., Ravitz, G., Shyu, M.-L., and Chen, S.-C., Effective feature space reduction with imbalanced data for semantic concept detection, in *Proceedings of IEEE International Conference on Sensor Networks, Ubiquitous, and Trustworthy Computing*, Taichung, Taiwan, R.O.C., pp. 262–269, 2008b.

Lindley, D., A statistical paradox, *Biometrika*, 44(1–2), 187–192, 1957.

Lu, Y., Cohen, I., Zhou, X. S., and Tian, Q. Feature selection using principal feature analysis. In *Proceedings of the International Conference on Multimedia*, Augsburg, Germany, pp. 301–304, 2007.

Lowe, D., Distinctive image features from scale-invariant keypoints, *International Journal of Computer Vision*, 2(60), 91–110, 2004.

Manning, D. C., Raghavan, P., and Schutze, H., Text classification and Naive Bayes, *Introduction to Information Retrieval*, Cambridge University Press, Cambridge, UK, pp. 253–287, 2008.

Naphade, M., Smith, J. R., Tesic, J., Chang, S.-F., Hsu, Wu., Kennedy, L., Hauptmann, A., and Curtis, J., Large-scale concept ontology for multimedia, *IEEE Multimedia Magazine*, 13(3), 86–91, 2006.

Quinlan, R., *C4.5: Programs for Machine Learning*, Morgan Kaufmann Publishers, San Francisco, CA, 1993.

Robnik-Sikonja, M. and Kononenko, I., An adaptation of relief for attribute estimation in regression, in *International Conference on Machine Learning*, Nashville, Tennessee, USA, pp. 296–304, 1997.

Saeys, Y., Inza, I., and Larranaga, P., A review of feature selection techniques in bioinformatics, *Bioinformatics*, 23(19), 2507–2517, 2007.

Shyu, M.-L., Xie, Z., Chen, M., and Chen, S.-C., Video semantic event/concept detection using a subspace-based multimedia data mining framework, *IEEE Transaction on Multimedia*, 10(2), 252–259, 2008.

Smeulders, A. W. M., Worring, M., Santini, S., Gupta, A., and Jain, R., Content-based image retrieval: The end of the early years, *IEEE Transaction Pattern Analysis and Machine Intelligence*, 22, 1349–1380, 2000.

Sun, Y., Iterative relief for feature weighting: Algorithms, theories, and applications, *IEEE Transactions on Pattern Analysis and Machine Intelligence*, 29(6), 1035–1051, 2007.

Witten, I. H. and Frank, E., *Data Mining: Practical Machine Learning Tools and Techniques*, 2nd ed. Morgan Kaufmann, San Francisco, CA, 2005.

Zhu, Q., Lin, L., and Shyu, M.-L., Feature selection using correlation and reliability based scoring metric for video semantic detection, In *Proceedings of IEEE International Conference on Semantic Computing*, Pittsburgh, PA, USA, pp. 462–469, 2010.

3 Improved Feature Vocabulary-Based Method for Image Categorization

Frank Y. Shih and Alexander Sheppard

CONTENTS

3.1 INTRODUCTION

With the enormous and growing amounts of captured digital images, there has been tremendous interest in image categorization. Image categorization aims to label or classify images into one of the predefined categories. It attempts to retrieve all the images from the same category as a given query image. The attributes of similarity vary from system to system, which are mostly based on color, texture, and shape features.

Image categorization has a wide range of applications, including image search (Sivic and Zisserman, 2003), event detection (surveillance) (Li et al., 2005), process control (via robots or autonomous vehicles) (Loncomilla and Ruiz-del-Solar, 2005), and human–computer interaction (Nakajima et al., 2000). The use in such applications is primarily facilitated by the rising number of consumer devices employing digital image sensors, with mobile phones being a prominent example. A second factor is the continuing leaps in the computer-processing power, allowing for faster execution on image-processing algorithms.

Preceding our discussion, it is important to call attention to the distinction between object categorization and other similar but different tasks, such as object recognition, content-based image retrieval, and object detection. Object categorization entails associating some object present in an image with the correct label for that object. Object recognition can be said to identify one specific object instance, such as a particular red Honda Civic as opposed to a car. Content-based image retrieval deals mainly with low-level image features, not objects, and therefore may be quite imprecise. Object detection often focuses on a single visual category as opposed to many.

The bag-of-feature and feature vocabulary-based approaches have been presented for image categorization due to their simplicity and competitive performance. Some modified versions have been subsequently proposed, incorporating the methods such as adapted vocabularies, fast indexing, and Gaussian mixture models. In this chapter, we propose an improvement of replacing the Harris-affine detection method of Csurka et al. (2004) by a random sampling procedure together with an increased number of sample points (Shih and Sheppard, 2011).

The rest of the chapter is organized as follows. Section 3.2 describes feature extraction and description. Section 3.3 discusses object categorization. In Section 3.4, we present the improved method based on random sampling. Experimental results are provided in Section 3.5. Finally, we draw conclusions in Section 3.6.

3.2 FEATURE EXTRACTION AND DESCRIPTION

Feature extraction is the method for locating points of interest in an image that can be added to a database to be searched later in order to identify objects. Feature description is the method for describing the neighborhood of the located points in the database. In comparing various feature extraction and description methods, Moreels and Perona (2007) identified two leading candidates for real-time processing: one is the affine-invariant interest point detector by Mikolajczyk and Schmid (2002) and the other is Lowe's scale-invariant features transform (2004).

The affine-invariant interest point detector is based on the Harris–Laplace detector (Mikolajczyk and Schmid, 2004; Lindeberg and Garding, 1994; Theodoridis and Koutroumbas, 2006), which is invariant only to scale change. This detector iteratively finds the scale-space extrema of the Laplacian-of-Gaussian (LoG) filter convolved with the image, identifies the maxima of the Harris corner measure taken at the selected scale, and repeats this process until two points are identical. This method is similar to Lowe's Difference-of-Gaussian (DoG) detection method (Lowe, 2004) since DoG is a discrete approximation to LoG. The Harris–Laplace detector is briefly described as follows:

1. Find the local extremum σ over scale of the LoG for the point $x^{(k)}$; otherwise, reject the point. The investigated range of scales is limited to $\sigma^{(k+1)} = t\sigma^{(k)}$ with $t \in [0.7, 1.4]$.
2. Detect the spatial location of a maximum of the Harris measure nearest to $x^{(k)}$ for the selected $\sigma^{(k+1)}$.
3. Go to step 1 if $\sigma^{(k+1)} \neq \sigma^{(k)}$ or $x^{(k+1)} \neq x^{(k)}$.

However, this process fails for significant affine transformations due to the mapping of circular neighborhoods to elliptical neighborhoods in the affine transformed image. A modified detector intends to iteratively transform the neighborhood of a point and its affine-transformed version until these two neighborhoods can be related by a pure rotation matrix, which does not affect the LoG extremum since the magnitude of LoG is rotation-invariant. The formulae from scale-space theory are

$$x'_L = M_L^{-1/2} x_L, \quad x'_R = M_R^{-1/2} x_R, \quad x'_L = R x'_R, \tag{3.1}$$

where M is the second moment matrix smoothed over the appropriate neighborhood and R is a pure rotation matrix. Provided that the necessary convergence criteria are satisfied, this implies that by iterating the application of the inverse second moment matrix over successive newly defined circular neighborhoods, it is possible to obtain a matrix U that will approach a fixed point, such that $x = (\mu^{-1/2})^{(k)} x$. Hence, μ must approach the identity matrix as k becomes large. Therefore, we have the following procedure:

1. Initialize $U^{(0)}$ to the identity matrix.
2. Normalize window $W(x_w) = I(x)$ centered on $U^{(k-1)} x_w^{(k-1)} = x^{(k-1)}$.
3. Select integration scale σ_I at point $x_w^{(k-1)}$.
4. Select differentiation scale $\sigma_D = s\sigma_I$, where s maximizes $\lambda_{\min}(\mu)/\lambda_{\max}(\mu)$ with $s \in [0.5, 0.75]$ and $\mu = \mu(x_w^{(k-1)}, \sigma_I, \sigma_D)$.
5. Detect spatial localization $x_w^{(k)}$ of a maximum of the Harris corner measure nearest to $x_w^{(k-1)}$ and compute the location of the interest point $x^{(k)}$.
6. Compute $\mu_i^{(k)} = \mu^{-1/2}(x_w^{(k)}, \sigma_I, \sigma_D)$.
7. Concatenate transformation $U^{(k)} = \mu_i^{(k)} \cdot U^{(k-1)}$ and normalize $U^{(k)}$ to $\lambda_{\max}(U^{(k)}) = 1$.
8. Go to step 2 if $1 - (\lambda_{\min}(\mu_i^{(k)})/\lambda_{\max}(\mu_i^{(k)})) \geq \varepsilon_C$.

Once interest points have been identified, it is necessary to compute their descriptors. A standard method for this step is the scale-invariant feature transform (SIFT), which is pervasive in almost all the bags of keypoints categorization methods. The description involves first orienting the point neighborhood with respect to the image gradient, thus achieving rotation invariance to representation, and then representing the gradients around the point as a 4×4 array of histograms. Values are weighted using a Gaussian window with σ equal to one half the length of the histogram array, which ensures that changes in values fall off smoothly.

The descriptor is more robust than simply using image matching by correlation, which is highly sensitive to image changes that occur under 3D viewpoint change. To minimize abrupt changes in value, the descriptor uses trilinear interpolation to distribute the value of each gradient sample into adjacent bins. This avoids boundary effects between samples as they shift from one histogram to another. Also, to minimize disturbances from large nonlinear brightness changes near the image patch, all lengths are capped at a maximum of 0.2 (a value found by experimentation) and the vector is then renormalized. The main result is that each point is represented by a $4 \times 4 \times 8 = 128$ dimensional feature vector.

3.3   OBJECT CATEGORIZATION

Csurka et al. (2004) proposed a simple, but effective, method for object categorization based on a bag of keypoints. Their method consists of the following steps:

1. Assign feature descriptors to a set of predetermined clusters (a vocabulary) with a vector quantization algorithm.
2. Construct a bag of keypoints, which counts the number of feature vectors assigned to each cluster.
3. Apply a multi-class classifier, treat each bag of keypoints as the feature vector, and thus determine which category or categories to assign the image.

In order to generate the set of predetermined clusters, the simple method of k-means clustering is used based on assigning points to their closest cluster centers and then recomputing the cluster centers. It is described as follows:

1. Choose initial estimates $\theta_j(0)$ arbitrarily for the θ_j cluster representatives, where $j = 1, 2, \ldots, m$.
2. While change still occurs, repeat:
   a.   For $i = 1, 2, \ldots, N$
      i.   Determine the closest representative θ_j for x_i.
      ii. Set $b(i) = j$, where b maps member vectors to clusters.
   b.   For $j = 1, 2, \ldots, m$

Determine θ_j as the mean of the vectors $x_i \in X$ with $b(i) = j$, where X is the complete set of vectors.

We can minimize the cost function:

$$J(\theta, U) = \sum_{i=1}^{N} \sum_{j=1}^{m} u_{ij} \left\| x_i - \theta_j \right\|^2 \tag{3.2}$$

It means that the cost function recovers the clusters that are as compact as possible, as measured by the squared Euclidean distance. This can be represented in terms of vector quantizers Q, probability density function $p(x)$, and vector quantizers R_j as

$$Q(x) = \theta_j \text{ iff } d(x, \theta_j) \leq d(x, \theta_k), \quad k = 1, 2, \ldots, m, \ k \neq j \tag{3.3}$$

$$\int_{R_j} d(x, \theta_j) p(x) \, dx = \min_y \int_{R_j} d(x, y) p(x) \, dx \tag{3.4}$$

The vector quantization method has a high impact on the time efficiency of constructing the histogram bins necessary for the bag-of-feature methods.

A disadvantage of the k-means algorithm is the lack of guarantee of convergence to a global minimum of the cost function, as opposed to a local minimum. Various strategies have been proposed to deal with this issue, such as initializing the cluster centers to the centers of a set of random partitions of the data, or using the optimization techniques such as genetic algorithms. However, these may be ineffective or may be costly in computation. Moreover, the bag-of-feature method is not primarily interested in the efficacy of clustering in terms of error, but rather in empirical discrimination during categorization. A second relevant feature of k-means is that the number of cluster centers is predetermined, which has a great effect on discriminative power up to a certain point. Beyond this point, results continue to improve to some extent, but it was found according to Csurka et al. (2004) that typically $k = 1000$ gives the best trade-off between computational speed and accuracy.

On completion of k-means, the set of clusters is formed into a histogram (the bag of keypoints) that counts the number of points assigned to each cluster. Each image can then be represented by such a histogram. A large database will contain many images, and hence many histograms will be generated. The histograms can be conceived as new feature vectors, which are fed into a classifier for training and testing. The classifiers may be the Naive Bayes Classifier or the support vector machine (SVM) with linear, quadratic, or cubic kernels.

In the proposed method, we use the random sampling method to produce improved categorization results over the keypoint-based detectors. We observe that using a large number of points in the initial clustering step tends to improve results. The scale-space extrema detected by the keypoint method are too few to ultimately compete with a larger random sample.

Nowak et al. (2006) identified four main implementation choices for bag-of-feature approaches, such as how to sample patches, which descriptor to use, how to quantify the resulting descriptor space distribution, and how to classify images based on the resulting global image descriptor. On the matter of the first implementation choice, random sampling on a pyramid in scale space is compared with Harris-affine and LoG detectors. Also, the size of the codebook (translating to the number of clusters if k-means were used) is examined. It is observed that as the codebook size increases, all approaches except random sampling begin to suffer from problems of dataset overfitting. Moreover, due to inherent limitations, Harris-affine and LoG cannot find sufficient points to compete with random sampling. Therefore, we propose an optimal categorization system using as large a keypoint sampling density as possible and a relatively large codebook size.

Notably, it is observed that although some methods of quantizing the data are helpful, the random construction of codebooks offers only relatively small disadvantages as compared with more sophisticated methods. In other words, the impact on bag-of-feature methods from the codebook construction algorithm (Csurka et al., 2004) is relatively slight as compared with sampling density, codebook size, and classifier design. The graphical representation of the variation of codebook construction algorithm is shown in Figure 3.1.

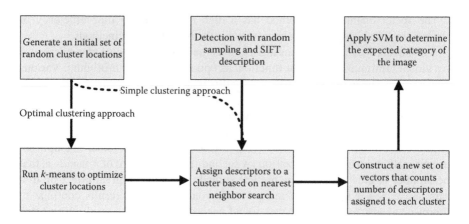

FIGURE 3.1 The graphical representation of the variation of codebook construction algorithm.

3.4 IMPROVED METHOD BASED ON RANDOM SAMPLING

To improve the method by Csurka et al. (2004), the optimized clustering approach should be taken as combined with the best possible detection method, random sampling, and a large number of sample points. In order to implement the description stage, the SIFT ++ program of Vedaldi (2005) was incorporated into the system. Vedaldi's bag-of-feature program was used to generate random keypoints although the remainder of the program, which employs a hierarchical clustering strategy as opposed to k-means, was removed. For lucid comparisons, our program is run on the same Caltech 4 dataset in (Csurka et al., 2004). This generates up to 5000 random keypoints per image. Five image categories are used: car side views, car rear views, faces, airplanes, and motorbikes. Some images are shown in Figure 3.2. The number of images used is 450 per category, forming a total of 2250 images.

3.4.1 K-MEANS CLUSTERING AND NEAREST NEIGHBOR QUANTIZATION

For clustering and vector quantization, the SciPy library of algorithms and mathematical tools are employed in the Python programming language. The PyML is an interactive object-oriented framework for machine learning written in Python. Once loaded into SciPy, a relatively small subsample (5000 keypoints per category) of the entire dataset is used in simple k-means to generate a codebook of optimized cluster locations. The 1000-D feature vector representing each image is initialized to zero. Each keypoint vector is then compared with each of the optimized cluster locations to quantize it to an appropriate location in the codebook. For each image, the cluster location selected is then incremented in the 1000-D feature vector representing the entire image. From this process, a complete set of 1000-D histogram vectors is generated, each providing a unique representation of one image.

FIGURE 3.2   Image categories selected from the Caltech 4 database.

3.4.2 NEAREST-NEIGHBOR CLASSIFICATION AS A BENCHMARK METHOD

The classifier step is critical in the bag-of-feature method. Two basic approaches are employed: Bayesian classifier and SVM, with the former serving as benchmark for the latter. The Bayesian classifier is based on minimizing the probability of error using Bayesian statistical prediction. Applying this to the bag-of-feature method yields

$$P(C_j \mid I_i) = P(C_j) \prod_{t=1}^{|V|} P(v_t \mid C_j)^{N(t,i)} \tag{3.5}$$

where C_j denotes the categories, I_i denotes images (corresponding to classes and vectors), v_t denotes the keypoint, and $N(t, i)$ is the number of times that the cluster occurs in an image.

However, in order to vary the results, we use the k-nearest neighbor algorithm in place of Bayesian classification. This method is a nonparametric method classifying each point in the testing data according to the classes of the k closest vectors in the training data. Motivation comes from Bayesian theory: Given a small region R in the neighborhood of a test point x, the probability P of finding a training point in R, and N observations, we have $K \approx NP$, where K is the number of training points in R. If the volume of R is small, the probability density of the data $p(x)$ on R will be approximately constant, or $P \approx p(x)V$. Removing P from the set of equations gives the density $p(x) \approx (K/NV)$. Applying the following Bayesian formula:

$$p(C_k \mid x) = \frac{p(x \mid C_k)p(C_k)}{p(x)} \tag{3.6}$$

with density for class k: $p(x \mid C_k) = K_k/N_k V$ and the class prior probabilities: $p(C_k) = N_k/N$ yields

$$p(C_k \mid x) = \frac{K_k}{K} \tag{3.7}$$

In other words, the class with the highest K_k value is optimal.

3.4.3 LINEAR SOFT-MARGIN SVMs

A sophisticated classifier is the SVM (Abe, 2005; Bishop, 2006; Cheng and Shih, 2007), a learning system that separates two classes of vectors with an optimal separating hyperplane. An important property of the SVM is that its hyperplane is one of maximal margins: the distance between vectors of each class is the largest that can exist given the dataset. There are several basic types of SVM, including the relatively simple linear maximal margin classifier, the linear soft margin classifier, and in the case where a nonlinear curve produces an optimal result; a nonlinear kernel function can be used to create a nonlinear classifier. In this chapter, we only consider the linear soft margin case because the linear maximal margin is too simple and using nonlinear kernels has been found to produce overfitting.

Given a set of training data $x_1, x_2, \ldots, x_k$ with $x_i \in R^n$ for $i = 1, 2, \ldots, k$. The maximal margin hyperplane must be represented by the equation: $\mathbf{w} \cdot \mathbf{x} + b = 0$, where $\mathbf{w} \in R^n$ and $b \in R$. It is assumed to be possible to divide the training data into two classes, represented by a mapping to a $y_i \in R^n$ such that

$$\begin{cases} \mathbf{w}^T x_i + b \geq 1 & \text{for } y_i = 1 \\ \mathbf{w}^T x_i + b < 1 & \text{for } y_i = -1 \end{cases} \tag{3.8}$$

which is equivalent to

$$y_i(\mathbf{w}^T x_i + b) \geq 1 \quad \text{for } i = 1, \ldots, M \tag{3.9}$$

Planes that satisfy this criterion are given by

$$(\mathbf{w}^T x_i + b) = c \quad \text{for } -1 < c < 1 \tag{3.10}$$

Assuming that each plane with $c = 1$ and -1 includes at least one data point, the plane with $c = 0$ will provide maximal margin. It can be easily shown that the distance between any data point x and the optimal hyperplane is given by

$$\frac{|D(\mathbf{x})|}{|\mathbf{w}|} = \frac{|\mathbf{w}^T \mathbf{x} + b|}{|\mathbf{w}|} \tag{3.11}$$

Therefore, the maximal margin is simply $2/|\mathbf{w}|$. This is maximized if the function $Q(\mathbf{w}) = |\mathbf{w}|^2/2$ is minimized with respect to the constraints given by the data. This can be reduced to an equivalent minimization problem:

$$Q(\mathbf{w}, b, \xi) \frac{1}{2} |\mathbf{w}|^2 + C \sum_{i=1}^{M} \xi_i^p \tag{3.12}$$

which in turn is reduced to a quadratic programming problem except that $C \geq \alpha_i$ is added as an additional constraint and $p = 1$ for present purposes. For the dataset, C is set to be 0.005.

3.4.4 Categorization with Multi-Class SVMs

There are several possible extensions of SVM to multi-class problems. Two most important extensions are *one-against-all* and *pairwise coupling*. The former is based on computing n decision functions, one for each class. If $D_i(x) = \mathbf{w}_i^T x + b_i$ is the decision function between two classes i and j, $R_i = \{x | D_i(x) > 0, i = n \text{ and } D_i(x) \leq 0, i \neq n\}$ are non-overlapping regions such that class n is favored above all others.

On the other hand, for $x \notin R_i$, $i = 1, 2, \ldots, n$, we use membership functions defined based on the distance from the optimal separating hyperplanes $D_j(x) = 0$. In the case $i = j$, we have

$$m_{ii}(x) = \begin{cases} 1 & \text{for } D_i(x) \geq 1 \\ D_i(x) & \text{otherwise} \end{cases} \tag{3.13}$$

and in the case $i \neq j$, we have

$$m_{ij}(x) = \begin{cases} 1 & \text{for } D_j(x) \leq -1 \\ -D_j(x) & \text{otherwise} \end{cases} \tag{3.14}$$

A test vector $\mathbf{x}$ is assigned to the class: $\arg \min_{i=1\ldots n} D_i(\mathbf{x})$ provided that there is only one minimum. This scheme resolves the potential existence of unclassifiable regions.

3.5   EXPERIMENTAL RESULTS

In this section, the results of using the proposed method on the Caltech 4 dataset are presented. The images previously known to be in a particular category are inputted to the multi-class classifier. The classifier then uses the 1000 data attributes of the histogram vector to come up with its own classification. The classification is then compared with the known ground truth value to produce error measures. Four major runs are performed on 2250 images. Each of them uses seven values for keypoints per image in the interval of 5–5000. Description of the four major runs is presented in Table 3.1. Figure 3.3 illustrates the graphical comparison of the results.

Details for each run are described as follows. In all cases where the SVM classifier is used on histogram vectors, the parameter C is set to be 0.005 although the results appear almost identical for $C = 1$. For the k-NN classifier, $k = 3$. For building the vocabulary using k-means, 5000 keypoints per category are aggregated into a set of 25,000, from which the 1000 cluster centers are extracted. A random subset of feature vectors is also tried for a vocabulary in run 4. Finally, for high numbers of points allowed, the available number of SIFT keypoints begins to decline. In

TABLE 3.1
Description of Four Major Runs

	Classifier	Vocabulary	Detector
Run 1	SVM	k-means	Random
Run 2	SVM	k-means	SIFT
Run 3	k-NN	k-means	Random
Run 4	SVM	Random	Random

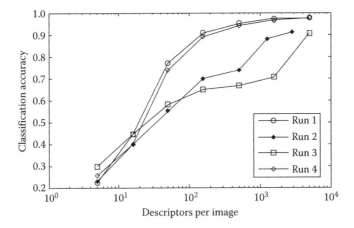

FIGURE 3.3 A graphical comparison of the results.

particular, for 1581 points the average actual number is decreased to 1272, and at 5000 it is only 2865.

Results show that the SVM outperforms *k*-NN classification in the bag-of-feature methods. However, the most important result is that random keypoint locations conclusively outperform the SIFT detector even with the same number of keypoints. Thus, the new method is superior to the traditional methods. In particular, its accuracy on Caltech 4 reaches as high as 97.6%, which is 1.5% higher than the previous result (Csurka et al., 2004) and more than 6% higher in this implementation as compared with SIFT detectors at maximum accuracy. Table 3.2 shows the classification rate for each run and the number of keypoints used. All steps of the classification process can be implemented efficiently. The total time to classify all objects in Figure 3.3 is <0.2 s on a 2 GHz Pentium 4 processor.

The performance of SIFT keypoints in this implementation is noticeably poor, with only 91.4% maximum accuracy. This may be due to the SIFT detector used as opposed to the newer Harris affine detector used. Another issue is that for high numbers of allowed points, SIFT cannot procure as many as a random sample. At 5000 points allowed, the actual mean number of SIFT keypoints found per image is only 2865. However, it appears that the large majority of the difference is due to the SIFT

TABLE 3.2

The Classification Rate for Each Run and the Number of Keypoints Used

Keypoints	5	16	50	158	500	1581	5000
Run 1	0.222	0.443	0.770	0.908	0.953	0.972	0.976
Run 2	0.229	0.400	0.552	0.697	0.739	0.880	0.914
Run 3	0.298	0.443	0.579	0.648	0.666	0.704	0.905
Run 4	0.255	0.400	0.739	0.892	0.942	0.965	0.974

detector drawing an inherently poorer sample since the difference in accuracy between 1581 and 5000 keypoints is a mere 0.4% in Run 1.

The positive effect of using k-means to optimize the vocabulary appears to be small but consistent for large numbers of points used. Vocabulary composition has a significant effect on classification accuracy: for large numbers of points, a completely random vocabulary appears to drop the accuracy over using a random sample of keypoints, and performing k-means improves it by up to 0.2%.

Finally, confusion matrices are shown below for each run at maximum accuracy. Their graphical plot with SVM using k-means and 5000 random keypoints in the order of cars_brad, cars_brad_bg, motorbikes_side, faces, and airplanes_side in Run 1 is shown in Figure 3.4.

Run 1 :   Run 2 :

$$
\begin{bmatrix}
95.6 & 0.4 & 1.8 & 0 & 1.6 \\
0 & 99.3 & 0.4 & 0 & 0 \\
1.1 & 0.2 & 97.6 & 0.2 & 0.2 \\
0.7 & 0 & 0 & 98.2 & 0.7 \\
2.7 & 0 & 0.2 & 1.6 & 97.3
\end{bmatrix}
\begin{bmatrix}
89.8 & 0.2 & 0 & 2.7 & 5.6 \\
0.2 & 95.3 & 4.0 & 0.2 & 3.6 \\
0.4 & 3.3 & 95.3 & 0 & 2.0 \\
2.4 & 0 & 0 & 91.1 & 4.0 \\
7.1 & 1.1 & 0.2 & 6.0 & 84.9
\end{bmatrix}
$$

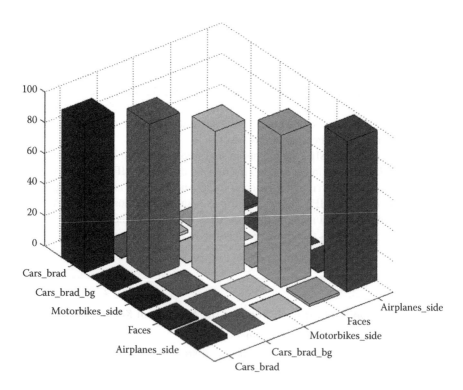

FIGURE 3.4 Graphical plot of the confusion matrix with SVM using k-means and 5000 random keypoints in Run 1.

Run 3 :   Run 4 :

$$
\begin{bmatrix}
93.6 & 5.6 & 4.0 & 2.2 & 14.7 \\
0 & 92.2 & 0 & 0.4 & 2.4 \\
4.9 & 2.2 & 95.8 & 0.9 & 6.2 \\
1.1 & 0 & 0 & 96.0 & 1.6 \\
0.4 & 0 & 0.2 & 0.4 & 74.9
\end{bmatrix}
\begin{bmatrix}
94.4 & 0.4 & 2.2 & 0 & 1.8 \\
0.2 & 99.1 & 0.2 & 0 & 0 \\
1.8 & 0.2 & 97.3 & 0 & 0 \\
0.7 & 0 & 0 & 98.7 & 0.9 \\
2.9 & 0.2 & 0.2 & 1.3 & 97.1
\end{bmatrix}
$$

3.6 CONCLUSIONS

The described new improvement is to replace the Harris-affine detection method by a random sampling procedure together with an increased number of sample points. We observe that using a large number of points in the initial clustering step tends to improve results. The scale-space extrema detected by the keypoint method are too few to ultimately compete with a larger random sample. Experimental results show that the classification of cluster histogram (bag of keypoints) vectors using random sampling and a large number of keypoints significantly outperforms the methods based on local features and the same number of keypoints. Mean classification accuracies exceed those of the latter methods by at least 6% in the proposed system, and exceed those using the k-NN classifier by at least 7%, giving an average of 97.6% accuracy on the Caltech 4 dataset. This value beats previously published results in (Csurka et al., 2004) by 1.5%.

REFERENCES

Abe, S., *Support Vector Machines for Pattern Classification*, Springer, New York, 2005.

Bishop, C., *Pattern Recognition and Machine Learning*, Springer, New York, 2006.

Cheng, S. and Shih, F. Y., An improved incremental training algorithm for support vector machines using active query, *Pattern Recogn*, 40(3), 964–971, 2007.

Csurka, G., Dance, C. R., Fan, L., Willamowski, J., and Bray, C., Visual categorization with bags of keypoints, *Proc ECCV Intl Workshop on Statistical Learning in Computer Vision*, pp. 1–24, 2004.

Li, Y., Atmosukarto, I., Kobashi, M., Yuen, J., and Shapiro, L. G., Object and event recognition for aerial surveillance, *Proc SPIE Optics and Photonics in Global Homeland Security*, 5781, Orlando, FL, pp.139–149, May 2005.

Lindeberg, T. and Garding, J., Shape-adapted smoothing in estimation of 3-D shape cues from affine deformations of local 2-D brightness structure, *Proc of European Conf on Computer Vision*, Stockholm, Sweden, pp. 389–400, May 1994.

Loncomilla, P. and Ruiz-del-Solar, J., Improving SIFT-based object recognition for robot applications, *Lect Notes Comput Sci*, 3617, 1084–1092, 2005.

Lowe, D., Distinctive image features from scale-invariant keypoints, *Intl J Comput Vis*, 60(2), 91–110, 2004.

Mikolajczyk, K. and Schmid, C., An affine invariant interest point detector, *Proc of European Conf on Computer Vision—Part I*, Copenhagen, Denmark, pp. 128–142, May 2002.

Mikolajczyk, K. and Schmid, C., Scale and affine invariant interest point detectors, *Intl J Comput Vis* 60(1), 63–86, 2004.

Moreels, P. and Perona, P., Evaluation of features detectors and descriptors based on 3D objects, *Intl J Comput Vis*, 73(3), 263–284, 2007.

Nakajima, C., Pontil, M., and Poggio, T., Object recognition and detection by a combination of support vector machine and rotation invariant phase only correlation, *Proc Intl. Conf Pattern Recognition*, vol. 4, Washington, DC, pp. 4787–4790, Sep. 2000.

Nowak, E., Jurie, F., and Triggs, B., Sampling strategies for bag-of-features image classification, *Proc of European Conf on Computer Vision*, Graz, Austria, pp. 490–503, May 2006.

Shih, F. Y. and Sheppard, A., An improved feature vocabulary based method for image categorization, *Pattern Recognit and Artificial Intelligence*, 25(3), 415–429, 2011.

Sivic, J. and Zisserman, A., Video google: A text retrieval approach to object matching in videos, *Proc Intl Conf on Computer Vision*, vol. 2, Nice, France, pp. 1470–1477, Oct. 2003.

Theodoridis, S. and Koutroumbas, K., *Pattern Recognition*, Academic Press, Orlando, FL, 2006.

Vedaldi, A., SIFT++: A lightweight C++ implementation of SIFT, 2005–7, http://vision.ucla.edu/~vedaldi/code/siftpp/siftpp.html.

Part II

Watermarking

4 Automatic Detection and Removal of Visible Image Watermarks

Hong-Ren Su, Ya-Yun Cheng, and Shang-Hong Lai

CONTENTS

4.1 INTRODUCTION

In recent years, widespread use of the Internet increased the exchange of information and knowledge. However, the accompanying copyright and ownership problems became very critical to the success of digital content distribution. In order to solve these problems for digital content, information-hiding techniques have become more and more important in many application areas. Electronic watermark, which is also called digital watermark, is a branch of information-hiding techniques and it is a common function in several famous photo-processing application programs. Digital watermarking can be roughly divided into two types—visible marking and invisible marking. The former is commonly used in academic theses and stock photos, and the latter is popular in commercial document sharing, copyright protection, and bill antifaking. Both these types have their own special requirements, such as easy embedding, copy prevention, hard removing, robustness against tempering, and preservation of original information.

The development of invisible watermarking technology has been almost two decades and there are two main approaches: one is the spatial-domain approach and the other is the frequency-domain approach. Van Schyndel et al. (1994) proposed an algorithm called Least-Significant Bits (LSB), which changes the least important bit to add the hiding information. Later, Pitas and Kaskalis (1995) and Pitas (1996) proposed a modified LSB method for information hiding, but the problem with this approach is that it is sensitive to image compression, noise adding, or other image attacking. The frequency-domain approach includes the discrete Cosine Transform (DCT)-based methods and discrete wavelet transform (DWT)-based methods. For example, Barni et al. (1998) proposed to embed a pseudo-random sequence of real numbers in a selected set of DCT coefficients, and Xia et al. (1998) proposed to add pseudo-random codes to the large coefficients at the high- and middle-frequency bands in the DWT of an image.

Sometimes watermarks need not be hidden in the digital contents, as some companies use visible watermarks, but most of the literature has focused on invisible (or transparent) watermarking. Visible watermarks overlaid on the original images or documents are strongly linked to the original watermarks and should be easy to recognize without blocking the original information. The critical point of the visible watermark is how to maintain the balance of watermark clearly visible yet difficult to move. Braudaway et al. (1996) used an analytic human perceptual models and varying pixel brightness to embed reasonably unobtrusive visible logos in color images, and an automatic way of adjusting the watermark intensity for a given image was presented in Rao et al. (1998) by using a texture-based human visual system (HVS) metric. Kankanhalli et al. (1999) worked in the DCT domain to classify DCT image blocks by image characteristics with the strength of the watermark in a block determined by its class. In Hu and Kwong (2001), they proposed a method to overcome the limitations of the DCT-domain-based methods and make the marking applied varied according to the features of the host images. The scaling factors are adaptively determined based on the luminance masking and local spatial characteristics of the host images.

Visible image watermark is commonly used to protect the copyright of documents and images. However, visible image watermark may cause problems for automatic understanding or analysis of the document or image contents, which is required for information or image retrieval. Thus, the visible watermark may need to be removed or modified, and the concept of reversible visible watermarks was proposed by IBM in 1997. The related literature has just rarely been reported. Pei and Zeng (2006) utilized independent component analysis to separate source images from watermarked and host images. Hu et al. (2006) presented a user-key-dependent removable visible watermarking system. Huang et al. (2007) proposed a copyright annotation scheme with both the visible watermarking and reversible data-hiding algorithms, and Liu and Tsai (2010) developed a lossless approach based on using deterministic one-to-one compound mappings of image pixel values for overlaying a variety of visible watermarks of arbitrary sizes on cover images.

In this chapter, we focus on the problems of detecting the visible watermarks from images and removing the watermarks to generate the watermark-free images. We will present our Fourier-based image alignment method with iterative refinement for the watermark detection. Subsequently, we describe how to remove the watermark from the image after the watermark is precisely detected. Some experimental results are given to show the performance of the proposed watermark detection and removal algorithm on real watermark documents and images.

4.2   VISIBLE IMAGE WATERMARK

Visible image watermark is commonly used to protect the copyright of documents and images. In general, most visible watermark embedding can be formulated as follows.

4.2.1   Watermark Embedding

The visible image watermark embedding can be modeled as follows by Hu et al. (2006):

$$I_t(x) = w_o(x)\, I_o(x) + w_w(x)\, I_w(x) \quad \text{for } x \in W \tag{4.1}$$

where W is the region of the watermark embedding, $I_o(x)$ is the original image, $I_w(x)$ is the watermark image, $w_o(x)$ and $w_w(x)$ are the associated weighting functions, and $I_t(x)$ is the watermark-embedded image. Note that $w_o(x) + w_w(x) = 1$. The relationship of an original watermark $I_w(y)$ and its transformed watermark $I_w(x)$ can be modeled by a geometric affine transform given by $x = Ay$, where A is the affine transform matrix.

Generally speaking, $w_o(x)$ and $w_w(x)$ are usually set to constants inside the watermark area and the transformation A is usually the rigid transform with scaling, rotation, and translation. Figure 4.1 shows the typical visible watermark images with

FIGURE 4.1   The watermark embedded images with (a) $w_o = 0.8$, (b) $w_o = 0.5$, and (c) $w_o = 0.3$.

different weighting w_o. If $w_o(x)$ and $w_w(x)$ are not constant, the most common setting is as follows:

$$w_o(x) = \frac{I_w(x)}{I_o(x) + I_w(x)}; \quad w_w(x) = \frac{I_o(x)}{I_o(x) + I_w(x)} \tag{4.2}$$

Combining Equations 4.1 and 4.2 leads to the following model:

$$I_t(x) = \left(\frac{2}{I_o(x) + I_w(x)} \right) I_o(x) I_w(x) \tag{4.3}$$

This equation shows that the embedded image $I_t(x)$ is proportional to the multiplication between $I_o(x)$ and $I_w(x)$. The main advantage of this embedding scheme is that no parameter is required to be adjusted.

For documents, the content can be segmented into two classes: that is, foreground and background. The foreground includes the texts or photos, and the background is always the white color area without anything, such as pdf or word files. In order to preserve the information for understanding, foreground will not be allowed to be modified. Thus the visible document watermarking can be defined as follows:

$$I_t(x) = \begin{cases} I_o(x) & x \in \text{foreground} \\ kI_w(w) & x \in \text{background} \cap \text{watermark region} \end{cases} \tag{4.4}$$

k is a scale factor for decreasing the intensity values of the watermark in the documents. Figure 4.2 depicts two examples of typical watermarked documents.

4.2.2 WATERMARK REMOVAL

Given two images, a watermark image and a watermarked document image, there are two steps in the watermark removal procedure. The first step is to detect the visible watermark from the watermarked image or document, and the second step is to remove the watermark to generate the watermark-free image. Figure 4.3 shows the system of the watermark removal. In the remainder of this chapter, we focus on giving a detailed description of watermark detection and water removal components. In the next section, we first review some previous works on Fourier-based image alignment methods, and then we present a novel Fourier-moment-based image alignment algorithm to precisely estimate the position, orientation, and scaling of the watermark from a watermarked image. Subsequently, we propose how to improve the image alignment accuracy under occlusion or outliers by introducing the distance weighting scheme in the proposed algorithm. In Section 4.4, we present how to remove watermark from a watermarked image after the watermark is precisely aligned. Finally, we show some experimental results, followed by some conclusions.

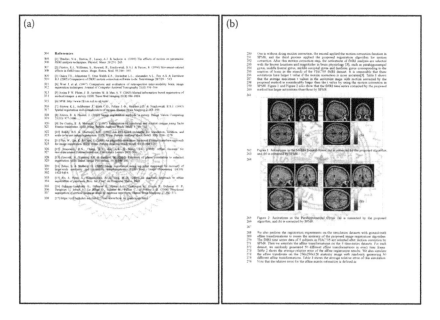

FIGURE 4.2 Two examples of watermarked documents: (a) watermarked document, (b) watermarked document containing images.

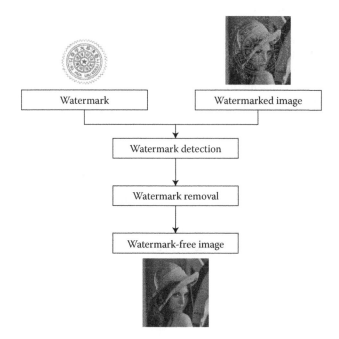

FIGURE 4.3 The system flow of watermark removal.

4.3   FOURIER-BASED IMAGE REGISTRATION

There have been several Fourier-based methods proposed for image registration in the past by De Castro and Morandi (1987), Bracewell et al. (1993), and Reddy and Chatterji (1996). All the Fourier-based methods have the advantages that it is robust to noise with low computational complexity and it can handle large motion. Unfortunately, the accuracy of the Fourier-based image registration methods is limited to the mapping of log-polar transform due to the spectrum interpolation error from the Cartesian plane to the log-polar plane for estimating rotation and scaling. The interpolation directly on the 2D Cartesian Fourier domain will result in error in the image registration. Thus, the log-polar transform is the primary challenge to improve the registration precision and the alignment range. Many modified Fourier-based image registration methods, like the log-polar method in Reddy and Chatterji (1996), unequally spaced Fourier transform by Averbuch et al. (2006), and multilayer fractional Fourier transform (MLFFT) by Pan et al. (2009), have been proposed to evaluate the log-polar transform more efficiently and reliably. MLFFT uses the fractional Fourier transform to create several spectrums with different resolutions from an image and sums them into one for the log-polar transform. This strategy makes the Fourier-based image registration more accurate than the other Fourier-based methods. However, MLFFT cannot go beyond the framework of the log-polar transform, and it just selects multiple spectrums from the fractional FFT to achieve better approximation.

In the next section, we will introduce a novel Fourier-based image registration algorithm without using the log-polar transform. This algorithm can be used in both rigid and affine registration, while the previous Fourier-based methods can only deal with the rigid alignment, which is composed of scaling, rotation, and translation. The proposed algorithm is based on the fact that the rigid transform between two images corresponds to a related rigid transform between their Fourier spectrums, whose energies are normally concentrated around the origin in the frequency domain. Thus, the moments for the corresponding Fourier spectrum distributions can be calculated as probability density functions. The novel registration algorithm is based on minimizing the relationship between the moments for the Fourier spectrums of the two images.

4.3.1   RIGID TRANSFORM RELATIONSHIP BETWEEN IMAGE AND FOURIER DOMAIN

Consider two images $g(x,y)$ and $h(x,y)$ and they are related by a rigid transformation, that is, $h(x,y) = g(r_0 \cos\theta_0 \, x + r_0 \sin\theta_0 \, y + c, -r_0 \sin\theta_0 \, x + r_0 \cos\theta_0 \, + f)$ with the rotation θ_0, scaling r_0 and translation vector $[\, c \, f \,]^t$. Assume the Fourier transforms of the image functions $g(x,y)$ and $h(x,y)$ be denoted by $G(u,v)$ and $H(u,v)$, respectively. Then, we can derive the following rigid relationship between the Fourier transforms $G(u,v)$ and $H(u,v)$ given as follows:

$$G(u,v) = \frac{1}{r_0^2} e^{\frac{i[(c\cos\theta_0 - f\sin\theta_0)u + (f\cos\theta_0 + c\sin\theta_0)v]}{r_0}} H\left(\frac{\cos\theta_0 u + \sin\theta_0 v}{r_0}, \frac{-\sin\theta_0 u + \cos\theta_0 v}{r_0} \right)$$

(4.5)

By letting $u' = (\cos\theta_0\, u + \sin\theta_0\, v)/r_0$ and $v' = (-\sin\theta_0\, u + \cos\theta_0\, v)/r_0$, we have the relationship $u = r_0\,(\cos\theta_0\, u' - \sin\theta_0\, v')$ and $v = r_0\,(\sin\theta_0\, u' + \cos\theta_0\, v')$. Taking the absolute values on both sides of Equation 4.5, we have the rigid transformation relationship between the spectrums $|G(u,v)|$ and $|H(u,v)|$ as follows:

$$|G(u,v)| = \frac{1}{|r_0^2|}|H(u',v')| \tag{4.6}$$

where

$$\begin{bmatrix} u \\ v \end{bmatrix} = \begin{bmatrix} r_0\cos\theta_0 & -r_0\sin\theta_0 \\ r_0\sin\theta_0 & r_0\cos\theta_0 \end{bmatrix}\begin{bmatrix} u' \\ v' \end{bmatrix} \tag{4.7}$$

Equations 4.6 and 4.7 show that the rigid transformation relationship in the amplitude of their Fourier spectrums only has two parameters, rotation θ_0 and scaling r_0. In order to evaluate them easily, Reddy and Chatterji (1996) proposed the log-polar transform from the Cartesian plane $|G(u,v)|$ and $|H(u,v)|$ into the log-polar plane $M_g(\log r, \theta)$ and $M_h(\log r, \theta)$ in Equation 4.6 by letting $u = r\cos\theta$ and $v = r\sin\theta$. Equation 4.6 can be rewritten as follows:

$$M_g(\log r,\theta) = \frac{1}{r_0^2} M_h(\log r - \log r_0, \theta - \theta_0) \tag{4.8}$$

Equation 4.8 shows that the problem for estimating rotation and scaling is converted into a translation-like problem by taking the log-polar transform. Thus, the rotation and scaling can be determined as follows:

$$(\log r_0, \theta_0) = \arg\max_{(\log r,\theta)} \text{real}\left(\text{IFT}\left\{ \frac{\text{FT}(M_g)\text{conj}(\text{FT}(M_h))}{|\text{FT}(M_g)\text{conj}(\text{FT}(M_h))|} \right\} \right) \tag{4.9}$$

where FT and IFT denotes the forward and inverse Fourier transform operators, respectively, and conj denotes the complex conjugate operator.

The rotation θ_0 and scaling r_0 can be first estimated based on Equation 4.9 from the two Fourier spectrums $\text{FT}(M_g)$ and $\text{FT}(M_h)$ through the log-polar transform, and the remaining translation vector (c,f) can also be computed by transforming image $g(x,y)$ with the transformation $g'(x,y) = g(r_0\cos\theta_0\, x + r_0\sin\theta_0\, y, - r_0\sin\theta_0\, x + r_0\cos\theta_0)$ and determining the translation between $g'(x,y)$ and $h(x,y)$ from their cross-power spectrum. To be more specific, the translation vector is determined as follows:

$$(c,f) = \arg\max_{(x,y)} \text{real}\left(\text{IFT}\left\{ \frac{G'(u,v)\, H^*(u,v)}{|G'(u,v)\, H^*(u,v)|} \right\} \right) \tag{4.10}$$

where $G'(u,v)$ is the Fourier transform of $g'(x,y)$, and $H^*(u,v)$ is the complex conjugate of $H(u,v)$.

4.3.2 MOMENT MATCHING APPROACH TO ESTIMATING RIGID TRANSFORM MATRIX

Owing to the interpolation error in the log-polar transform, we introduce the moment matching approach to estimate the rotation and scaling of the rigid transform in Equation 4.6 instead of using the log-polar transform. Let the Fourier spectrums of the 2D image functions $f_1(x,y)$ and $f_2(x,y)$ be denoted by $F_1(u,v)$ and $F_2(u,v)$, respectively. If $f_1(x,y)$ and $f_2(x,y)$ are related by a rigid transformation, then their Fourier spectrums are also related by the corresponding rigid transform, that is, $|F_1(u,v)| = |F_2(u',v')|/r_0^2$ with the relation between (u, v) and (u', v') given in Equation 4.7. We rewrite Equation 4.7 by $a = r_0 \cos \theta_0$ and $b = r_0 \sin \theta_0$ as follows:

$$\begin{bmatrix} u \\ v \end{bmatrix} = \begin{bmatrix} a & -b \\ b & a \end{bmatrix} \begin{bmatrix} u' \\ v' \end{bmatrix} \tag{4.11}$$

To determine the rigid transform parameters, a and b, we employ the moment matching technique to the Fourier spectrums $F_1(u,v)$ and $F_2(u,v)$. The $(i+j)$th-order moment for the Fourier spectrum $|F_k(u,v)|$, $k = 1$ or 2, is defined as

$$m_{i,j}^{(k)} = \iint u^i v^j |F_k(u,v)| \, du \, dv \tag{4.12}$$

By coordinate substitution, we can derive the following equation:

$$m_{i,j}^{(1)} = \iint u^i v^j |F_1(u,v)| \, du \, dv \tag{4.13}$$

$$m_{i,j}^{(2)} = \iint (au' - bv')^i (bu' + av')^j |F_2(u',v')| \, du' dv' \tag{4.14}$$

Thus, we have the following relationship between the first-order moments of the two Fourier spectrums from Equations 4.13 and 4.14.

$$\begin{bmatrix} m_{1,0}^{(1)} \\ m_{0,1}^{(1)} \end{bmatrix} = \begin{bmatrix} a & -b \\ b & a \end{bmatrix} \begin{bmatrix} m_{1,0}^{(2)} \\ m_{0,1}^{(2)} \end{bmatrix} \tag{4.15}$$

The 2D rigid transform parameters (a,b) can be estimated by minimizing the errors associated with the constraints in Equation 4.15 in a least-squares estimation framework as follows:

$$a = \frac{m_{1,0}^{(1)} m_{1,0}^{(2)} + m_{0,1}^{(1)} m_{0,1}^{(2)}}{\left(m_{1,0}^{(2)} \right)^2 + \left(m_{0,1}^{(2)} \right)^2} \tag{4.16}$$

$$b = \frac{m_{0,1}^{(1)}m_{1,0}^{(2)} - m_{1,0}^{(1)}m_{0,1}^{(2)}}{\left(m_{1,0}^{(2)}\right)^2 + \left(m_{0,1}^{(2)}\right)^2} \tag{4.17}$$

4.3.3 A NOVEL FOURIER-MOMENT-BASED IMAGE REGISTRATION ALGORITHM

In this section, we summarize the novel Fourier-moment-based image registration algorithm. The detailed procedure is given as follows:

1. Compute the discrete Fourier transforms of two images $h(x,y)$ and $g(x,y)$ via FFT as Equation 4.6.
2. Compute the first-order moments for the Fourier spectrums $|H(u,v)|$ and $|G(u, v)|$.
3. Determine the rigid transform parameters (a,b) in Equations 4.16 and 4.17 by minimizing the least-square errors associated with the moment matching constraints given in Equation 4.15.
4. Transform the image $g(x,y)$ with the rigid transform associated with the estimated parameters (a,b) and the transformed data are denoted by $g'(x,y)$.
5. Determine the translation from the cross-power spectrum of $g'(x,y)$ and $h(x,y)$ based on Equation 4.10.

Compared with the previous Fourier-based image registration methods, the proposed Fourier-moment-based image registration approach has less time complexity and can provide higher accuracy, because it does not require the log-polar transform, which involves a computationally costly interpolation procedure on the log-polar plane and introduces errors in the interpolation.

4.4 ITERATIVE REFINEMENT PROCESS BY A DISTANCE WEIGHTING

In the previous section, we present a novel Fourier-moment-based image registration algorithm based on matching the moments with the Fourier spectrums of the image functions. To further improve the accuracy of the proposed image registration algorithm, especially to cope with the occlusion problem, we propose an iterative refinement process by introducing a distance weighting scheme into images, which is detailed in the following sections.

4.4.1 BINARY EDGE IMAGE DATA

Let a point set $p(x,y) \in E$, where E is the set of the Canny edge points, be extracted from an image $h(x,y)$. Then the image $h(x,y)$ is transformed into a binary image $B(x,y)$ with values of the pixels set to 1 if they are edge points, and set to 0 otherwise.

$$B(x, y) = \begin{cases} 1 & (x,y) \in E \\ 0 & (x,y) \notin E \end{cases} \tag{4.18}$$

4.4.2   DISTANCE WEIGHTING

The idea of improving the accuracy of the proposed Fourier-moment-based image registration algorithm under the occlusion or outlier problems is to assign an appropriate weight to each point p in the binary image such that the points without correspondences have very small weights and the points with proper correspondences have high weights in the binary image. In the previous definition of the binary image B given in Equation 4.18, the function has binary values to indicate the presence of data points. Since the point sets may contain some noise variation, we introduce a distance weighting to reduce the influence of the points without proper correspondences.

The distance for one data point $p \in E_1$ in the binary image $B_1(x,y)$ to the other point set E_2 in the $B_2(x,y)$ is defined as

$$d(p,E_2) = \min_{q \in E_2} \|p - q\| \tag{4.19}$$

Note that the distance for all data points in E_1 to E_2 can be efficiently computed by using the distance transform. Then, we define the weighting for each data point in E_1 to E_2 as follows:

$$w(p,E_2) = \frac{\sigma^2}{\sigma^2 + d^2(p,E_2)} \tag{4.20}$$

To improve the correctness of the point correspondences, we employ both Euclidean and gradient distances to compute their corresponding weighting functions given in Equation 4.20. Thus, the total weighting $w_t(p, E_2)$ is determined by the Euclidean weighting $w_e(p, E_2)$ and the gradient distance weighting $w_g(p, E_2)$, and they are defined as follows:

$$w_t(p,E_2) = w_e(p,E_2)w_g(p,E_2) = \frac{\sigma_e^2}{\sigma_e^2 + d_e^2(p,E_2)} \frac{\sigma_g^2}{\sigma_g^2 + d_g^2(p,E_2)} \tag{4.21}$$

where σ_e^2 and σ_g^2 are the parameters for the weighting function.

Thus, the binary image B_1 can be transformed into weighted image function Bw_1 as follows:

$$Bw_1(x,y) = \begin{cases} w_t((x,y),E_2) & (x,y) \in E_1 \\ 0 & (x,y) \notin E_1 \end{cases} \tag{4.22}$$

Similarly, the weight image function Bw_2 is defined as follows:

$$Bw_2(x,y) = \begin{cases} w_t((x,y),E_1) & (x,y) \in E_2 \\ 0 & (x,y) \notin E_2 \end{cases} \tag{4.23}$$

In our algorithm, we first apply the proposed Fourier-moment-based image registration to the binary image B_1 and B_2 without using the distance weighting. The estimated rigid transform is applied to all points in E_1 to update the point set E_1. Then, the distance weighting is computed to produce the weighted image function Bw_1 and Bw_2 as given in Equations 4.22 and 4.23. The proposed Fourier-moment-based image registration algorithm is applied to find the rigid transformation between Bw_1 and Bw_2, and the estimation result is used to refine the rigid registration. This refinement process is repeated several times until convergence.

The iterative registration is very crucial to the robustness of the registration. If the first step cannot provide robust registration, then the iterative refinement will not converge to the correct registration results. The novel Fourier-based registration algorithm is robust even without the iterative refinement. It is because the Fourier transform of the image brings most of the energy to the low-frequency region in the Fourier domain, thus making the registration determined from the moments of the Fourier coefficients robust against noise which is normally considered to be mostly distributed in the high-frequency domain.

4.4.3 Fourier-Based Image Registration Algorithm with Iterative Refinement

In this section, we summarize the Fourier-moment-based image registration algorithm with iterative refinement. The detailed procedure is given as follows:

1. Generate the binary images B_1 and B_2 from the two images $h(x)$ and $g(x)$ as Equation 4.18.
2. Compute the discrete Fourier transforms of B_1 and B_2 via FFT.
3. Compute the first-order moments for the Fourier spectrums of B_1 and B_2 from Equations 4.13 and 4.14.
4. Determine the rigid parameters by minimizing the least-square errors associated with the moment matching constraints given in Equation 4.15.
5. Transform B_1 with the rigid transform associated with the estimated parameters (a,b) only and the transformed data are denoted by B_1'.
6. Determine the translation vector t between B_1' and B_2 via the cross-power spectrum method given in Equation 4.10.
7. Shift the map B_1' with the translation vector t and compute the distance weighting from Equation 4.22 to form the weighting function Bw_1.
8. Repeat step 2 through step 7 with the binary image replaced by the weighting functions computed in the previous step until the changes in the rigid transformation parameters are within a small threshold.

4.5 VISIBLE WATERMARK REMOVAL

According to the general model of visible watermark embedding, the production of watermarked text images and pictures is different; therefore, the methods of removing the watermarks to generate the watermark-free text and picture images are also different. We will describe the watermark removal methods for these two types of images subsequently in this section.

4.5.1 WATERMARK REMOVAL FOR PICTURES

Equation 4.1 shows the general model of watermark embedding. We consider the most common situation that $w_o(x)$ and $w_w(x)$ are set to a constant in all x positions belonging to the watermark area. If $w_o(x)$ equals to c, then the model is transformed as follows:

$$I_t(x) = cI_o(x) + (1 - c) I_w(x) \qquad (4.24)$$

Equation 4.24 can be rewritten as follows:

$$I_t(x) - I_w(x) = c (I_o(x) - I_w(x)) \qquad (4.25)$$

In a natural picture, it is very common that the intensity values of adjacent pixels are very close to each other. Thus, a small patch with watermarked area and its neighbors can be considered as having the similar intensity value in the watermark-free picture. Let I_t be the watermarked area and its neighbors be I_o. Figure 4.4 shows the intensity relationship between watermarked area and its neighboring region. After detecting the location and its geometric transformation of the watermark I_w in the watermarked image, we select the entire small patches with watermarked area and its neighbors, we can acquire many patch sets $S = \{S_1, S_2, S_3, \ldots, S_n\}$ and $S_n = \{I_o, I_t, I_w\}$. Then c can be solved by minimizing the least-square errors in Equation 4.25 between $I_t - I_w$ and $I_o - I_w$ using the entire patch S.

FIGURE 4.4   Intensity relationship between watermarked area and the neighboring region.

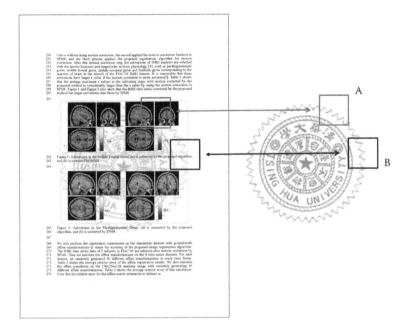

FIGURE 4.5 Detection of the watermark in the watermarked document image.

4.5.2 WATERMARK REMOVAL FOR TEXT IMAGES

The image in the watermarked text region is not changed by the watermark, which is visible only in the background of the document image and has smaller intensity values than the document image. After detecting the location and estimating the geometric transformation of the watermark I_w in the watermarked document image, we use the corresponding sliding windows both in watermarked document and in watermark. If the areas in the corresponding sliding windows are similar, such as the B area in Figure 4.5, intensity thresholding is simply applied to distinguish text and watermark regions based on their intensity distribution. If the areas in the corresponding sliding windows are very different, such as the A area in Figure 4.5, do nothing. A small sliding window is better to recover the watermark-free document image.

4.6 EXPERIMENTS AND DISCUSSIONS

There are two types of watermark removal experiments to demonstrate the effectiveness of the proposed algorithm in this section; namely, the picture and document image watermark removal. In the picture watermark removal experiments, the color Lena image (512×512) is used as the original picture with the watermark of Tsing Hua logo (146×139) imposed with random rigid geometric transform and different blending weight c values as shown in Figure 4.6. The documents are produced by ourselves and the same watermark of Tsing Hua logo is added with random rigid geometric transforms. The watermarked images and documents are printed by HP Color LaserJet CP2020 Series PCL6 and then the printed watermarked images and

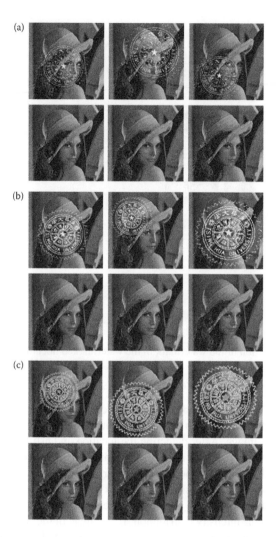

FIGURE 4.6 First row: watermarked images with random watermark positions and scales. The blending weight c value is set to (a) 0.8, (b) 0.5, and (c) 0.3. Second row: the corresponding watermark-removed images by using the proposed algorithm.

documents are scanned by HP Scanjet for practical watermark removal from paper-copy documents.

When there is no noise or picture in the watermarked document image, the watermark removal can be simply accomplished by intensity thresholding if the watermark can be precisely detected and aligned from the image. However, there could be considerable noise corruption and image degradation after image scanning in practice. Figure 4.7 depicts an example of the document images before and after paper scanning. It is obvious that the image quality is greatly degraded after paper scanning. Figure 4.8 shows an example of the watermark removal result on a scanned document image containing pictures and texts by using the proposed watermark removal

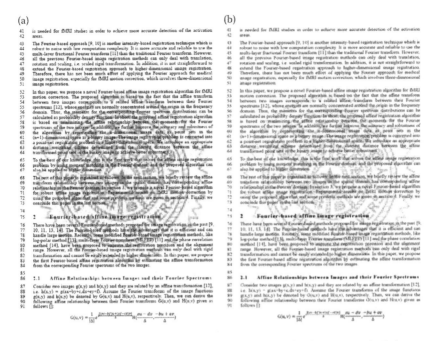

FIGURE 4.7 (a) The ideal watermarked document and (b) the watermarked document image after paper scanning.

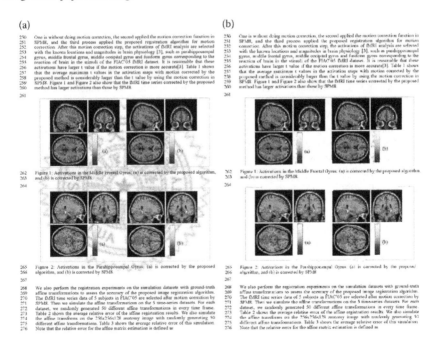

FIGURE 4.8 (a) A scanned watermarked document image and (b) the watermark-removed document image by using the proposed algorithm.

algorithm. It is clear that the proposed algorithm provides a satisfactory watermark removal result in this experiment.

4.7 CONCLUSIONS

In this chapter, we presented a watermark removal algorithm to remove visible watermarks from picture or text images. Our algorithm is based on first applying a robust Fourier-moment-based image registration algorithm for watermark detection, followed by a watermark removal process for picture and text images. Experimental results demonstrate satisfactory watermark removal results for both the picture and document images. In the future, we aim to extend this robust Fourier-moment-based image registration algorithm to overcome more challenging watermark-embedding problems.

REFERENCES

Averbuch, A., Coifman, R. R., Donoho, D. L., Elad, M., and Israeli, M., Fast and accurate polar Fourier transform, *Appl. Comput. Harmon. Anal.*, 21, 145–167, 2006.

Barni, M., Bartolini, F., Cappellini, V., and Piva, A., A DCT-domain system for robust image watermarking, *Signal Process.*, 66(3), 357–372, 1998.

Bracewell, R. N., Chang, K. Y., Jha, A. K., and Wang, Y. H., Affine theorem for two-dimensional Fourier transform, *Electron. Lett.*, 29(3), 304–309, 1993.

Braudaway, G. W., Magerlein, K. A., and Mintzer, F., Protecting publicly-available images with a visible image watermark, in *Optical Security and Counterfeit Deterrence Techniques*, Vol. 2659, R. L. van Renesse (ed.), San Jose, CA: ISandT and SPIE, pp. 126–133, 1996.

De Castro, E. and Morandi, C., Registration of translated and rotated images using finite Fourier transforms, *IEEE Trans. Pattern Anal. Mach. Intell.*, 3, 700–703, 1987.

Hu, Y. and Kwong, S., Wavelet domain adaptive visible watermarking, *IET Electron. Lett.*, 37(20), 1219–1220, September 2001.

Hu, Y. J., Kwong, S., and Huang, J., An algorithm for removable visible watermarking, *IEEE Trans. Circuits Syst. Video Technol.*, 16(1), 129–133, 2006.

Huang, H. C., Chen, T. W., Pan, J. S., and Ho, J. H., Copyright protection and annotation with reversible data hiding and adaptive visible watermarking, *Second International Conference on Innovative Computing, Information and Control*, pp. 292–292, September 2007.

Kankanhalli, M. S., Rajmohan, and Ramakrishnan, K. R., Adaptive visible watermarking of images, in *Proceedings of the IEEE Conference Multimedia Computing and Systems*, vol. 1, pp. 568–573, 1999.

Liu, T. Y. and Tsai, W. H., Generic lossless visible watermarking—A new approach, *IEEE Trans. Image Process.*, 19(5), 1224–1235, 2010.

Pan, W., Qin, K., and Chen, Y., An adaptable-multilayer fractional Fourier transform approach for image registration, *IEEE Trans. Pattern Anal. Mach. Intell.*, 31(3), 400–413, 2009.

Pei, S. C. and Zeng, Y. C., A novel image recovery algorithm for visible watermarked images, *IEEE Trans. Inf. Forensics Sec.*, 1(4), 543–550, 2006.

Pitas, I., A method for watermark casting on digital images, in *Proceedings of the IEEE International Conference on Image Processing*, vol. 3, pp. 215–218, 1996.

Pitas, I. and Kaskalis, P. H., Applying signatures on digital images, in *Proceedings of the IEEE Workshop Nonlinear Image and Signal Processing*, Neos Marmaros, Greece, pp. 460–463, June 1995.

Rao, A. R., Braudaway, G. W., and Mintzer, F. C., Automatic visible watermarking of images, *Proceedings of the SPIE, Optical Security and Counterfeit Deterrence Techniques II*, vol. 3314, pp. 110–121, 1998.

Reddy, B. S. and Chatterji, B. N., An FFT-based technique for translation, rotation, and scale-invariant image registration, *IEEE Trans. Pattern Anal. Mach. Intell.*, 5(8), 1266–1270, 1996.

Van Schyndel, R. G., Tirkel, A. Z., and Osborne, C. F., A digital watermark, in *Proceedings of the IEEE International Conference on Image Processing*, vol. 2, pp. 86–90, November 1994.

Xia, X. G., Boncelet, C., and Arce, G., Wavelet transform based watermark for digital images, *Opt. Express*, 3(12), 497–511, 1998.

5 Digital Watermarking Based on Chaotic Map and Reference Register

Yi-Ta Wu and Frank Y. Shih

CONTENTS

5.1 INTRODUCTION

Since multimedia technologies have been becoming increasingly sophisticated in the rapidly growing Internet applications, data security, including copyright protection and data integrity detection, has raised tremendous concerns. One solution to achieve data security is the digital watermarking technology that embeds hidden information or secret data in a host image (Berghel and O'Gorman, 1996; Eggers and Girod, 2002; Wu, 2005). It serves as a suitable tool to identify the source, creator, owner, distributor, or authorized consumer of a document or an image. It can also be used to detect whether a document or an image is illegally distributed or modified.

There are two domain-based watermarking techniques: one in the spatial domain and the other in the frequency domain. In the spatial domain (Voyatzis and Pitas, 1996; Nikolaidis and Pitas, 1998; Wong, 1998; Celik et al., 2002; Mukherjee et al., 2004), we insert watermarks into a host image by changing the gray levels of certain pixels. The embedding capacity may be large, but the hidden information could be easily detected by means of computer analysis. In the frequency domain (Cox et al., 1997; Lin and Chen, 2000; Shih and Wu, 2003, 2005; Wang et al., 2003; Miller et al., 2004; Wu and Shih, 2004, 2006; Zhao et al., 2004), we insert watermarks into frequency coefficients of the image transformed mainly by *Discrete Fourier Transform*, *Discrete Cosine Transform* (DCT), or *Discrete Wavelet Transform*. The hidden information is in general difficult to detect, but we cannot embed a large volume of watermarks in the frequency domain due to significant image distortion.

In general, the frequency-domain watermarking would be robust since the embedded watermarks are spread out all over the spatial extent of an image (Cox et al., 1997). If the watermarks are embedded into locations of large absolute values (we name "*significant coefficients*") of the transformed image, the watermarking technique would become more robust. Unfortunately, the transformed images mostly contain only a few significant coefficients, so the watermarking capacity is limited.

Zhao et al. (2004) presented a wavelet-domain watermarking algorithm based on chaotic map. They divided an image into a set of 8×8 blocks, and selected the first 256 blocks for embedding watermarks. Miller et al. (2004) proposed a watermarking algorithm using informed coding and embedding. They could embed 1380 bits of data in an image of size 240×368; that is, an embedding capacity of 0.015625 bits/ pixel, which still leaves room for improvement.

In this chapter, we propose the strategy of enlarging the watermarking capacity by breaking local spatial similarity in an image to generate more significant coefficients of the transformed image (Wu and Shih, 2007). The rest is organized as follows. We present the block-based chaotic map (BBCM) in Section 5.2. We describe the reference register (RR) in Section 5.3. We explain the proposed watermarking algorithm in Section 5.4. We provide experimental results in Section 5.5. Finally, we draw conclusions in Section 5.6.

5.2 BLOCK-BASED CHAOTIC MAP

In principle, the amount of significant coefficients of a transformed image is limited due to local image similarity. We notice that if the pixel locations are rearranged by destroying the image similarity, the number of significant coefficients in the transformed image would be increased. Figures 5.1a–c show an image, its pixel values, and its transformed image by DCT, respectively. Figures 5.1d–f show the relocated image (i.e., the noisy image), its pixel values, and its transformed image by DCT, respectively. In this example, we define the absolute value >70 to be significant, and there are totally 14 significant coefficients in Figure 5.1f, but only two significant coefficients in Figure 5.1c. In general, the higher degree of noises an image has, the larger amount of significant coefficients its transformed image possesses. We adopt the chaotic map in this chapter to mess up pixels for obtaining noisy images.

In mathematics and physics, chaos theory deals with the behavior of certain nonlinear dynamic systems that under certain conditions exhibit a phenomenon known as *chaos*, which is characterized by a sensitivity to initial conditions. The chaotic maps can be considered as a tool to relocate the pixels of an image. Voyatzis and Pitas (1996) presented a well-known chaotic system called "toral automorphism" to a squared image, which first rearranges the locations of an image and then embeds the watermark into the spatial domain of the relocated image. The chaotic map for changing pixel (x, y) to (x', y') can be represented by

$$\begin{bmatrix} x' \\ y' \end{bmatrix} = \begin{bmatrix} 1 & 1 \\ l & l+1 \end{bmatrix} \begin{bmatrix} x \\ y \end{bmatrix} \mod N \tag{5.1}$$

(a)   (b)   (c)

215	201	177	145	111	79	55	41
215	201	177	145	111	79	55	41
215	201	177	145	111	79	55	41
215	201	177	145	111	79	55	41
215	201	177	145	111	79	55	41
215	201	177	145	111	79	55	41
215	201	177	145	111	79	55	41
215	201	177	145	111	79	55	41

1024.0	499.4	0.0	1.7	0.0	1.2	0.0	1.4
0.0	0.0	0.0	0.0	0.0	0.0	0.0	0.0
0.0	0.0	0.0	0.0	0.0	0.0	0.0	0.0
0.0	0.0	0.0	0.0	0.0	0.0	0.0	0.0
0.0	0.0	0.0	0.0	0.0	0.0	0.0	0.0
0.0	0.0	0.0	0.0	0.0	0.0	0.0	0.0
0.0	0.0	0.0	0.0	0.0	0.0	0.0	0.0
0.0	0.0	0.0	0.0	0.0	0.0	0.0	0.0

(d)   (e)   (f)

215	201	177	145	111	79	55	41
201	177	145	111	79	55	41	215
177	145	111	79	55	41	215	201
145	111	79	55	41	215	201	177
111	79	55	41	215	201	177	145
79	55	41	215	201	177	145	111
55	41	215	201	177	145	111	79
41	215	201	177	145	111	79	55

1024.0	0.0	0.0	0.0	0.0	0.0	0.0	0.0
0.0	110.8	244.0	−10.0	35.3	−2.6	10.5	−0.5
0.0	244.0	−116.7	−138.2	0.0	−27.5	0.0	−6.4
0.0	−10.0	−138.2	99.0	72.2	0.2	13.3	0.3
0.0	35.3	0.0	72.2	−94.0	−48.2	0.0	−7.0
0.0	−2.6	−27.5	0.2	−48.2	92.3	29.9	0.4
0.0	10.5	0.0	13.3	0.0	29.9	−91.3	−15.7
0.0	−0.5	−6.4	0.3	−7.0	0.4	−15.7	91.9

FIGURE 5.1 An example of increasing the number of significant coefficients. (a) and (b) An image and its gray values, (c) the DCT coefficients, (d) and (e) the relocated image and its gray values, (f) the DCT coefficients.

where $\det\left(\begin{bmatrix} 1 & 1 \\ l & l+1 \end{bmatrix}\right) = 1$ or -1; l denotes an integer; N denotes the width of a square image. Let $l = 2$ and $N = 101$ in the case of Figure 5.2a of size 101×101. When we use Equation 5.1 to rearrange the image, it is obvious that the reconstructed image will be the same as the original image after 17 iterations. Figures 5.2b1–b17 are the iterative noisy images after applying the chaotic map.

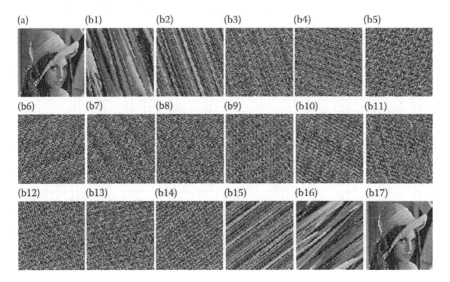

FIGURE 5.2 An example of chaotic map. (a) The original image and (b1–b17) the relocated images in each iteration.

In order to obtain proper coefficients from a transformed image for embedding and extracting watermarks, the concept of RR is developed. Its ability is based on the local image similarity. Therefore, we create the BBCM to partially break the local image similarity in the unit size of block (i.e., a set of connected pixels) instead of pixel. In principle, the bigger the block size is, the larger the local similarity is.

5.3　REFERENCE REGISTER

In this section, we present the *intersection-based pixels collection* (IBPC) method to compute the RR efficiently. The IBPC labels image pixels using two symbols alternatively, and then collects the same symbol to form two subimages. The RR is designed to locate significant DCT coefficients, where watermarks can be embedded. Figures 5.3a–c show three collection approaches in horizontal, vertical, and diagonal directions, respectively. The two subimages are constructed as Figures 5.3d and e. Afterwards, one is used as the RR for indicating significant coefficients, and the other as the container for embedding watermarks.

Note that the pair of subimages generated possess local spatial similarity. From experiments, we observe that the diagonal direction produces better similarity for watermarking. Next, we design a table for nonuniform luminance quantization. It is well known that human perception is more sensitive to low frequencies and less sensitive to high frequencies. Let $QTable(i, j)$ denote the quantization table, where $0 \leq i, j \leq 7$. After applying scale-up (or scale-down) by *quality factor* (QF), we obtain the new quantization table NewTable (i, j) as

$$NewTable(i, j) = \begin{cases} QTable(i, j) \times \dfrac{50}{QF} & \text{if } QF < 50 \\ QTable(i, j) \times (2 - 0.02 \times QF) & \text{otherwise} \end{cases} \quad (5.2)$$

(a)

A	A	A	A	A	A	A	A	A	A	A	A	A	A	A	A
B	B	B	B	B	B	B	B	B	B	B	B	B	B	B	B
A	A	A	A	A	A	A	A	A	A	A	A	A	A	A	A
B	B	B	B	B	B	B	B	B	B	B	B	B	B	B	B
A	A	A	A	A	A	A	A	A	A	A	A	A	A	A	A
B	B	B	B	B	B	B	B	B	B	B	B	B	B	B	B
A	A	A	A	A	A	A	A	A	A	A	A	A	A	A	A
B	B	B	B	B	B	B	B	B	B	B	B	B	B	B	B

(b)

A	B	A	B	A	B	A	B	A	B	A	B	A	B	A	B
A	B	A	B	A	B	A	B	A	B	A	B	A	B	A	B
A	B	A	B	A	B	A	B	A	B	A	B	A	B	A	B
A	B	A	B	A	B	A	B	A	B	A	B	A	B	A	B
A	B	A	B	A	B	A	B	A	B	A	B	A	B	A	B
A	B	A	B	A	B	A	B	A	B	A	B	A	B	A	B
A	B	A	B	A	B	A	B	A	B	A	B	A	B	A	B
A	B	A	B	A	B	A	B	A	B	A	B	A	B	A	B

(c)

A	B	A	B	A	B	A	B	A	B	A	B	A	B	A	B
B	A	B	A	B	A	B	A	B	A	B	A	B	A	B	A
A	B	A	B	A	B	A	B	A	B	A	B	A	B	A	B
B	A	B	A	B	A	B	A	B	A	B	A	B	A	B	A
A	B	A	B	A	B	A	B	A	B	A	B	A	B	A	B
B	A	B	A	B	A	B	A	B	A	B	A	B	A	B	A
A	B	A	B	A	B	A	B	A	B	A	B	A	B	A	B
B	A	B	A	B	A	B	A	B	A	B	A	B	A	B	A

(d)

A	A	A	A	A	A	A	A
A	A	A	A	A	A	A	A
A	A	A	A	A	A	A	A
A	A	A	A	A	A	A	A
A	A	A	A	A	A	A	A
A	A	A	A	A	A	A	A
A	A	A	A	A	A	A	A
A	A	A	A	A	A	A	A

(e)

B	B	B	B	B	B	B	B
B	B	B	B	B	B	B	B
B	B	B	B	B	B	B	B
B	B	B	B	B	B	B	B
B	B	B	B	B	B	B	B
B	B	B	B	B	B	B	B
B	B	B	B	B	B	B	B
B	B	B	B	B	B	B	B

FIGURE 5.3　(a), (b), and (c) respectively show three approaches of collecting the pixels with same label in horizontal, vertical, and diagonal directions, (d) and (e) the generated two subimages.

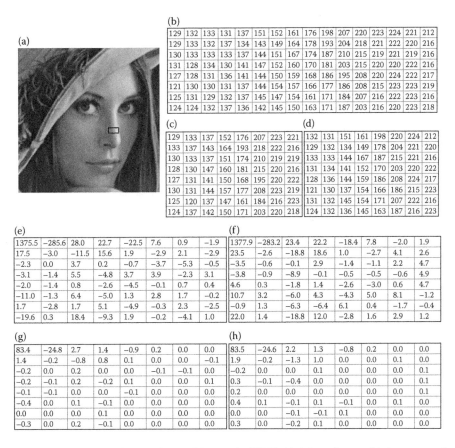

FIGURE 5.4   The similarity of two subimages by IBPC. (a) A Lena image, (b) the extracted 16×8 image, (c) and (d) the pair of extracted 8×8 subimages, (e) and (f) the 8×8 DCT coefficients, (g) and (h) the quantized images.

Figure 5.4a shows the Lena image with a 16×8 region being cropped as in Figure 5.4b. Figures 5.4c and d show the pair of subimages obtained by the diagonal IBPC. Figures 5.4e and f are their respective DCT images. Figures 5.4g and h show the results after dividing Figures 5.4e and f by the quantization table with QF = 50, respectively. We observe that the two results are very similar.

5.4   PROPOSED WATERMARKING ALGORITHM

The proposed algorithm consists of two procedures: watermark embedding and watermark extracting. During the watermark-embedding procedure, the block-based pixels of an original image are first relocated by BBCM to obtain a noisy image. Second, the noisy image is separated into 16×8 subimages. Third, a pair of 8×8 containers and an RR are obtained based on IBPC from each 16×8 subimage. Fourth, a DCT and a scale-down procedure are applied on the container and RR. Fifth, the proper coefficients are located by the RR, and the watermarks are embedded into the corresponding locations in the container to obtain the frequency-domain water-marked image. After performing the scale-up procedure and an inverse DCT on the 8×8 frequency-domain watermarked image, we obtain the 8×8 spatial-domain watermarked image and form a new 16×8 watermarked image by combining with the original 8×8 RR. Then, the watermarked noisy image is obtained by collecting all the 16×8 watermarked images. Finally, the watermarked image is obtained by continuously performing the BBCM on the watermarked noisy image. The embedding procedure is described below, and its flowchart is shown in Figure 5.5.

The Embedding Procedure:

1. Rearrange a host image H of size $n \times m$ to H^C by BBCM.
2. Divide a host image H^C into a set of subimages, $H_{(i,j)}^{16 \times 8}$, of size 16×8, where $1 \le i \le \lfloor n/16 \rfloor$ and $1 \le j \le \lfloor m/8 \rfloor$.
3. Build $\mathrm{HA}_{(i,j)}^{8 \times 8}$ and $\mathrm{HB}_{(i,j)}^{8 \times 8}$ from each subimage $H_{(i,j)}^{16 \times 8}$ by diagonal IBPC.
4. Obtain $\mathrm{DA}_{(i,j)}^{8 \times 8}$ and $\mathrm{DB}_{(i,j)}^{8 \times 8}$ from $\mathrm{HA}_{(i,j)}^{8 \times 8}$ and $\mathrm{HB}_{(i,j)}^{8 \times 8}$ by DCT, respectively.
5. Obtain $\mathrm{QA}_{(i,j)}^{8 \times 8}$ and $\mathrm{QB}_{(i,j)}^{8 \times 8}$ from dividing $\mathrm{DA}_{(i,j)}^{8 \times 8}$ and $\mathrm{DB}_{(i,j)}^{8 \times 8}$ by the JPEG quantization table, respectively.
6. Determine the proper positions of significant coefficients in $\mathrm{QB}_{(i,j)}^{8 \times 8}$. Let $V_{(k,l)}^C$ and $V_{(k,l)}^R$ denote the values of $\mathrm{QA}_{(i,j)}^{8 \times 8}$ and $\mathrm{QB}_{(i,j)}^{8 \times 8}$, respectively. Let $S_{(k,l)}^C$ denote the result after embedding the message. Let α be a weighting factor and $\mathrm{RR_{Th}}$ be the threshold for significant coefficients. If the embedding message is "1" and $V_{(k,l)}^R \ge \mathrm{RR_{Th}}$, we obtain $S_{(k,l)}^C = V_{(k,l)}^C + \alpha V_{(k,l)}^R$; otherwise, we set $S_{(k,l)}^C = V_{(k,l)}^C$. The 8×8 $S_{(k,l)}^C$'s are collected to form the corresponding $E_{(i,j)}^{8 \times 8}$. In general, a bigger α will produce the higher robustness of watermarking.
7. Obtain $M_{(i,j)}^{8 \times 8}$ from multiplying $E_{(i,j)}^{8 \times 8}$ by the JPEG quantization table.
8. Obtain $I_{(i,j)}^{8 \times 8}$ by applying the *Inverse Discrete Cosine Transformation* (IDCT) on $M_{(i,j)}^{8 \times 8}$.
9. Reconstruct $C_{(i,j)}^{16 \times 8}$ by combining $I_{(i,j)}^{8 \times 8}$ and $\mathrm{HB}_{(i,j)}^{8 \times 8}$ using diagonal IBPC.
10. Obtain the output watermarked image O^C by collecting all $C_{(i,j)}^{16 \times 8}$'s.
11. Obtain the final watermarked image O by continuously applying BBCM.

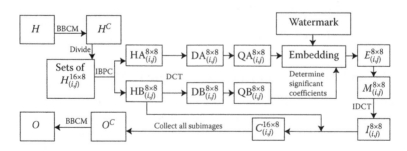

FIGURE 5.5 The embedding procedure.

After receiving the watermarked image O, we extract the watermark information. During the watermark-extracting procedure, the pair of 8×8 frequency-domain containers and RR are obtained by using the same procedure as in the embedding procedure. Similarly, the locations for hiding the watermarks are selected by the RR, and the watermarks are obtained from the corresponding locations in the container. Meanwhile, after removing the watermark from the container, we reconstruct the original 8×8 image by the scale-up procedure and the inverse DCT. Finally, the original image is obtained by combining all the 16×8 recovered image following the BBCM procedure. The extracting procedure is presented below, and its flowchart is shown in Figure 5.6.

The Extracting Procedure:

1. Rearrange the watermarked image O of size $n \times m$ to O^C by BBCM.
2. Divide O^C into a set of subimages, $O_{(i,j)}^{16 \times 8}$, of size 16×8.
3. Build $\text{OA}_{(i,j)}^{8 \times 8}$ and $\text{OB}_{(i,j)}^{8 \times 8}$ from each subimage $O_{(i,j)}^{16 \times 8}$ by IBPC.
4. Obtain $\text{TA}_{(i,j)}^{8 \times 8}$ and $\text{TB}_{(i,j)}^{8 \times 8}$ from $\text{OA}_{(i,j)}^{8 \times 8}$ and $\text{OB}_{(i,j)}^{8 \times 8}$ by DCT, respectively.
5. Obtain $\text{RA}_{(i,j)}^{8 \times 8}$ and $\text{RB}_{(i,j)}^{8 \times 8}$ from dividing $\text{TA}_{(i,j)}^{8 \times 8}$ and $\text{TB}_{(i,j)}^{8 \times 8}$ by the JPEG quantization table, respectively.
6. Determine the extracting location by checking the significant coefficients of $\text{RB}_{(i,j)}^{8 \times 8}$. Let $W_{(k,l)}^C$ and $W_{(k,l)}^R$ denote the values of $\text{RA}_{(i,j)}^{8 \times 8}$ and $\text{RB}_{(i,j)}^{8 \times 8}$, respectively. Let w denote the embedded message. If $(W_{(k,l)}^C - W_{(k,l)}^R) \geq (\alpha/2)W_{(k,l)}^R$, w is set to be 1; otherwise, w is 0.
7. Collect all the subwatermarks to obtain the watermark.
8. Let $F_{(k,l)}^C$ denote the result after the additional amount is removed. If w is 1, we calculate $F_{(k,l)}^C = W_{(k,l)}^C - \alpha W_{(k,l)}^R$; otherwise, we set $F_{(k,l)}^C = W_{(k,l)}^C$. The 8×8 $F_{(k,l)}^C$'s are collected to form the corresponding $U_{(i,j)}^{8 \times 8}$.
9. Obtain $\overline{M}_{(i,j)}^{8 \times 8}$ from multiplying $U_{(i,j)}^{8 \times 8}$ by the JPEG quantization table.
10. Obtain $\overline{I}_{(i,j)}^{8 \times 8}$ by applying the IDCT on $\overline{M}_{(i,j)}^{8 \times 8}$.
11. Reconstruct $\overline{C}_{(i,j)}^{16 \times 8}$ by combining $\overline{I}_{(i,j)}^{8 \times 8}$ and $\text{OB}_{(i,j)}^{8 \times 8}$ using diagonal IBPC.
12. Obtain the recovered image RC^C by collecting all $\overline{C}_{(i,j)}^{16 \times 8}$'s.
13. Obtain the final recovered image RC by continuously applying BBCM.

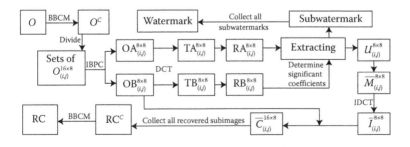

FIGURE 5.6   The extracting procedure.

5.5 EXPERIMENTAL RESULTS

We conducted experiments using 200 images and provide comparisons with the "iciens" algorithm by Miller et al. (2004). Figure 5.7a shows the relocated image of Figure 5.2a after performing seven iterations of relocation using BBCM, where the block size is 2×2 and $l = 2$. We obtain Figure 5.7b from the small region in (a), and generate (c) and (d) by diagonal IBPC. Figures 5.7e and f are, respectively, obtained by DCT followed by a division of the quantization table, where $QF = 80$. Note that, Figure 5.7e is used as the container for watermark embedding, and (f) is used as the RR as colored in gray. We embed the watermark as "11111111," so that the original image could be distorted to a large degree. Figure 5.7g provides the result in the frequency domain after embedding watermarks, where $\alpha = 0.5$ and $RR_{Th} = 9$. Figure 5.7h shows the watermarked results in the spatial domain.

In the extracting procedure, our goal is to not only correctly obtain the embedded message, but also successfully recover the original image from the watermarked one.

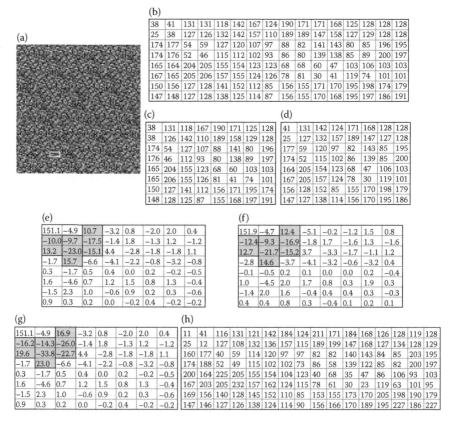

FIGURE 5.7 The embedding strategy. (a) A relocated image, (b) a 16×8 subimage, (c) and (d) a pair of extracted results using IBPC, (e) a container, (f) a RR, (g) the result after embedding watermarks, and (h) the watermarked result.

An example of the watermarked result of size 16×8 is shown in Figure 5.7h. We separate it into two subimages using IBPC as shown in Figures 5.8a and b. Note that, since Figure 5.8b is the same as Figure 5.7d, we generate the same RR. After obtaining Figure 5.8c by DCT, we can extract the watermark as "11111111" and obtain Figure 5.8d. Figure 5.8e shows the recovered result, which is exactly the same as Figure 5.7b.

Miller et al. (2004) stated that their watermarking scheme is robust if the *message error rate* (MER) is lower than 20%. We tested our proposed algorithm under JPEG compression. Figure 5.9a is the original Lena image. After continuously performing nine times of BBCM with block size of 2×2, $l = 2$, $\alpha = 0.2$, and $RR_{Th} = 1$, we obtain Figure 5.9b. Figure 5.9c is the watermarked image. Figure 5.9d is the image attacked by JPEG compression using QF = 20. Figure 5.9e is the reconstructed image after the embedded watermarks are removed. The result shows that the proposed watermarking technique obtains MER = 0.19 which is <20%, and the capacity of 0.116 bits/pixel with PSNR = 35.24 which is much higher than the "iriens" algorithm (Miller et al., 2004) which achieves the capacity of 0.015625 bits/pixel with PSNR = 37.85.

(a)

11	116	121	184	211	184	126	119
12	108	136	115	199	168	134	129
160	40	114	97	82	140	84	203
188	49	102	73	58	122	82	197
200	225	155	104	40	35	86	93
203	232	162	115	61	23	63	95
169	140	145	110	153	173	205	190
146	126	124	90	166	189	227	227

(b)

41	131	142	124	171	168	128	128
25	127	132	157	189	147	127	128
177	59	120	97	82	143	85	195
174	52	115	102	86	139	85	200
164	205	154	123	68	47	106	103
167	205	157	124	78	30	119	101
156	128	152	85	155	170	198	179
147	127	138	114	156	170	195	186

(c)

151.0	-5.0	16.8	-3.1	0.8	-2.0	2.0	0.4
-16.2	-14.3	-26.1	-1.3	1.8	-1.3	1.2	-1.2
19.5	-33.9	-22.7	4.4	-2.7	-1.8	-1.8	1.1
-1.7	23.0	-6.6	-4.1	-2.3	-0.8	-3.2	-0.8
0.3	-1.7	0.5	0.4	0.1	0.2	-0.2	-0.5
1.6	-4.6	0.7	1.2	1.5	0.8	1.3	-0.3
-1.5	2.3	1.0	-0.6	0.9	0.2	0.3	-0.6
0.9	0.3	0.2	0.0	-0.2	0.4	-0.2	-0.2

(d)

151.0	-5.0	10.6	-3.1	0.8	-2.0	2.0	0.4
-10.0	-9.7	-17.6	-1.3	1.8	-1.3	1.2	-1.2
13.2	-23.0	-15.1	4.4	-2.7	-1.8	-1.8	1.1
-1.7	15.7	-6.6	-4.1	-2.3	-0.8	-3.2	-0.8
0.3	-1.7	0.5	0.4	0.1	0.2	-0.2	-0.5
1.6	-4.6	0.7	1.2	1.5	0.8	1.3	-0.3
-1.5	2.3	1.0	-0.6	0.9	0.2	0.3	-0.6
0.9	0.3	0.2	0.0	-0.2	0.4	-0.2	-0.2

(e)

38	41	131	131	118	142	167	124	190	171	171	168	125	128	128	128
25	38	127	126	132	142	157	110	189	189	147	158	127	129	128	128
174	177	54	59	127	120	107	97	88	82	141	143	80	85	196	195
174	176	52	46	115	112	102	93	86	80	139	138	85	89	200	197
165	164	204	205	155	154	123	123	68	68	60	47	103	106	103	103
167	165	205	206	157	155	124	126	78	81	30	41	119	74	101	101
150	156	127	128	141	152	112	85	156	155	171	170	195	198	174	179
147	148	127	128	138	125	114	87	156	155	170	168	195	197	186	191

FIGURE 5.8 The extracting strategy. (a) and (b) a pair of extracted 8×8 results by IBPC, (c) an 8×8 watermarked image by DCT, (d) the result after removing the additional amount, and (e) a reconstructed 16×8 image.

FIGURE 5.9   The robustness experiment using JPEG compression. (a) A 202×202 Lena image, (b) the relocated image, (c) the watermarked image, (d) the image attacked by the JPEG compression, and (e) the reconstructed image.

Figures 5.10 and 5.11 show the experiments under low-pass filter and Gaussian noise attacks, respectively. We use the same parameters (block size of 2×2, $l = 2$, $\alpha = 0.2$, and $RR_{Th} = 1$) and a 3×3 low-pass filter with all ones, and obtain that MER is 16% and the capacity is 0.117 bits/pixel with PSNR = 34.91 in Figure 5.10, and the "iciens" algorithm achieves the capacity of 0.015625 bits/pixel with PSNR = 37.71. We use the same parameters and a Gaussian noise with the *standard deviation* (SD) 500 and mean 0 to be added into the watermarked image, and obtain MER is 23% and the capacity is 0.124 bits/pixel with PSNR = 34.58 in Figure 5.11, and the "iciens" algorithm achieves the capacity of 0.015625 bits/pixel with PSNR = 37.08.

Although our algorithm achieves a slightly lower PSNR than "iriens" developed by Miller et al. (2004), the watermark capacity derived by our algorithm is much higher than "iriens." The "iriens" watermarking pursues the robustness by encoding each bit of watermark into several bits using the informed coding and embedding. Our algorithm provides a different perspective of performing the high-capacity watermarking by increasing significant coefficients.

Figure 5.12 shows the effect under JPEG compression with different QFs. For simplicity, we use the symbol "*" to represent our proposed algorithm and "▲" to represent the "iciens" algorithm by Miller et al. (2004). Note that for display convenience, the x-coordinates of JPEG QF are marked in a decreasing order. It is clear that our algorithm produces a slightly higher MER when QF is larger than 40, but a much lower MER when QF is smaller than 40.

FIGURE 5.10 The robustness experiment using low-pass filter. (a) A 202×202 image, (b) the relocated image, (c) the watermarked image, (d) the image attacked by a low-pass filter, and (e) the reconstructed image.

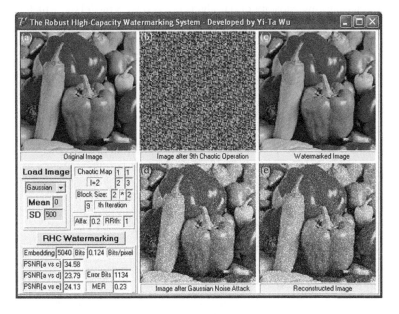

FIGURE 5.11 The robustness experiment using Gaussian noise. (a) A 202×202 image, (b) the relocated image, (c) the watermarked image, (d) the image attacked by Gaussian noise, and (e) the reconstructed image.

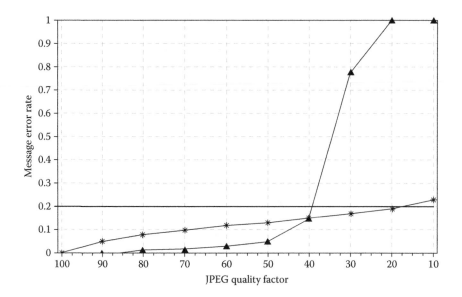

FIGURE 5.12   Robustness versus JPEG compression.

Figure 5.13 shows the case of Gaussian noise with mean 0 and different SDs. The parameters used are block size of 2×2, $l = 2$, $\alpha = 0.2$, and $RR_{Th} = 1$. The result shows that our algorithm outperforms iciens in terms of low MER. Moreover, the average watermarking capacity of our algorithm is 0.1238 bits/pixel, which is much higher than 0.015625 bits/pixel in iciens.

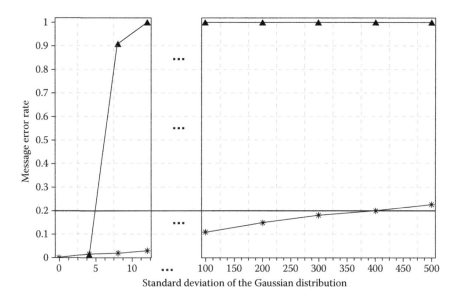

FIGURE 5.13   Robustness versus Gaussian noise.

TABLE 5.1
Effect of RR$_{Th}$

RR$_{Th}$	Capacity (bits)	Capacity (bits/pixel)	JPEG (QF = 20)		Gaussian Noise (SD = 500, mean = 0)	
			Error Bits	MER	Error Bits	MER
10	810	0.0199	104	0.12	140	0.17
9	924	0.0226	120	0.13	170	0.18
8	1061	0.0260	139	0.13	195	0.18
7	1228	0.0301	163	0.13	228	0.19
6	1439	0.0353	197	0.14	271	0.19
5	1715	0.0420	243	0.14	329	0.19
4	2087	0.0511	304	0.15	403	0.19
3	2617	0.0641	404	0.15	522	0.20
2	3442	0.0844	577	0.18	712	0.21
1	5050	0.1238	968	0.19	1116	0.22

Table 5.1 shows effects of threshold RR$_{Th}$ under JPEG compression with QF = 20 and under Gaussian noise with SD = 500 and mean = 0. We observe that our algorithm can enlarge the capacity (i.e., >0.12 bits/pixel), as well as maintain a low MER (i.e., <0.22).

5.6 CONCLUSIONS

The described new watermarking method is based on the BBCM and the RR. The watermarks are embedded into the DCT coefficients for the robust purpose. In order to increase the watermarking capacity, the noisy image is generated based on the BBCM. The watermark-embedding and -extracting procedures are conducted on the DCT coefficients located by the RR. We demonstrate the superiority of the proposed scheme that can enlarge the watermarking capacity as well as maintain a low MER. Experimental results show that the proposed algorithm works accurately under some image-processing distortions, such as the JPEG compression, Gaussian noise, and low-pass filter.

REFERENCES

Berghel, H. and O'Gorman, L., Protecting ownership rights through digital watermarking, *IEEE Comput. Mag.*, 29(7), 101–103, 1996.

Celik, M. U., Sharma, G., Saber, E., and Tekalp, A. M., Hierarchical watermarking for secure image authentication with localization, *IEEE Trans. Image Process.*, 11(6), 585–595, 2002.

Cox, I. J., Kilian, J., Leighton, T., and Shamoon, T., Secure spread spectrum watermarking for multimedia, *IEEE Trans. Image Process.*, 6(12), 1673–1687, 1997.

Eggers, J. and Girod, B., *Informed Watermarking*, Kluwer: Academic, 2002.

Lin, S. D. and Chen, C.-F., A robust DCT-based watermarking for copyright protection, *IEEE Trans. Consum. Electron.*, 46(3), 415–421, 2000.

Miller, M. L., Doerr, G. J., and Cox, I. J., Applying informed coding and embedding to design a robust high-capacity watermark, *IEEE Trans. Image Process.*, 13(6), 792–807, 2004.

Mukherjee, D. P., Maitra, S., and Acton, S. T., Spatial domain digital watermarking of multimedia objects for buyer authentication, *IEEE Trans. Multimed.*, 6(1), 1–15, 2004.

Nikolaidis, N. and Pitas, I., Robust image watermarking in the spatial domain, *Signal Process.*, 66(3), 385–403, 1998.

Shih, F. Y. and Wu, Y. T., Combinational image watermarking in the spatial and frequency domains, *Pattern Recognit.*, 36(4), 969–975, 2003.

Shih, F. Y. and Wu, Y., Enhancement of image watermark retrieval based on genetic algorithm, *J. Vis. Commun. Image Represent.*, 16, 115–133, 2005.

Voyatzis, G. and Pitas, I., Applications of toral automorphisms in image watermarking, in *Proceedings of the IEEE International Conference on Image Processing*, Vol. 2, Lausanne, Switzerland, pp. 237–240, September 1996.

Wang, H., Chen, H., and Ke, D., Watermark hiding technique based on chaotic map, in *Proceedings of the IEEE International Conference on Neural Networks and Signal Processing*, Nanjing, China, pp. 1505–1508, December 2003.

Wong, P. W., A public key watermark for image verification and authentication, in *Proceedings of the IEEE International Conference Image Processing*, Chicago, IL, pp. 425–429, 1998.

Wu, Y. T., *Multimedia Security, Morphological Processing, and Applications*, PhD dissertation, New Jersey Institute Technology, Newark, 2005.

Wu, Y. T. and Shih, F. Y., An adjusted-purpose digital watermarking technique, *Pattern Recognit.*, 37(12), 2349–2359, 2004.

Wu, Y. T. and Shih, F. Y., Genetic algorithm based methodology for breaking the steganalytic systems, *IEEE Trans. Syst. Man Cybern. B*, 36(1), 24–31, 2006.

Wu, Y. and Shih, F. Y., Digital watermarking based on chaotic map and reference register, *Pattern Recognit.*, 40 (12), 3753–3763, 2007.

Zhao, D., Chen, G., and Liu, W., A chaos-based robust wavelet-domain watermarking algorithm, *Chaos Solitons Fractals*, 22, 47–54, 2004.

6 Pseudo-Random Pixel Rearrangement Algorithm Based on Gaussian Integers for Image Watermarking

Aleksey Koval, Frank Y. Shih, and Boris S. Verkhovsky

CONTENTS

6.1 INTRODUCTION

Steganography is a process of hiding information in a medium in such a manner that no one except for the anticipated recipient knows of its existence (Shih, 2008). The history of steganography can be traced back to around 440 B.C.E., where the Greek historian Herodotus described in his writings about two events: one used wax to cover secret messages, and the other used shaved heads. With the explosion of Internet as a carrier for various digital media, many new directions of this state of the art emerged.

A notable application of steganography is watermarking of digital images, which is a useful tool for identifying the source, creator, owner, distributor, or authorized consumer of a document or an image. It has become very easy nowadays to copy or distribute digital images (whether copyrighted or not). A watermark is a pattern of bits inserted into a digital media for copyright protection (Berghel and O'Gorman, 1996). There are two kinds of watermarks: visible and hidden. A good visible

99

watermark must be difficult for an unauthorized person to remove and can resist falsification. Since it is relatively easy to embed a pattern or a logo into a host image, the authorized person must make sure that the visible watermark was indeed the one inserted by the author. In contrast, a hidden watermark is embedded into a host image by some sophisticated algorithm and is invisible to the naked eye. It could, however, be extracted by a computer.

There are many innovating watermarking algorithms and many more get published everyday (such as Al-Qaheri et al., 2010; Huang et al., 2010; Lin and Shiu, 2010; Yamamoto and Iwakiri, 2010). In many image watermarking algorithms, for example (Dawei et al., 2004; Wu and Shih, 2007; Yantao et al., 2008; Ye, 2010), it is required to rearrange the pixels as a part of watermarking process. Randomness is desired during this step. Modular arithmetic and, specifically, the integer exponentiation modulo prime numbers are widely used in modern cryptographic algorithms. One important property of integer exponentiation modulo prime is that it generates a sequence of integers that looks very much like a sequence of random numbers. This is a property that is desirable for pixel rearrangement algorithms. In this chapter, the rearrangement step of watermarking algorithms is revisited and a different universal method for doing it is described. It is easy to replace the rearrangement step in Dawei et al. (2004), Wu and Shih (2007), Yantao et al. (2008), and Ye (2010). Moreover, this method can be used with most image watermarking algorithms to enhance them.

One can look at Gaussian integers as an extension of real integers into two dimensions. They exhibit similar properties as regular integers but have some notable differences that could be exploited in various fields such as cryptography (El-Kassar et al., 2001, 2005; Elkamchouchi et al., 2002; Verkhovsky and Koval, 2008). One important difference is that they have a larger order for the same prime size, which provides the increased security.

Wu and Shih (2007) and Yantao et al. (2008) Arnold's cat map (Arnold and Avez, 1968) was used to rearrange pixels for improving the performance of watermarking techniques. Here a replacement is described, namely, a novel pixel rearrangement algorithm based on Gaussian integers, to rearrange pixels in an image (Koval et al., 2010). It is demonstrated that the new algorithm is superior to Arnold's cat map in both time complexity and security. This technique is not a watermarking algorithm by itself but rather a universal enhancement to any existing watermarking algorithms. The technique tends to increase robustness to noise by uniformly distributing noise throughout the image. The increase in robustness depends on the watermarking algorithm enhanced by the technique.

6.2   OVERVIEW TO GAUSSIAN INTEGERS

A Gaussian integer is a complex number: $Z[i] = \{a + bi : a,b \in \mathbb{Z}\}$, where both a and b are integers. Gaussian integers, with ordinary addition and multiplication of complex numbers, form an integral domain. The norm of a Gaussian integer is a natural number, defined as $|a + bi| = a^2 + b^2$.

The prime elements of $Z[i]$ are also known as Gaussian primes, and can be divided into two subgroups. One subgroup consists of primes $P = (p, 0)$, where p is a real prime and $(p \bmod 4) = 3$, referred to as Blum primes. The second subgroup consists of primes $P = (a, b)$, where $|P|$ is a real prime and $|P| \bmod 4 = 1$ (real non-Blum prime), referred to as non-Blum Gaussian primes.

There is one-to-one relationship between groups modulo real non-Blum primes and non-Blum Gaussian primes. Owing to this fact, the algorithms based on such Gaussian integers do not offer any advantages over real integers. Hence, a subset of $Z[i]$: real primes $p:(p \bmod 4) = 3$ or Blum primes should be considered. This allows the following definition of modulo (mod) operation for Gaussian primes to be

$$A \bmod p = (a \bmod p) + (b \bmod p)i. \tag{6.1}$$

In this chapter, Gaussian integers are denoted with capital letters and real integers with lower case letters. Also, vector notation for Gaussian integers is used (i.e., $G = (a,b)$ is equivalent to $G = a + bi$).

The order for Gaussian integers is defined in the same way as for real integers. k is called an *order* of a Gaussian integer H if $H^k = (1,0) \pmod p$. This is equivalent to stating that

$$\text{ord}(H) \bmod p = k. \tag{6.2}$$

The ideas in Karatsuba and Ofman (1962) could be applied to speed up multiplication of Gaussian integers. It takes three multiplications to multiply two different Gaussian integers. Below is the algorithm inspired by (Karatsuba and Ofman, 1962).

Algorithm 1: Gaussian Integer Multiplication

Given: Gaussian integers (a,b) and (c,d).
Output: Gaussian integer $(x,y) = (a,b)(c,d)$.

$$v_1 := (a + b)(c + d); \quad v_2 := ac; \quad v_3 := bd; \tag{6.3}$$

$$x = v_2 - v_3; \quad y = v_1 - v_2 - v_3; \tag{6.4}$$

It takes two multiplications to square a Gaussian integer:

$$(a,b)^2 = ((a + b)(a - b), ab + ab). \tag{6.5}$$

For the main algorithm described in this chapter, the exponentiation of Gaussian integers modulo prime has to be performed. To do this, any one of the many published real integer exponentiation algorithms that have multiplication and square modulo prime as the basic step could be used. The only thing required to do is to replace real integers with Gaussian integers and use Algorithm 1 for multiplication and Equation 6.5 for squaring. As an example, the widely used left-to-right binary exponentiation algorithm (a form of *Square-and-Multiply* exponentiation algorithm) described in Menezes et al. (1997, Section 14.6) will be adopted.

Algorithm 2: Left-to-Right Binary Exponentiation Algorithm for Gaussian Integers

Given: Gaussian integer $G = (a,b)$, prime p, exponent $e = (e_t\, e_{t-1} \ldots e_1\, e_0)_2$
Output: Gaussian integer $A = G^e \bmod p$

```
1. A:=(1,0)
2. for i:=t downto 0
3.    A:=A² mod p
4. if (eᵢ=1)
5. A:=AG mod p
6. endif
7. endfor
8. return(A)
```

For squaring, on line 3 of Algorithm 2, one can use Equation 6.5 and for multiplication operation Algorithm 1 could be used. There are as many iterations of the main loop as there are bits in the exponent e. One squaring operation is performed for each iteration. Additionally, one multiplication is performed if the current e_i bit is 1. Assuming that e is random, the number of 1's equals to the number of 0's in e on average. Thus, the average number of integer multiplications in Algorithm 2 is

$$3.5 \lfloor \log_2(e) \rfloor. \tag{6.6}$$

The example below demonstrates the steps of Algorithm 2.

Example 1: Illustration of Algorithm 2

Suppose $(1,2)^{357} \bmod 23$ has to be computed. First note that $e = 357_{10} = 101100101_2$. Set $A := (1,0)$ (line 1 of the Algorithm 2). After this, iterate through the bits of $e = 101100101_2$:

1. $i=8$, $e_8=1$. Computing $A=(1,0)^2 \bmod 23=(1,0)$. Since $e_8=1$, compute $A=AG \bmod p=(1,0)(1,2) \bmod 23=(1,2)$.
2. $i=7$, $e_7=0$. Computing $A=(1,2)^2 \bmod 23=(20,4)$.
3. $i=6$, $e_6=1$. Computing $A=(20,4)^2 \bmod 23=(16,22)$. Since $e_6=1$, compute $A=AG \bmod p=(16,22)(1,2) \bmod 23=(18,8)$.
4. $i=5$, $e_5=1$. Computing $A=(18,8)^2 \bmod 23=(7,12)$. Since $e_5=1$, compute $A=AG \bmod p=(7,12)(1,2) \bmod 23=(6,3)$.
5. $i=4$, $e_4=0$. Computing $A=(6,3)^2 \bmod 23=(4,13)$.
6. $i=3$, $e_3=0$. Computing $A=(4,13)^2 \bmod 23=(8,12)$.
7. $i=2$, $e_2=1$. Computing $A=(8,12)^2 \bmod 23=(12,8)$. Since $e_2=1$, compute $A=AG \bmod p=(12,8)(1,2) \bmod 23=(19,9)$.
8. $i=1$, $e_1=0$. Computing $A=(19,9)^2 \bmod 23=(4,20)$.
9. $i=0$, $e_0=1$. Computing $A=(4,20)^2 \bmod 23=(7,22)$. Since $e_0=1$, compute $A=AG \bmod p=(7,22)(1,2) \bmod 23=(9,13)$.

Therefore, $(1,2)^{357} \bmod 23 = (9,13)$

For the main algorithm described in this chapter, it is necessary to find a special Gaussian integer called a generator, as defined below.

Definition 1: Gaussian Integer Generator

A Gaussian integer G is a generator for a Blum Gaussian prime p iff ord(G) = $p^2 - 1$ (mod p).

Generators are easy to find. To find a generator for the Gaussian integer group modulo prime p, the following algorithm is sufficient (improvement is possible for certain cases (Verkhovsky and Sadik, 2009)):

Algorithm 3: A Simple Algorithm for Finding Gaussian Integer Generators

1. Factor $p^2 - 1$:

$$p^2 - 1 = (f_1)^{e_1} (f_2)^{e_2} \ldots (f_k)^{e_k}. \tag{6.7}$$

2. Select a Gaussian integer $G = (a, b)$ such that $a \neq 0$, $b \neq 0$, and $a^2 \neq b^2$ (mod p).
3. For each factor f_i of $p^2 - 1$, compute

$$B_i = G^{(p^2-1)/f_i} \bmod p. \tag{6.8}$$

If any of $B_i = (1,0)$ mod p, then G is not a generator and go to Step 2. Otherwise, G is a generator. The example below demonstrates the steps of Algorithm 3.

Example 2: Illustration of Algorithm 3 (Finding of a Gaussian Generator)

Suppose a Gaussian generator for the prime $p = 23$ has to be found. At first, factor $p^2 - 1$:

$$p^2 - 1 = 23^2 - 1 = 528 = 2^4 \cdot 3 \cdot 11. \tag{6.9}$$

After that, try $G = (1,2)$. To see if G is a generator, compute:

$$(1,2)^{528/2} \bmod 23 = (22,0). \tag{6.10}$$

Since $(22,0)$ is not $(1,0)$, try the next factor of 528: integer 11.

$$i(1,2)^{528/11} \bmod 23 = (2,0). \tag{6.11}$$

Since $(2,0)$ is not $(1,0)$, we obtain that $(1,2)$ is a generator.

6.3   PIXEL REARRANGEMENT ALGORITHM

Algorithm 4: Pixel Rearrangement Based on Gaussian Integers

Given: Image $I = (x,y)$ of size $m \times n$;
Output: Image $I' = (x',y')$ of size $m \times n$;

```
 1. Generate a prime p > max(m,n)and p mod 4=3.
 2. Find a Gaussian integer generator G=(a,b)mod p using
    Algorithm 3.
 3. Generate a random number s, such that 0 < s < p²-1.
 4. S=(sₓ,sᵧ):=Gˢ mod p      (6.12)
 5. while (sₓ≤m or sᵧ≤n)
 6. S:=SG mod p
 7. end-while
 8. C=(c₁,c₂):=S
 9. for i=1 to m
10.   for j=1 to n
11. I'{c₁,c₂}:=I{i,j}         (6.13)
12. C:=CG mod p               (6.14)
13.     while c₁>m or c₂>m
14. C:=CG mod p               (6.15)
15.     end-while
16.   end-for
17. end-for
```

Note that the last value of $C = (c_1, c_2)$ needs to be saved in order to rearrange back the pixels. Without the value of C, pixels could be rearranged back; however, it would require additional computation.

Algorithm 5: Reverse of Algorithm 4

```
 1. Cᵣ :=C                          (6.16)
 2. for i=m downto 1
 3.   for j=n downto 1
 4. I{i,j}:=I'{ c₁, c₂}             (6.17)
 5. Cᵣ:=Cᵣ G⁻¹ mod p               (6.18)
 6.     while (c₁> m or c₂> m)
 7. Cᵣ :=CᵣG⁻¹ mod p               (6.19)
 8.     end-while
 9.   end-for
10. end-for
```

The time complexity of Algorithms 4 and 5 can be defined in terms of p. The most computationally expensive operations of the algorithm are Equations 6.12, 6.14, and 6.18. Suppose that u is the time spent to multiply two integers of size p. Assuming that the Square-and-Multiply algorithm is used for exponentiation and Algorithm 1 is used to multiply two Gaussian integers, the time complexity of Equation 6.12 is approximately:

$$3.5u \log_2 (p^2 - 1) \approx 7u \log_2 p. \tag{6.20}$$

Because the order of Gaussian integers is $p^2 - 1$, in Step 4 of Algorithm 4, $p^2 - 1$ multiplications are performed. Therefore, the number of multiplications required is

$$O(3u(p^2 - 1)) = O(3up^2). \tag{6.21}$$

The total time complexity of Algorithm 4 is

$$O(3u(p^2 - 1) + 7u \log_2 p) = O(up^2). \tag{6.22}$$

The complexity of integer multiplication u depends on the size of p. For small p, the most efficient algorithm is the naive multiplication with time complexity of $O(l^2)$, where $l = \log_2 p$ is the size of p in bits. For a larger p, the multiplication algorithm in Karatsuba and Ofman (1962) is faster than the naive method. The time complexity of Karatsuba multiplication is $O(3l^{1.585})$. For an even larger p, Toom-Cook (or Toom-3) algorithm is more efficient with a time complexity of $O(n^{1.465})$ (Knuth, 1998). The thresholds for the size of p vary widely with implementation details. However, it is reasonable to assume that most images would not be sufficiently large for Toom-Cook or Karatsuba multiplication. Therefore, it can be assumed that the naive multiplication method can be used and Equation 6.22 becomes

$$O(up^2) = O[(p \log_2 p)^2]. \tag{6.23}$$

which is the time complexity of Algorithm 4.

To minimize the time complexity, it is reasonable to select p close to max(m,n). If p is selected in such a way, then the time complexity in terms of image size is

$$O\{[\max(m,n) \log_2 (\max (m,n))]^2\}. \tag{6.24}$$

The rearrangement algorithm described above is universal and can be used for many purposes. It can be applied for image watermarking as follows.

Algorithm 6: Watermarking with Pixel Rearrangement Based on Gaussian Integers

1. Rearrange the image using Algorithm 4.
2. Apply the desired watermarking technique to the resulting rearranged image from Step 1.
3. Apply Algorithm 5 to the resulting image from Step 2.

Algorithm 7: Extraction of the Watermark Applied with Algorithm 4

1. Rearrange the image using Algorithm 4.
2. Extract the watermark using the watermarking extraction technique in Algorithm 5.

Note that in Algorithm 5, depending on the watermarking technique, it may be possible to extract watermark and perform rearrangement on the watermark rather than on the image.

6.4 PROOF OF ALGORITHM VALIDITY

The validity of the algorithms arises from the properties of the Gaussian integer group. In this section, these properties are going to be described and proved. For any two complex numbers A and B, it is true that $|AB| = |A||B|$. Gaussian integer is a special kind of complex number, so it is true for Gaussian integers also. When a Gaussian integer C is multiplied by itself modulo p, in turn, the norm of C mod p gets multiplied by itself also. This means that $|C^i|$mod p ($i = 1, 2, \ldots$) will cycle with a period of ord($|C|$)mod p, as illustrated in Tables 6.1 and 6.2.

In addition, $C^{\text{ord}(|C|)}$ is a Gaussian integer with norm equal to 1 mod p. In fact, the Gaussian integers $U: |U| = 1$ mod p form a cyclic subgroup with an order ($p + 1$). This subgroup will be referred to as a *Norm 1* subgroup. Moreover, the order of any Gaussian integer C is a product of ord($|C|$) and ord($|U|$), where $U = C^{\text{ord}(|C|)}$ mod p. From this, the algorithms for finding Gaussian generators to use for discrete logarithm-based cryptography are derived.

Lemma 1

If C is a complex number and p is a prime, then $|C^n| = |C|^n$ mod p.

Proof:

For any complex number, it is true that $|C^n| = |C|^n$; therefore, $|C^n| = |C|^n$ mod p.
Q.E.D.

Lemma 2

If C is a Gaussian integer and p is a Blum prime, then

- ord(C)mod p is divisible by ord($|C|$)mod p
- if $C^{\text{ord}(|C|)} = U$ mod p, then $|U| = 1$mod p
- if $U = C^{\text{ord}(|C|)}$ mod p, then ord(C)mod p is divisible by ord(U)mod p

Proof:

Suppose that ord(C)mod p is not divisible by ord($|C|$)mod p. This means that $|C^{\text{ord}(C)}|$ mod p is not equal to 1, but $C^{\text{ord}(C)} = (1,0)$. This is a contradiction. $|U|$ must equal to 1 mod p because $|C^n| = |C|^n$ mod p and, in this case, $n = $ ord($|C|$). If ord(C) mod p is not divisible by ord(U), then $C^{\text{ord}(C)}$ would not equal to (1,0), so ord(C) must be divisible by ord(U).
Q.E.D.

TABLE 6.1

Repeating Norm Examples for Prime $p = 7$

Power:	1	2	3	4	5	6	7	8	9	10	11	12	13	14	15	16
Norm:	2	4	**1**	2	4	**1**	2	4	**1**	2	4	**1**	2	4	**1**	4
	(1,6) [2]	(0,5) [4]	**(5,5) [1]**	(3,0) [2]	(3,4) [4]	**(0,1) [1]**	(1,1) [2]	(2,0) [4]	**(2,5) [1]**	(0,3) [2]	(3,3) [4]	**(6,0) [1]**	(6,1) [2]	(0,2) [4]	**(2,2) [1]**	(4,0) [2]
	(1,1) [2]	(0,2) [4]	**(5,2) [1]**	(3,0) [2]	(3,3) [4]	**(0,6) [1]**	(1,6) [2]	(2,0) [4]	**(2,2) [1]**	(0,4) [2]	(3,4) [4]	**(6,0) [1]**	(6,6) [2]	(0,5) [4]	**(2,5) [1]**	(4,0) [2]
Norm:	3	2	6	4	5	**1**	3	2	6	4	5	**1**	3	2	6	4
	(3,1) [3]	(1,6) [2]	(4,5) [6]	(0,5) [4]	(2,1) [5]	**(5,5) [1]**	(3,6) [3]	(3,0) [2]	(2,3) [6]	(3,4) [4]	(5,1) [5]	**(0,1) [1]**	(6,3) [3]	(1,1) [2]	(2,4) [6]	(2,0) [4]
	(4,6) [3]	(1,6) [2]	(3,2) [6]	(0,5) [4]	(5,6) [5]	**(5,5) [1]**	(4,1) [3]	(3,0) [2]	(5,4) [6]	(3,4) [4]	(2,6) [5]	**(0,1) [1]**	(1,4) [3]	(1,1) [2]	(5,3) [6]	(2,0) [4]

Note: The cases of Norm = 1 are in bold face.

TABLE 6.2

Repeating Norm Examples for Prime $p = 11$

Power:	1	2	3	4	5	6	7	8	9	10	11	12	13	14	15
Norm:	2	4	8	5	10	9	7	3	6	**1**	2	4	8	5	10
	(3,2) [2]	(5,1) [4]	(2,2) [8]	(2,10) [5]	(8,1) [10]	(0,8) [9]	(6,2) [7]	(3,7) [3]	(6,5) [6]	**(8,5) [1]**	(3,9) [2]	(2,0) [4]	(6,4) [8]	(10,2) [5]	(4,4) [10]
	(10,1) [2]	(0,9) [4]	(2,2) [8]	(7,0) [5]	(4,7) [10]	(0,8) [9]	(3,3) [7]	(5,0) [3]	(6,5) [6]	**(0,1) [1]**	(10,1) [2]	(2,0) [4]	(9,2) [8]	(0,7) [5]	(4,4) [10]
Norm:	3	9	5	4	**1**	3	9	5	4	**1**	3	9	5	4	**1**
	(3,4) [3]	(4,2) [9]	(4,0) [5]	(1,5) [4]	**(5,8) [1]**	(5,0) [3]	(4,9) [9]	(9,10) [5]	(9,0) [4]	**(5,3) [1]**	(3,7) [3]	(3,0) [9]	(9,1) [5]	(1,6) [4]	**(1,0) [1]**
	(7,8) [3]	(7,2) [9]	(0,4) [5]	(1,6) [4]	**(3,6) [1]**	(6,0) [3]	(9,4) [9]	(9,1) [5]	(0,2) [4]	**(6,3) [1]**	(7,3) [3]	(3,0) [9]	(10,2) [5]	(10,6) [4]	**(0,1) [1]**

Note: The cases of Norm = 1 are in bold face.

Lemma 3

If U is a Gaussian Integer, p is a Blum prime, and $|U| = 1 \bmod p$, then

1. The maximum order of U is $(p + 1)$, and
2. ord$(U) \bmod p$ must divide $(p + 1)$.

Proof:

1. Any Gaussian integer A taken to the power $(p + 1) \bmod p$ is in the form of $(c,0)$.

    In this case, $U^{p+1} \bmod p$ could be one of either $(1,0)$ or $(-1,0)$ because $|U| = 1 \bmod p$. Since $p + 1$ is divisible by 4 for all Blum primes, $U^{(p+1)/4}$ is a Gaussian integer of norm 1 and is a root of degree of U^{p+1}. For $(-1,0)$, all roots of degree 4 have a norm equal to $-1 \bmod p$. This means that $U^{(p+1)}$ must equal to $(1, 0) \bmod p$.
2. If $p + 1$ is not divisible by ord(U), then $U^{(p}+1)$ would not be equal to $(1,0)$, so $p + 1$ must be divisible by ord(U).

Q.E.D.

Lemma 4

If C is a Gaussian integer and p is a Blum prime, then ord$(C) = $ ord$(|C|)$ord$(U) \bmod p$, where $|C^n| = |C|^n \bmod p$.

Proof:

ord(C) must be divisible by ord$(|C|)$ and ord(U), so ord$(C) = n*$ord$(|C|)$ord(U), where n is an integer. In addition, $C^{\text{ord}(|C|)}$ord$^{(U)} = U^{\text{ord}(U)} = (1,0)$. Consequently, n must be equal to 1.
Q.E.D.

6.5 CRYPTOIMMUNITY OF THE REARRANGEMENT ALGORITHM

From the properties of the Gaussian integer group, it could be estimated how hard it is for an adversary to obtain the original image from the rearranged one. The less an adversary knows about the algorithm and parameters, the harder it is to determine the original arrangement. It is reasonable to look at the following three cases:

Case 1. The adversary knows nothing about the rearrangement algorithm used, but one suspects that some kind of algorithm has been used. In this case, it is extremely hard for the adversary to figure out the original arrangement because there are too many possibilities. That is, there are $n!$ possible permutations, where n is the number of pixels in the image.

Case 2. The adversary knows that Algorithm 4 was used, but one does not know the parameters such as prime p, generator G, or private key s. In this case, the number of possible permutations for an image I of size $m \times n$ is

$$(p^2 - 1)[\varphi(p^2 - 1)]. \tag{6.25}$$

where φ is Euler's totient function (Abramowitz and Stegun, 1964).

Equation 6.25 does not include the complexity of guessing p. The reason for this is that it is too computationally expensive to use a large p (refer to Equation 6.23). For efficiency, p should be close to the image size. The prime p in Equation 6.25 can be selected in such a way that $\varphi(p - 1)$ is maximized. To do this, a prime with large prime divisors of $p + 1$ and $p - 1$ could be selected. For example,

$$p + 1 = s_1 q_1 \tag{6.26}$$

and

$$p - 1 = s_2 q_2, \tag{6.27}$$

where s_1 and s_2 are small integers, and q_1 and q_2 are primes close to p in size. In this case,

$$\varphi(p^2 - 1) = \varphi((p - 1)(p + 1)) = \varphi(s_1 s_2)(q_1 - 1)(q_2 - 1) \tag{6.28}$$

and

$$o(\varphi(p^2 - 1)) = o((q_1 - 1)(q_2 - 1)) = o(q_1 q_2) = o(p^2). \tag{6.29}$$

Consequently, the approximate computational complexity is

$$o((p^2 - 1)[\varphi(p^2 - 1)]) = o(p^4) = o(\max{(m,n)^4}). \tag{6.30}$$

Case 3. The adversary knows Algorithm 4 used, prime p, and a generator G. In this case, the number of possible permutations is limited to

$$p^2 - 1. \tag{6.31}$$

While it may be unreasonable to assume that the adversary would not know Algorithm 4, there is no reason to make a prime p and a generator G known. Therefore, case 2 may be the most reasonable security estimate.

If increased protection is desired, Algorithm 4 could be applied several times on the same image. Suppose that Algorithm 4 was applied t times on image I of size $m \times n$. In this case, the number of possible permutations is

$$o(\max{(m,n)}^{4t}), \tag{6.32}$$

while the time to compute the rearranged image would still be reasonable and be on the same order in terms of image size:

$$O\{t[\max{(m,n)}\log_2 \max{(m,n)}]^2\} = O\{[\max{(m,n)}\log_2 \max{(m,n)}]^2\}. \tag{6.33}$$

Therefore, one can achieve the desired level of security by increasing the time it takes to rearrange the image somewhat. Multiple rearrangements could provide a desirable and practical trade-off.

6.6   COMPARISON TO ARNOLD'S CAT MAP CHAOS TRANSFORMATION

The Arnold's cat map transformation variation used in Wu and Shih (2007) is defined as

$$\begin{bmatrix} x' \\ y' \end{bmatrix} = \begin{bmatrix} 1 & 1 \\ l & l+1 \end{bmatrix} \begin{bmatrix} x \\ y \end{bmatrix} \bmod N, \tag{6.34}$$

where N is the width of the square image. The possible values of l in Equation 6.34 are $l: 1 < l < N-2$. Therefore, the number of the transformations required is $O(N)$. It is reasonable to assume that N is small enough to call for naive multiplication algorithms. Thus, the multiplication time complexity is

$$O(\log_2^2 N), \tag{6.35}$$

which has to be performed for every pixel (i.e., N^2 times). Therefore, the time complexity of Arnold's Cat Map is

$$O(N^3 \log_2^2 N). \tag{6.36}$$

Equation 6.26 should be compared with Equation 6.24, assuming $N \approx \max(m,n)$. It is obvious that the computational complexity of Algorithm 4 described by Equation 6.24 is much better than that of Arnold's Cat map described by Equation 6.36.

As far as security is concerned, it is obvious that there are only $o(N)$ possible permutations because $l: 1 < l < N-2$. It is much smaller than $o(\max(m,n)^4)$ for Algorithm 4.

Another important advantage of Algorithm 4 is that the transformed image does not have any visible patterns. After rearrangement with this algorithm, the resulting

FIGURE 6.1 Image rearranged by and Algorithm 4 Arnold's Cat map side-by-side. A is the original image, B is the rearranged image by Algorithm 4, and C1–C7 are the steps of Arnold's Cat map rearrangement.

image looks like random noise. The transformation with Arnold's Cat map, on the other hand, preserves visible patterns. Figure 6.1 clearly illustrates this point. At every step of Arnold's Cat map transformation, C1–C7 patterns are clearly visible. The image B, on the other hand, looks like random noise. Consequently, Algorithm 4, when used for watermarking, is far superior to Arnold's Cat map in terms of security and computational time.

6.7 AN EXAMPLE IN IMAGE WATERMARKING

Algorithm 4 can be used with general watermarking techniques. The following example illustrates its use of applying LSB substitution for watermark. Even though this technique does not provide a robust watermark, the use of rearrangement does improve the security by making the watermark virtually undetectable. When pixel rearrangement is used and the adversary looks at the last two bits of the watermarked image, all one sees is random noise. The only way to see the watermark is to rearrange the pixels.

FIGURE 6.2   (a) The original Cameraman image, (b) the two most significant bits of Lena as the watermark, (c) the rearranged image of Cameraman using Algorithm 4, (d) the water-marked image of the rearranged image using LSB substitution, (e) the rearranged back of the preceding watermarked image using Algorithm 5, (f) the extracted 2 bits of LSB, and (g) the rearranged back of the preceding extracted image using Algorithm 5.

Figure 6.2 illustrates the advantages of using the rearrangement algorithm for image watermarking, where (a) is the original Cameraman image, (b) is the two most significant bits of the Lena image to be used as the watermark, (c) is the rearranged image of Cameraman using Algorithm 4, (d) is the watermarked image of the rear-ranged image using LSB substitution, (e) is the rearranged back of the preceding watermarked image using Algorithm 5, (f) is the extracted 2 bits of LSB, and (g) is the rearranged back of the preceding extracted image using Algorithm 5. Note that image (g) is exactly the same as the original watermark (b).

If the watermarking is performed without rearrangement, then the hidden water-mark is easily detectible. By using the described algorithms, it is impossible to see the original watermark in image (f), which is random noise just like images (c) and (d). It is fairly difficult for the adversary to extract the original watermark, even though one knows that the watermark is hidden there. With sequential applications of Algorithm 4, the security could be enhanced to an arbitrary level, making water-mark practically impossible to reconstruct for the adversary.

6.8 CONCLUSIONS

The described new method of rearranging image pixels for watermarking is based on the properties of Gaussian integers. It results in such a more random-looking image transformation that significantly improves the security of the embedded watermark. Moreover, its speed is much faster as compared with the Arnold cat map. The described algorithm is an easy-to-implement practical technique that would enhance the security of any watermarking algorithm. It is flexible enough to offer variable levels of security.

REFERENCES

Abramowitz, M. and Stegun, I. A. *Handbook of Mathematical Functions*. New York: Dover, 1964.

Al-Qaheri, H., Mustafi, A., and Banerjee, S. Digital watermarking using ant colony optimization in fractional Fourier domain. *J. Inf. Hiding Multimedia Signal Process*, 1(3), 220–240, 2010.

Arnold, V. I. and Avez, A. *Ergodic Problems in Classical Mechanics*. New York: Benjamin, 1968.

Berghel, H. and O'Gorman, L. Protecting ownership rights through digital watermarking. *IEEE Comput. Mag.*, 29(7), 101–103, 1996.

Dawei, Z., Guanrong, C., and Wenbo, L. A chaos-based Robust wavelet-domain watermarking algorithm. *Chaos Solitons Fractals*, 22(1), 47–54, 2004.

El-Kassar, A., Rizk, M., Mirza, N., and Awad, Y. El-Gamal public-key cryptosystem in the domain of Gaussian integers. *Int. J. Appl. Math.*, 7(4), 405–412, 2001.

El-Kassar, A. N., Haraty, R. A., Awad, Y. A., and Debnath, N. C. Modified Rsa in the domains of Gaussian integers and polynomials over finite fields. In *Proceedings of the ISCA 18th International Conference on Computer Applications in Industry and Engineering*, Dascalu, S. (ed.), pp. 298–303. HI, USA: ISCA, 2005.

Elkamchouchi, H., Elshenawy, K., and Shaban, H. Extended Rsa cryptosystem and digital signature schemes in the domain of Gaussian integers. In *Proceedings of the 8th International Conference on Communication Systems*, pp. 91–95, IEEE Computer Society, 2002.

Huang, H.-C., Chen, Y.-H., and Abraham, A. Optimized watermarking using swarm-based bacterial foraging. *J. Inf. Hiding Multimedia Signal Process*, 1(1), 51–58, 2010.

Karatsuba, A. and Ofman, Y. Multiplication of many-digital numbers by automatic computers. *Proceedings of the USSR Academy of Sciences*, Vol. 145, pp. 293–94, 1962.

Knuth, D. E. *The Art of Computer Programming*. 3rd ed., Vol. 2, Redwood City, CA: Addison-Wesley, 1998.

Koval, A., Shih, F. Y., and Verkhovsky, B. S. A pseudo-random pixel rearrangement algorithm based on Gaussian integers for image watermarking. *J. Inf. Hiding Multimedia Signal Process*, 2(1), 60–70, 2010.

Lin, C.-C. and Shiu, P.-F. Highcapacity data hiding scheme for Dct-based images. *J. Inf. Hiding Multimedia Signal Process*, 1(3), 220–240, 2010.

Menezes, A. J., Oorschot, P. C. V., and Vanstone, S. A. *Handbook of Applied Cryptography*, Boca Raton, FL, USA: CRC Press, 1997.

Shih, F. Y. *Digital Watermarking and Steganography: Fundamentals and Techniques*. Boca Raton, FL, USA: Taylor & Francis Group, CRC Press, Inc., 2008.

Verkhovsky, B. and Koval, A. Cryptosystem based on extraction of square roots of complex integers. In *Fifth International Conference on Information Technology: New Generations (ITNG 2008)*, pp. 1190–1191. Las Vegas, NV, USA: IEEE Computer Society, 2008.

Verkhovsky, B. S. and Sadik, S. Accelerated search for Gaussian generator based on triple prime integers. *J. Comput. Sci.*, 5(9), 614–618, 2009.

Wu, Y. and Shih, F. Y. Digital watermarking based on chaotic map and reference register. *Pattern Recognit.*, 40(12), 3753–3763, 2007.

Yamamoto, K. and Iwakiri, M. Real-time audio watermarking based on characteristics of Pcm in digital instrument. *J. Inf. Hiding Multimedia Signal Process*, 1(2), 59–71, 2010.

Yantao, Z., Yunfei, M., and Zhiquan, L. A Robust Chaos-based Dct-domain watermarking algorithm. In *Proceedings of the 2008 International Conference on Computer Science and Software Engineering*, Yanshan University, China, pp. 935–938, IEEE Computer Society, 2008.

Ye, G. Image Scrambling encryption algorithm of pixel bit based on chaos map. *Pattern Recognit. Lett.*, 31(5), 347–354, 2010.

7 Reversible Data-Hiding Techniques for Digital Images

*Zhi-Hui Wang, Ming-Ting Sun,
and Chin-Chen Chang*

CONTENTS

7.1 INTRODUCTION

Data hiding for digital images is the science of embedding data into digital images by making slight changes to the original images (also known as the cover images). The changes are invisible to the naked eye, so that no one would recognize the existence of the secret messages except for the sender and the recipient. Reversible (also called invertible or lossless) data-hiding techniques refer to the schemes that the cover image can be recovered losslessly after extracting the data from the stego image (i.e., the image with data embedded). To the best of our knowledge, there were

TABLE 7.1

Approximate Numbers of Publications on Reversible Data Hiding from 1997 to April 2011

Year	1997	1998	1999	2000	2001	2002	2003	2004	2005	2006	2007	2008	2009	2010	2011
Number of Publications	1	0	0	1	4	8	11	12	11	15	18	25	34	24	5

more than 160 papers published on the subject of reversible data-hiding techniques between 1997 and April 2011, as shown in Table 7.1. Reversible data-hiding techniques have attracted extensive research interest due to the fact that a recoverable cover image is more suitable for sensitive applications, such as military and medical applications. As an example of a medical application, a doctor may wish to embed a patient's private or secret information into the patient's medical images without degrading the images.

The basic requirements of a reversible data-hiding scheme are visual quality of the stego image, hiding capacity, and robustness (Chang and Kieu, 2010). A data-hiding scheme with good visual quality (also called low image distortion) is more secure than one with high distortion because it is less likely that it will attract the attention of potential adversaries. Researchers usually use peak signal-to-noise ratio (PSNR) to measure the difference in visual quality between the cover image and the stego image:

$$PSNR(dB) = 10\log_{10}\left(\frac{255^2}{MSE}\right), \tag{7.1}$$

$$MSE = \frac{1}{M \times N}\sum_{i=1}^{M}\sum_{j=1}^{N}(x(i,j) - x'(i,j))^2, \tag{7.2}$$

where M and N are the width and height of the cover image in pixels, respectively. $x(i,j)$ and $x'(i,j)$ denote the intensities of the pixels located at coordinates (i,j) in both images. A data-hiding scheme with a high-hiding capacity is better than one with low-hiding capacity. Hiding capacity is usually measured in bits per pixel (bpp), which is calculated by $S/(M \times N)$, where S is the number of secret bits embedded into the cover image. In terms of robustness, reversible data-hiding techniques can be divided into two main categories, that is, fragile and semifragile (Caldelli et al., 2010). A fragile technique is one in which the embedded data may not be extracted correctly when the stego image has undergone some modification, such as a JPEG compression. In a semifragile technique, the embedded data are able to survive a possible modification of the stego image. The importance of robustness varies from application to application. This factor is particularly important in watermarking applications, since the watermark must be able to survive attacks to achieve copyright protection or source tracing. Visual quality, hiding capacity, and robustness are three conflicting factors in a data-hiding scheme. Thus, according to users' requirements and the application domains, different trade-offs among these three factors are incorporated into the designs of different data-hiding techniques.

At present, fragile reversible data-hiding techniques can be conducted in three domains, that is, the spatial domain, the transformed domain, and the compressed domain. Semifragile data-hiding techniques can be conducted only in the spatial and transformed domains, because high-level information about the structure of the data stream usually is not available in compressed streams with embedded secret data. In the spatial domain, the values of the pixels of the cover image are altered directly to embed the data. In the transformed domain, the cover image should be preprocessed by a transform, such as the integer wavelet transform (IWT), the discrete cosine transform (DCT), the discrete wavelets transform, or the discrete Fourier transform, to get the frequency coefficients. Then, the frequency coefficients are modified slightly to embed data, and the stego image can be obtained by using the modified frequency coefficients. In the compression domain, the compression code is altered to embed the data.

Most of the published papers related to the reversible data-hiding schemes address fragile techniques because it is difficult to recover the cover image losslessly if the stego image undergoes a lossy modification, such as JPEG compression. A good fragile reversible data-hiding technique should provide high-hiding capacity and low distortion simultaneously. So, the main purpose of fragile techniques is to provide a secure channel to protect communication among legitimate users.

In this chapter, an overview of existing fragile reversible data-hiding schemes is provided. In particular, we first discuss basic schemes in each category mentioned above to provide basic information in this field for the reader. We also used examples to illustrate some complicated schemes, describe the history of the development of the techniques, and show the experimental results of some schemes in each category.

7.2 SPATIAL DOMAIN REVERSIBLE DATA-HIDING TECHNIQUES

From the aspect of embedding strategy, fragile techniques in the spatial domain can be subdivided into two categories, that is, difference expansion (DE)-based techniques and histogram shifting (HS)-based techniques.

7.2.1 DE-Based Techniques

DE refers to expanding the difference between the neighboring pixels to hide secret bits. In the decoding phase, the original difference value can be calculated from the expanded difference value to achieve the reversibility of the cover image. In this section, first, we describe the two most important and basic DE-based reversible data-hiding methods, that is, Tian's method (Tian, 2003) and Thodi and Rodriguez's method (Thodi and Rodriguez, 2007), which provide some basic concepts of DE-based techniques. Then, we review briefly the development history of DE-based techniques, and provide some conclusions based on the experimental results obtained with some DE-based methods.

7.2.1.1   Tian's DE Scheme

Given the average value and difference value of a number pair, the exact values of the two numbers in this number pair can be calculated. Also an integer number is unchanged if it is modulo by two after adding a least significant bit. Tian utilized the above two characteristics to propose the DE scheme. The first characteristic is used to achieve reversibility, and the second characteristic is used to embed the secret bit. The detailed procedure of Tian's scheme is presented as follows.

In the data-embedding phase, given a pixel pair with pixel values (x, y) in a grayscale image, where $0 \leq x, y \leq 255$, the procedure starts with calculating the average l and the difference h,

$$l = \left\lfloor \frac{x + y}{2} \right\rfloor, \quad h = x - y, \tag{7.3}$$

where $\lfloor \ \rfloor$ is the floor function. Assume the datum to be embedded into this pixel pair is bit b, where b is either 0 or 1, the new difference value h' is obtained as follows:

$$h' = 2 \times h + b. \tag{7.4}$$

Then, the stego pixel pair (x', y') can be computed by

$$x' = l + \left\lfloor \frac{h' + 1}{2} \right\rfloor, \quad y' = l - \left\lfloor \frac{h'}{2} \right\rfloor. \tag{7.5}$$

In the data-extracting phase and the pixel pair-recovering phase, the embedded datum bit b is extracted by

$$b = (x' - y') \bmod 2. \tag{7.6}$$

The pixel pair (x, y) can be recovered by

$$h = \left\lfloor \frac{h'}{2} \right\rfloor, \quad x = \left\lfloor \frac{x' + y'}{2} \right\rfloor + \left\lfloor \frac{h + 1}{2} \right\rfloor, \quad y = \left\lfloor \frac{x' + y'}{2} \right\rfloor - \left\lfloor \frac{h}{2} \right\rfloor. \tag{7.7}$$

It is easy to prove that the cover image can be recovered losslessly by considering even and odd values of x and y. Since the first step of Tian's scheme is the same as the procedure of horizontal wavelet transform (HWT), it can also be classified into one of the transformed domain schemes.

7.2.1.2   Thodi and Rodriguez's Prediction Error Expansion (PEE) Scheme

Since a number pair is the smallest unit in the DE scheme, Tian used two adjacent pixels as a number pair to embed one secret bit. Thodi and Rodriguez found that for every pixel except those pixels located in the first row and the first column in the image, a predicted value can be calculated by its top, top-left, and left pixels. The pixel value and its predicted value can form a new kind of number pair to embed one secret bit. In this way, it is possible to increase the hiding capacity from one bit per pixel pair to almost one bit per pixel. The detailed embedding procedure is discussed as follows. Given a grayscale cover pixel with pixel value x, where $0 \leq x \leq 255$, the three-neighbor context of x, as shown in Figure 7.1, is chosen to compute the predicted value p by

$$p = f(x_1, x_2, x_3) = \begin{cases} \max(x_1, x_3), & \text{if } x_2 < \min(x_1, x_3), \\ \min(x_1, x_3), & \text{if } x_2 \geq \max(x_1, x_3), \\ x_1 + x_3 - x_2, & \text{otherwise.} \end{cases} \qquad (7.8)$$

The modified predicted value p' of x used to embed the secret bit is an even number computed by

$$p' = 2 \times \left\lfloor \frac{p}{2} \right\rfloor. \qquad (7.9)$$

The prediction error e between p' and x is obtained by

$$e = x - p'. \qquad (7.10)$$

Next, the PEE operation to embed a secret bit b into x and get stego pixel x' is as follows:

$$x' = p' + 2 \times e + b. \qquad (7.11)$$

In the secret data extracting and pixel-recovering phases, the embedded secret bit is extracted by

$$b = x' \bmod 2. \qquad (7.12)$$

To recover the original cover image, the receiver can compute the predicted value p by using Equation 7.8. Then, the modified predicted value p' used in the

x_2	x_3
x_1	x

FIGURE 7.1   The three-neighbor context of pixel x.

data-embedding phase can be calculated by Equation 7.9 and the original cover pixel x is recovered by

$$x = p' + \left\lfloor \frac{x' - p'}{2} \right\rfloor. \qquad (7.13)$$

7.2.1.3   History of the Development of DE-Based Techniques

Tian's DE-based method (Tian, 2003) was the first reported DE-based reversible data-hiding scheme. This scheme acquires space for data hiding by exploring the redundancies in the content of the image. It embeds one bit of information at a time into each cover pixel pair. Thus, the maximum hiding capacity is 0.5 bpp for one level embedding (i.e., when the hiding process is conducted on the cover image once). To increase the hiding capacity, Alattar extended Tian's method to embed two bits at a time into a triplet of pixels (Alattar, 2003). Experimental results show that the hiding capacity of Alattar's method is greater than that of Tian's method, but it distorts the stego image to a greater extent. To further improve the performance of such DE-based methods, Alattar extended his own method (Alattar, 2003) to hide a triplet of bits in the DE for four adjoining pixels (Alattar, 2004a). Finally, Alattar generalized all the above methods to conceal $n - 1$ secret bits into a vector of n cover pixels. So, the hiding capacity of Alattar's method is, at most, $(n - 1)/n$ bpp (Alattar, 2004b). All the above DE-based techniques have the same problem, that is, the overflow and underflow problem after the expansion. The overflow and underflow problem is that some pixel values in the stego image may become larger than 255 or smaller than 0 after the expansion. To overcome this problem, DE-based techniques that were developed later used a location map to indicate the expandable and unexpandable pixels. If the value of a pixel exceeds 255 or is <0 after embedding the secret data, then "1" is recorded at the corresponding location in the location map, and this pixel is called an unexpandable pixel; otherwise, "0" is written at the corresponding location in the location map, and its corresponding pixel is called an expandable pixel. The location map represents overhead information and reduces the hiding capacity significantly. Weng et al. (2008) proposed a companding-based reversible data-hiding scheme. The procedure utilizes a companding function to convert the larger differences into smaller values, which can be used to improve the hiding capacity by increasing the number of expandable triplets while reducing the overhead caused by the location map. As a result, the hiding capacity of Attlar's DE of triplets scheme can be increased by this method. Lin et al. (2008c) proposed a location-map-free, DE-based technique. They divided the cover image into overlapping blocks and recorded the smallest value among all the differences between the minimum and maximum pixel values in every block. The smallest value is used to determine whether the block is embeddable or not. The maximum hiding capacity of a single level in this scheme is not more than 0.496 bpp.

Thodi and Rodriguez (2007) proposed PEE, another DE-based data-hiding scheme in which the secret bit is embedded by expanding the difference between the pixel value and its predicted value from neighboring pixels. Thodi and Rodriguez's method embeds one data bit into one cover pixel at a time, making the hiding capacity around 1 bpp. They also proposed two additional techniques to enhance the visual quality at low-embedding capacities. The major problem of Thodi and Rodriguez's method is the high distortion (PSNR <30 dB) at high-hiding capacities. Thodi and Rodriguez's PEE method is one of the most important DE-based methods because almost all new schemes proposed later used the basic idea of PEE, which uses information from neighboring pixels to compute a predicted value. Later, Lin and Hsueh (2008a) proposed a three-pixel block difference scheme. It embeds two secret bits into two differences—between the first and the second pixel as well as between the second and the third pixel—at a time. In this kind of three-pixel block, only the central pixel must be modified to embed two bits, so this scheme can achieve an average-hiding capacity of about 2 bpp. Lu and Chang (2008) proposed a nibble-based data-hiding scheme. Every pixel in the cover image is divided into two nibbles, and each neighboring nibble pair can be used to hide a secret bit. Lee et al. (2008) proposed an adaptive, block-based, lossless data-hiding technique. The cover image is partitioned into several nonoverlapping blocks. The proposed method conceals more secret bits in the smoother block to reduce the distortion of the stego image. Kim et al. (2008) proposed a DE transform scheme. This scheme improves two aspects of the original DE-based techniques, the size of the location map and the expandability of the pixel pair, by exploiting a new threshold. If the difference value of a pixel pair is smaller than half of the threshold, and the average value of this pixel pair is between the threshold and 255 minus the threshold, then this pixel pair is an absolute expandable pixel pair. This threshold can simplify the location map by eliminating the ambiguity in the decoding phase. Beginning in 2009, DE-based methods have been combined with the HS method to be discussed in the next subsection to get better hiding capacity and low distortion. Table 7.2 presents the

TABLE 7.2

Experimental Results of DE Techniques for Test Image Lena with One Level Embedding

Method	PSNR (dB)	Capacity (bpp)	Size of Location Map (bit)	Compression Technique
Tian (2003)	34.35	0.4996	$(M \times N)/2$	JBIG2 or run-length coding
Alattar (2003)	34.80	0.7228	N/A	N/A
Alattar (2004a)	≈35.00	≈0.3815	N/A	N/A
Alattar (2004b)	28.45	0.7449	$(M \times N)/K$	N/A
Lin et al. (2008c)	35.75	0.4951	0	N/A
Thodi and Rodriguez (2007)	28.13	0.9983	N/A	JBIG2
Lin et al. (2008a)	30.00	1.1800	N/A	N/A
Lu and Chang (2008)	38.09	0.9978	N/A	Prediction by partial matching (PPM)
Lee et al. (2008)	34.32	1.1000	N/A	N/A
Kim et al. (2008)	≈39.00	≈0.4000	Simplified	N/A

performances of the DE techniques mentioned above. In Table 7.2, M and N are the width and height of the cover image, respectively. K is the size of the vector in Alattar's method (Alattar, 2004b). The symbol "≈" means "is approximately equal to." The compression techniques in the last column are used to compress the location map in different papers losslessly. We did not list the work of Weng et al. (2008) because they did not test their method on Lena.

7.2.2 HS-Based Techniques

In this section, first, we use an example to briefly illustrate Ni et al.'s (2006) reversible data-hiding scheme based on HS. The development history of HS-based techniques and some experimental results are then introduced in Subsection 7.2.2.2.

7.2.2.1 Ni et al.'s HS Scheme

Ni et al.'s reversible data-hiding scheme utilizes the histogram of the cover image to embed secret data into the peak point pixels and reduce distortion by shifting a part of the histogram slightly. An example to illustrate Ni et al.'s scheme is presented in this section. The pixel values of the cover image in which secret data were embedded by Ni et al.'s method are shown in Figure 7.2.

In the process, first, Ni et al. scanned all pixel values in the cover image to generate a histogram of the image. Then, they found a peak point p_point and zero point z_point pair in the histogram. The p_point is the pixel value having the highest number of occurrences in the histogram of the image, and the z_point is the pixel value of the zero point at which there are no occurrence or the least occurrences, as shown in Figure 7.3. If there are multiple values with the same highest number of occurrences, or the least occurrences, the closest peak point and zero point pair can be chosen. The histogram of the 512×512-pixel Baboon is shown in Figure 7.4.

Next, the part of the histogram between the peak point and the zero point is shifted from the peak point in the direction of the zero point by one unit, which means all pixel values lying between the peak point and the zero point are increased by 1 if the pixel value of the zero point is greater than the pixel value of the peak point, as shown in Figures 7.5 and 7.6.

Finally, the secret data are embedded into the pixel with the value of the peak point. If the secret bit is "0," the pixel of the peak point is not changed, but, if the secret bit is "1," the pixel value of the peak point is increased (or reduced depending

2	5	3	1	1
4	2	5	1	1
2	3	4	4	1
3	3	1	2	5
3	3	3	3	2

FIGURE 7.2 Pixel values of the example cover image.

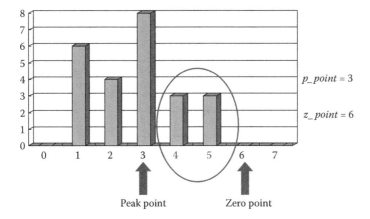

FIGURE 7.3 Image histogram, peak point, and zero point of Figure 7.2.

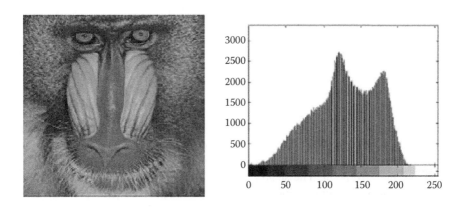

FIGURE 7.4 Baboon and its image histogram.

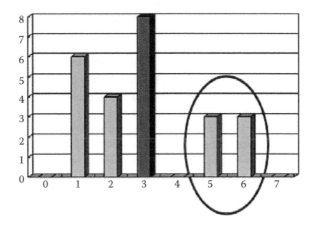

FIGURE 7.5 HS between peak and zero points.

2	6	3	1	1
5	2	6	1	1
2	3	5	5	1
3	3	1	2	6
3	3	3	3	2

FIGURE 7.6 Pixel values after HS.

2	6	4	1	1
5	2	6	1	1
2	4	5	5	1
3	3	1	2	6
4	4	3	4	2

FIGURE 7.7 Pixel values of the stego image.

on the HS direction) by 1. In our example, assuming that the secret data to be embedded are "11001101," the pixel values of the stego image are provided in Figure 7.7.

In the phase in which the secret data are extracted, the receiver gets the peak point and the zero point. The stego image is scanned in the same order as in the embedding phase. In our example, if the stego pixel equals the peak point, a secret bit "0" is extracted, and the recovered pixel remains unchanged. If the stego pixel equals one more than the peak point, a secret bit "1" is extracted, and the recovered pixel is restored by subtracting 1 from the current stego pixel. The cover image is restored by shifting all pixels lying between the peak point and the zero point to the left by one unit.

7.2.2.2 Development History of the HS-Based Techniques

The hiding capacity of a HS-based, reversible data-hiding method is primarily determined by the peak height of the histogram of the corresponding cover image. In 2007, Fallahpour and Sedaaghi used peak point and zero point pairs of the blocks of the image to hide the secret data instead of using the peak point and zero point pairs of the whole image as was done in Ni et al.'s scheme (Fallahpour and Sedaaghi, 2007). Fallahpour and Sedaaghi's method increases the hiding capacity by finding more peak point and zero point pairs in one image, but the distortion also is enlarged because more pixels are modified. Lin et al. (2008b) proposed a difference image-based data-hiding scheme to improve the hiding capacity of Ni et al.'s method while maintaining the low distortion of the stego image. The difference image is formed by the difference between two adjacent pixels in an image. In terms of an image, there

TABLE 7.3

Experimental Results of HS Techniques for Test Image Lena (Sized 512 × 512) with One Level Embedding

Method	PSNR (dB)	Capacity (bpp)
Ni et al. (2006)	48.2	0.0208
Fallahpour and Sedaaghi (2007)	47.29	0.0529
Lin et al. (2008b)	48.67	0.2493
Tai et al. (2009)	48.32	0.0854
Tsai et al. (2009)	≈48.6	0.3000
Hong et al. (2010)	≈51.2	0.3000
Luo et al. (2011)	49.68	0.1010

is a large probability that adjacent pixels have similar pixel values, so the peak point in the difference image histogram is ~0 and has much a higher value than the corresponding value in the cover image itself. Hence, the hiding capacity can be increased by using this characteristic. In the above HS-based data-hiding schemes, a side communication channel for pairs of peak and zero points is provided. Tai et al. (2009) introduced a binary tree that predetermines the multipeak points used to embed secret data. Tai et al. also used the difference histogram of the adjacent pixel pair to increase the hiding capacity. Tsai et al. proposed another HS-based, reversible data-hiding scheme at about the same time. It is known that the prediction error histogram has a higher peak height than the image histogram itself due to the similarity of the neighboring pixels. So, Tsai et al. embedded secret data by modifying the prediction error to achieve a greater hiding capacity. All test images in Tsai et al.'s method were medical images. Their experimental results showed that a better quality stego image (by about 1.5 dB) was obtained compared to that of Ni et al.'s scheme when the same quantity of secret data was embedded. Hong et al. (2010) presented an improved version of Tsai et al.'s scheme. They used the orthogonal projection technique to estimate the optimal weights of a linear predictor to increase the prediction accuracy and the hiding capacity. Luo et al. (2011) further improved the HS data-hiding techniques by utilizing prediction methods in the blocks of the cover image. The histogram is constructed by block differences with the reference of their integer medians, which is used to recover the cover image at the receiver end. Table 7.3 presents experimental results of the above-mentioned methods.

7.3 TRANSFORMED DOMAIN REVERSIBLE DATA-HIDING TECHNIQUES

In this section, first, Kamstra and Heijmans' reversible data-hiding scheme based on HWT is presented in detail (Kamstra and Heijmans, 2005). Then, we provide a detailed introduction to Chan et al.'s Haar digital wavelet transform (HDWT)-based reversible data-hiding method (Chan et al., 2009). The section concludes with a brief review of other works that deal with fragile reversible data hiding in the transformed domain.

7.3.1 Kamstra and Heijmans' Scheme

Kamstra and Heijmans (2005) transformed the cover image into a low-passed image L and a high-passed image H using the HWT as was done in Tian's method, and they embedded the secret data into the high-passed image H. In Tian's method, the small values in H is likely to be expandable. Kamstra and Heijmans exploited this fact by using L to predict the coefficients of H, both at the encoding side and the decoding side, since the coefficients of L are not changed during the data-hiding procedure. The content of the location map can be changed by using the prediction result. The main purpose of Kamstra and Heijmans' method was trying to have the 0's and 1's in the location map changed the order so that the 0's and 1's are more grouped together, thereby allowing strong compression, where 0 indicates the expandable pixel pair and 1 indicates the unexpandable pair. The strong compression of the location map could save more embeddable space for secret data. This goal is achieved by trying to perform the embedding according to the order indicated by the variance of L, since a small variance in L indicates a small value in H which is more likely to be expandable. The maximum hiding capacity that can be achieved by Kamstra and Heijmans' method is around 0.5 bpp. The details of this method are presented below.

The first step of Kamstra and Heijmans' scheme is to use the HWT algorithm to transform the cover image into a low-passed image, L, and a high-passed image, H, as presented in Figure 7.8.

The second step is to calculate the local variance $v_{(i, j)}$ of coefficient $L(i, j)$ by using Equation 7.14:

$$v_{(i,j)} = \frac{1}{\left|W(i,j)\right|} \sum_{i',j' \in W(i,j)} (L(i',j') - aveL(i,j))^2, \tag{7.14}$$

where $W(i, j)$ is a window surrounding $L(i, j)$ in L, as shown in Figure 7.9; $L(i', j')$ is a value in the $W(i, j)$; $|W(i, j)|$ is the number of coefficients in the window $W(i, j)$

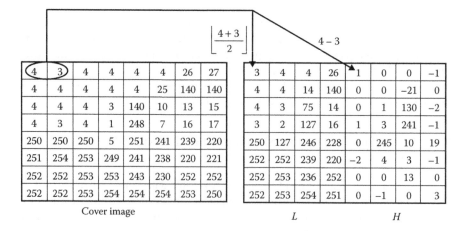

FIGURE 7.8   The HWT of the cover image.

3	4	4	26	1	0	0	−1
4	4	14	140	0	0	−21	0
4	3	75	14	0	1	130	−2
3	2	127	16	1	3	241	−1
250	127	246	228	0	245	10	19
252	252	239	220	−2	4	3	−1
252	253	236	252	0	0	13	0
252	253	254	251	0	−1	0	3

FIGURE 7.9 The window in the L image.

(i.e., $|W(i, j)| = 5$ in this case); and $aveL(i, j)$ is the average of *pixel values* inside $W(i, j)$. The calculated results of local variance except edge pixels are shown in Figure 7.10.

The third step is to sort all the local variances $v_{(i, j)}$ in L. The way that data are embedded in Kamstra and Heijmans' method is to utilize the difference between two neighboring pixels, just as was done in Tian's method. The only difference between the two methods is the order in which the neighboring pixel pairs were chosen. Tian's method chooses pixel pairs from left to right and from up to down. Kamstra and Heijmans' method utilizes the fact that the $H(i, j)$ with smaller $v_{(i, j)}$ is more likely to be expandable than the one that has larger $v_{(i, j)}$. So, the data-embedding order of Kamstra and Heijmans' scheme follows the order of the pixel pair's local variance $v_{(i, j)}$. In our example, the result of sorting the local variances $v_{(i, j)}$ is $\{v_{(1,6)}, v_{(1,1)}, v_{(2,1)}, v_{(2,2)}, v_{(1,5)}, v_{(2,5)}, v_{(1,2)}, v_{(2,6)}, v_{(2,3)}, v_{(1,4)}, v_{(2,4)}, v_{(1,3)}\}$. Assuming that the secret bits to be embedded in these 12 pixel pairs are 111111111111 (since an embedded bit of "0" will not change the data, here we use an all "1" example to show how the data change

	j			
x	x	x	x
x	17	2844	x
x	812	2259	x
x	3710	7820	x
x	9765	2972	x
x	2394	118	x
x	44	59	x
x	x	x	x

$\rightarrow i$

FIGURE 7.10 The local variance except for edge pixels in L.

4	3	4	4	4	4	26	27
4	4	5	4	−6	34	140	140
4	4	5	2	206	−55	13	15
4	3	6	−1	369	−114	16	17
250	250	373	−118	257	236	239	220
251	254	257	248	243	236	220	221
252	252	254	253	250	223	252	252
252	252	253	254	254	254	253	250

FIGURE 7.11   The results of expanding pixel pairs.

4	3	4	4	4	4	26	27
4	4	5	4	4	25	140	140
4	4	5	2	140	10	13	15
4	3	4	1	248	7	16	17
250	250	250	5	251	241	239	220
251	254	253	249	242	236	220	221
252	252	253	253	250	223	252	252
252	252	253	254	254	254	253	250

FIGURE 7.12   The embedding result of Kamstra and Heijmans' method.

with the embedding), the embedding results are shown in Figure 7.11. Seven pixels pairs in the small square in Figure 7.11 have pixel values either <0 or >255, so these seven pixel pairs are unexpandable. A location map is used in both Tian's method and Kamstra and Heijmans' method to separate expandable pixel pairs from unexpandable pixel pairs. If a pixel pair (x, y) is expandable, then the corresponding bit $loc_{(i, j)}$ in the location map is set to 0; otherwise, $loc_{(i, j)}$ is set to 1, where (x, y) is the pixel pair corresponding to the local variances $v_{(i, j)}$ located at (i, j), and $loc_{(i, j)}$ is the location bit of the pixel pair (x, y).

As a result, in our example, the location map obtained by using Tian's method is 010111111000, while the location map obtained by using Kamstra and Heijmans' method is 000001111111 (which will produce fewer bits after compression since the "0"s and "1"s are more grouped together).

The rest of the embedding procedure used in Kamstra and Heijmans' method is the embedding of secret data into the expandable pixel pairs by using Tian's method. Assuming that the secret data to be embedded into the five expandable pixel pairs in our example are 11001, the embedded pixel values are shown in Figure 7.12.

7.3.2   CHAN ET AL.'S HDWT-BASED SCHEME

Chan et al.'s (2009) scheme compresses the least significant bits of the coefficients of the high-frequency band by using Huffman coding and then uses the saved space to embed the secret data. The embedding process replaces the least significant bits of the coefficients in the high-frequency band with the bits of the compressed data and the secret data. The embedding procedure of Chan et al.'s scheme is introduced in detail below.

First, a cover image is transformed into a frequency domain cover image by HDWT decomposition. The frequency domain cover image consists of four different frequency bands, that is, LL, LH, HL, and HH, each of which is 1/4 of the size of the original image. LL represents the low-frequency band, LH and HL are the middle-frequency bands, and HH is the high-frequency band. The coefficients in LL are more important. If coefficients in LL are changed, the content of cover image c will be changed more significantly. The human eye is not sensitive to the change of coefficients in the HH band. To achieve high similarity between the stego image and the cover image, Chan et al. hide the secret data only in HH. The coefficients of LL, LH, HL, and HH in one block of the image "Lena" are shown in Figure 7.13.

Let S be a sequence of binary secret data to be embedded into the cover image, which is comprised of $M \times N$ pixels. Hence, each of LL, LH, HL, and HH consists of $(M/2) \times (N/2)$ pixels. The coefficients in HH are separated into sign, integer, and decimal segments, which are individually recorded by the sign matrix HHS, the integer matrix HHI, and the decimal matrix HHD, respectively. They can be encoded effectively by the Huffman coding method because there is significant redundancy. Chan et al. (2009) used only the k least significant bits of the HHI coefficients for Huffman coding and data embedding, where k is a constant determined by the user. The Huffman coding compresses the bits and creates space for embedding the secret data. The larger k will give more embedding capacity, but the resulting image will have

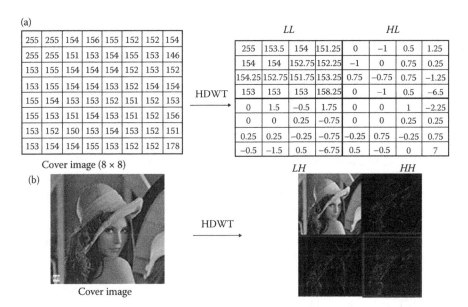

FIGURE 7.13 The decomposition of a cover image by HDWT—(a) the coefficients of the frequency bands of one block in image "Lena." (b) The visual image of the frequency bands.

poorer quality, and vice versa. The secret data and the compression code are embedded into the least significant bits of the coefficients in *HHI*. After data embedding in *HHI*, reversing the frequency-domain image back to the spatial-domain image may cause underflow and overflow problems, since the substitution of the least significant bits may cause some pixel values to be >255 or <0. To solve this problem, Chan et al. defined two thresholds, T_l and T_u. If any coefficient of *LL* is less than T_l or greater than T_u, the corresponding coefficient in *HHI* cannot be used to hide any data. In general, as described below, four steps are required to hide secret data in Chan et al.'s scheme:

Step 1: Scan each integer *HHI*[*i,j*], where $1 \leq i \leq M/2$ and $1 \leq j \leq N/2$. The *k* least significant bits of *HHI*[*i,j*] are copied to a set *ED* if $T_1 \leq LL[i,j] \leq T_u$.

Step 2: Huffman coding is used to compress *ED*. The generated compression codes are concatenated into one binary string *CD*.

Step 3: The proposed scheme links *CD* with secret data *S* into a binary string *HD*. The number of bits of secret data can be embedded equals the number of bits which is saved from Huffman coding, and so, *HD* has *k* bits. To prevent unauthorized people from figuring out any information of the secret data, Chan et al. encrypt the bit string *HD* to generate a sequence of encrypted binary string *HD'*, At last, *HD'* is used to replace the *k* least significant bits of *HHI*[*i,j*], where $1 \leq i \leq M/2$, $1 \leq j \leq N/2$, and $T_l \leq LL[i,j] \leq T_u$.

Step 4: *HHS*, *HHD*, and the new *HHI* are combined into a new *HH*. Finally, the stego image is produced from *LL*, *LH*, and *HL*, and the new *HH* by the reverse procedure of HDWT.

The data extraction and cover image recovery phase is the inverse process of the data-embedding procedure. First, the stego image is transformed into four frequency bands by using HDWT. Second, *HD'* is extracted from the *k* least significant bits in the *HHI* image of the *HH* band. *HD'* is decrypted to obtain the binary string *HD*. *HD* are then Huffman decoded to extract the original *k* least significant bits of the *HHI* coefficients and the following secret data. Then, the cover image is recovered by using the inverse procedure of HDWT.

7.3.3 OTHER TRANSFORMED DOMAIN WORKS

Xuan et al. (2002) proposed a reversible data-hiding method based on IWT. In their method, the grayscale cover image is divided into several bit planes. The bias between the number of zeroes and ones in a bit plane indicates the redundancy, which implies that one may compress bits in order to create space to hide data. The IWT was used in this method to eliminate the redundancy so that additional data could be embedded. Yang et al. (2004) proposed a histogram modification-based, reversible data-hiding scheme in the 8×8 integer DCT domain. This scheme utilized the high energy concentrating property of integer DCT and the histogram modification was used to embed the data into the integer DCT coefficients. Xuan et al. (2005) proposed another scheme using IWT and a threshold-embedding technique. Different embedding rules were applied to high-frequency wavelet coefficients, that is, depending upon whether the absolute value of the coefficient is less than, equal to, or greater than the threshold value.

TABLE 7.4

Experimental Results of Transformed Domain Techniques for Test Image Lena with One Level Embedding

Method	PSNR (dB)	Capacity (bpp)
Xuan et al. (2002)	36.64	0.3262
Yang et al. (2004)	44.98	0.1608
Xuan et al. (2005)	48.63	0.0985
Lee et al. (2007)	44.87	0.3000
Chan et al. (2009)	37.23	0.3687
Chang et al. (2010)	36.24	0.4610

In the preprocessing phase, histogram modification is used to narrow the histogram of the cover image to prevent the overflow and underflow problem. A new integer-to-IWT-based, reversible data-hiding technique was proposed by Lee et al. (2007). This scheme divided the cover image into nonoverlapping blocks and embedded data into the high-frequency wavelet coefficients of each block, since coefficients in the high-frequency band incorporate less energy than other bands of an image. Block-based embedding makes the size of the side information required to reconstruct the cover image small in proportion to the total embedding capacity. Since the hiding capacity includes not only the secret data but also side information, the reduction of side information can increase the hiding capacity of the secret data. Chang et al. (2010) proposed another HDWT-based, high-hiding capacity, reversible data-hiding scheme. This scheme utilized an adaptive arithmetic coding method to encode the HDWT coefficients in the high-frequency band. Chang et al.'s method embedded both compressed data and secret data into the high-frequency band of the cover image. They also substituted the least significant bits of the coefficients in the high-frequency band with the bits in the compression data and the secret data in the embedding procedure. Chang et al.'s method is superior to Chan et al.'s method because it uses a more efficient compression method.

The experimental results of the above-mentioned methods are shown in Table 7.4. The performance is evaluated by the hiding capacity and the quality of the stego image (i.e., PSNR). Yang et al.'s experimental results, shown in Table 7.4, were obtained from a test based on a 256×256 Lena image, whereas a 512×512 Lena image was used in the other schemes.

7.4 COMPRESSION DOMAIN REVERSIBLE DATA-HIDING TECHNIQUES

Currently, images are usually stored and compressed by lossy or lossless compression techniques for convenience and efficiency. Hence, some reversible data-hiding schemes based on compression codes have been proposed in the last few years. In this section, first, we introduce a simple and basic method. Then, other works on reversible data hiding in the compression domain are introduced briefly. The last section shows the comparison and analysis of these works.

7.4.1 Chang et al.'s Codebook Clustering-Based Scheme

Chang et al. (2007b) proposed a reversible data-hiding method for vector quantization (VQ)-compressed images to obtain higher-embedding capacity. In their method, the original codebook C with G codewords was redesigned to a new codebook C' with $G' = [G - 2^{s-1}/2^{s-1} \times 3] \times 2^{s-1} \times 3$ codewords based on the length of the secret data, where s denotes the length of the secret data to be embedded in a VQ index. The redundant 2^{s-1} codewords are used as indicators. If we want to embed one-bit secret data in an index table, codebook C' is divided into three separated clusters, that is, C_1, C_2, and C_3, with the same size $m = \lfloor (G - 1)/3 \rfloor$. The secret bit is embedded in the index by transforming the indices of C_2 into the corresponding indices in the other two clusters C_1 and C_3 according to the value of the secret bit. If the secret bit to be embedded is 0 and the current index belongs to cluster C_2, the current index will be transformed to the corresponding index in cluster C_1; if the secret bit is 1 and the current index belongs to cluster C_2, then the current index will be transformed into the corresponding index in cluster C_3. If the current index belongs to cluster C_1 or C_3, it cannot be used to embed any secret data. To avoid confusion in the extracting phase, the authors modified an original index in C_1 directly to an index in C_2, and used an indicator to mark the transformation of an original index in C_3 as "indicator ∥" corresponding index in C_2. The symbol "∥" represents a concatenation operation.

Figure 7.14 shows an example of Chang et al.'s method. In this figure, one secret bit is embedded into indices in index table T. The original codebook C contains 16 codewords, and the modified codebook C' contains 15 codewords; the codeword with index 0 is used as an indicator. The underlined index in index table T' means that there are no secret data embedded and that the original index is the corresponding index in C_1. The index with the symbol "∥" in index table T' means that there are no secret data embedded and that the original index is the corresponding index in C_3.

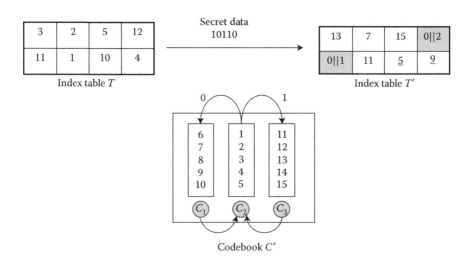

FIGURE 7.14 Example of Chang et al.'s method for embedding one bit of secret data.

However, in this case, the index following indicator index 0 sized $\lceil \log_2 G \rceil$ (=4 in this example) bits can be represented by only $\lceil \log_2 g \rceil$ (=3 in this example) bits, where g (=5 in this example) is the size of cluster C_2.

As shown in Figure 7.14, the first index in index table T is 3, which belongs to C_2, so it can be used to embed one secret bit. Then, the first secret bit 1 is embedded into index 3 by transforming index 3 to index 13, which belongs to C_3, so index 13 is put into the corresponding position in stego index table T'. The second index in index table T is 2, which belongs to C_2, so it also can be used to embed one secret bit. Then, the second secret bit 0 is embedded in index 2 by transforming it to index 7, which is the corresponding index of index 2 and which belongs to C_1. Index 7 is put into the second position in stego index table T'. The third index, index 5, performs the same embedding procedure as the second index to embed the third secret bit 1. In index table T, the fourth index 12 belongs to C_3, so it cannot be used to embed any secret bits. Index 12 is transformed as "0‖2" in index table T', where 0 is the indicator, and 2 is the corresponding index of index 12 in C_2. Index 10 in index table T belongs to C_1, which means it is also an unembeddable index. Then, index 10 is transformed to 5 directly in index table T'.

This method can also be expanded to enhance the embedding capacity. Figure 7.15 shows an example of the method for embedding two-bit secret data. In this example, the original codebook sized $G = 32$ is modified to a codebook sized $G' = 30$ and two codewords with index $i_0 = 0$ and index $i_1 = 31$. The new codebook G' is divided into six clusters. The size g of each cluster equals 5. If the current index belongs to cluster C_1 or C_2, two secret bits can be embedded in it. Otherwise, the current index is unembeddable. The embedding rule is similar to the method shown in Figure 7.14, except that the authors used indicator i_0 to mark an index belonging to cluster C_5 or C_6 and used indicator i_1 to mark the following two cases. The first case is an index belonging to cluster C_1, which is used to embed two bits of secret data,

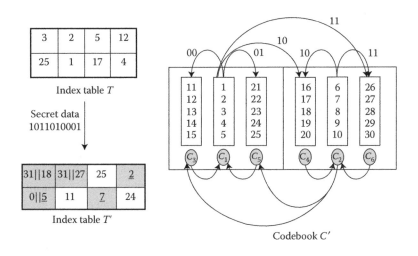

FIGURE 7.15 Example of Chang et al.'s method for embedding two bits of secret data.

$(10)_2$ and $(11)_2$. The second case is an index belonging to cluster C_2, which is used to embed two bits of secret data $(00)_2$ and $(01)_2$.

Chang et al. (2007a) proposed another reversible data-hiding scheme for VQ-compressed images. This scheme utilized a declustering strategy and the similarity property of adjacent areas in a natural image to achieve higher-embedding capacity than other schemes. However, one could argue that VQ is a lossy compression technique. It may not be suitable to use VQ with a reversible data-hiding technique where most of the applications require the lossless recovery of the original image. Also, VQ is rather specific. Most images are compressed with standard compression techniques such as JPEG instead of VQ.

7.4.2 Other Compression Domain Works

Chang and Lu (2006) proposed a reversible data-hiding scheme based on side-match vector quantization (SMVQ), which conceals the secret data in the indices of the compressed image with low distortion of the stego image. The secret data are embedded by separating the state codebook, which is generated in the SMVQ compression procedure, into two categories. Each category represents bit "0" or bit "1," respectively. Reversibility is achieved by using the prediction idea in the state codebook. Given the unchanged first row and first column blocks of the cover images, the decoder can generate the same state codebook for the rest of the blocks as the encoder. In the same year, Chang and Lin (2006) proposed another SMVQ-based reversible data-hiding scheme. The difference between Chang and Lu's method and Chang and Lin's methods is the way of finding two codewords in the state codebook to represent different bit values. Different from Chang and Lu's separating the state codebook into two parts, Chang and Lin used Linde-Buzo-Gray algorithm (Linde et al., 1980) to cluster all codewords in the codebook such that the codewords in the same cluster are similar. Assume the current encoding block X can be encoded into codeword C_0 by using VQ technique and C_0 exists in the state codebook of X, then, C_0 and its closest codeword C_1' which does not belong to the state codebook of X in the same cluster were used to represent bit "0" or bit "1," respectively. If C_0 does not appear in the state codebook of X or it cannot find its closest codeword C_0' in the same cluster (due to the codewords in the same cluster are in the state codebook or only one codeword being in the cluster), then, the current encoding block cannot be used to embed any secret data. Chang and Lin's method is inefficient when the size of the state codebook becomes large. Chang and Lin (2007) proposed a new reversible data-hiding scheme based on SMVQ by using the declustering strategy. The codebook is partitioned into two groups, and any codeword in one group has one and only one corresponding codeword in the other group. The Euclidean distance between each pair of codewords is as large as possible, which is the declustering strategy. The secret bit is embedded by choosing a different codeword in this dissimilar codeword pair. The use of declustering reduces the number of prediction errors significantly by using side match; however, it enlarges the distortion between the stego image and the cover image. Lee et al. (2010) proposed a reversible data-hiding scheme based on the distribution of the transformed image obtained by applying SMVQ. The size of the state codebook is identical to the size of the VQ codebook in Lee et al.'s scheme. They separated the transformed indices into three portions. The first portion contained the smallest values, which are utilized to hide secret data and reconstruct the cover indices. The second portion had indices with middle values, which are used to compress the code stream and to reconstruct the cover indices. The third portion had the largest values, which are only used to recover the cover indices. Because most transformed indices are distributed around zero, Lee et al.'s scheme can achieve high-hiding capacity.

Yang et al. (2005) proposed a reversible data-hiding scheme in the VQ compression domain based on the fast correlation VQ technique. The fast correlation-based VQ technique is a lossy compression method for the VQ index table (Lu and Sun, 2000). The main idea of this method is to modify each embeddable block X to the best matched codeword selected from c_a, c_b, c_c, c_d for c_m, where c_m is the mean codeword of c_a, c_b, c_c, c_d, that is, $c_m = (c_a + c_b + c_c + c_d)/4$ and c_a, c_b, c_c, and c_d are the four nearest adjacent blocks of X. Chang et al. (2007b) proposed a high-capacity reversible data-hiding scheme based on VQ. The idea of this scheme originated from the concept of codebook clustering proposed by Jo and Kim (2002). First, a codebook is sorted by the frequencies of the occurrences of codewords in the preprocessing phase. Then, they divide the sorted codebook into three clusters in accordance with the corresponding frequencies. The highest-frequency cluster is used to hide the secret data, and the others are used only for recovering the distortion of the VQ index table. Chang et al. (2007a) proposed another scheme that improved the hiding capacity significantly. It was based on the declustering for the VQ-compressed codes. The hiding capacity of this scheme is influenced by three factors, that is, the image context, the size of the codebook, and the number of the declustered groups. A smooth image can embed more secret data than a complicated image, and a larger codebook results in greater hiding capacity since both increase the possibility of generating large groups in the embedding phase. While a lower number of declustered groups would enhance the embedding capacity, a larger number of declustered groups would decrease the possibility of generating large groups. Chang et al. (2009) proposed a novel reversible data-hiding method for VQ indices based on locally adaptive coding. In this scheme, secret data are embedded into VQ indices in an index table during the process of compressing the index table in the block-by-block manner by using the locally adaptive data compression method. There are two different encoding types in the locally adaptive coding method. The encoding type is chosen to encode the index block according to the secret bit to be embedded. Lu et al. (2009) proposed a lossless data-hiding scheme, based on image VQ index residual value coding to improve the performance of Yang et al.'s (2005) method. This scheme uses the relationship between the neighboring four indices of the current index and their mean value to hide the secret bit. An index is encoded into the index difference appended with flag bits according to the secret bit and the context. Finally, the index difference is encoded by the proper bit stream to reduce the bit rate. Yang and Lin (2009) presented an improved method based on Chang et al.'s (2007b) method. Yang and Lin sorted the VQ codebook by referred counts and divided it into 2^s clusters, in which s denotes the size of the secret data embedded into each VQ index. This scheme achieves greater hiding capacity than Chang et al.'s scheme because half of the clusters can be used to embed secret data rather than the one-third of the clusters that are used in Chang et al.'s scheme. Wang and Lu (2009) proposed another lossless data-hiding scheme based on the joint neighboring coding of the VQ index table. According to an initial key and the secret data to be embedded, different adjacent indices may be chosen to perform the joint neighboring coding for each index. This scheme also uses the neighboring four indices of the current index to embed the secret data. The only difference between this method and Lu et al.'s method (2009) is the way of finding the adjacent index among those four indices to encode the current index into the difference and the flag bits.

Different from using visual quality and hiding capacity to evaluate the reversible data-hiding schemes in the spatial domain and transformed domain, compression ratio and hiding capacity are used to evaluate a reversible data-hiding scheme in the compression domain. Compression ratio represents the size of the bit stream after embedding the secret data into the compression code. A smaller value of compression ratio is better than a larger value when using the same compression technique. The unit of the compression ratio is bpp, which is the average number of bits used to represent one pixel after data embedding in the compression codes. We use bits as the new unit of the hiding capacity in this subsection to avoid confusion with the units of the compression ratio. Table 7.5 shows the performance of the reversible data-hiding schemes based on SMVQ, and Table 7.6 shows the performance based on VQ.

TABLE 7.5

Experimental Results of Compression Domain Techniques Based on SMVQ for Test Image Lena (512 × 512) with One Level Embedding

Method	Compression Ratio (bits/pixel)	Capacity (bits)
Chang and Lu (2006)	0.34	14,703
Chang and Lin (2006)	0.44	16,129
Chang and Lin (2007)	0.66	55,186
Lee et al. (2010)	0.34	31,308

TABLE 7.6

Experimental Results of Compression Domain Techniques Based on VQ for Test Image Lena with One Level Embedding (Codebook Size of 512)

Method	Compression Ratio (bits/pixel)	Capacity (bits)
Chang et al. (2007b)	0.60	11,313
Chang et al. (2007a)	N/A	36,288
Chang et al. (2009)	0.53	16,384
Lu et al. (2009)	0.635	15,319
Wang and Lu (2009)	0.641	32,004

7.5 CONCLUSIONS

The design philosophy behind the fragile reversible data-hiding methods has two main points, that is, (1) invariance of the local region in the image and (2) prediction by using the surrounding information. There are three common properties among all reversible data-hiding methods. The first property is that all these schemes are expandable to multilevel data-embedding schemes. The second property is the trade-off between the hiding capacity and the visual quality of the stego image. The third property is that these methods must utilize the surrounding information to guarantee reversibility. For reversible data hiding, there are several future research directions that could be considered: (1) the design of an adaptable reversible technique that has a hiding capacity and visual quality control mechanism to satisfy different require-ments of various applications easily; (2) the design of a reversible data-hiding scheme for medical images. The difference between medical images and other images is the region of interest (ROI), which is the most important part for doctors. It is desirable to develop a data-hiding scheme that is reversible in ROI and irreversible in other parts to increase the hiding capacity; and (3) more information is needed concerning the use of pixels to hide secret information, such as how best to use the frequency of pixel values, their magnitudes, their relationships with neighboring pixels, and their invariance in other regions.

REFERENCES

Alattar, A. M., Reversible watermark using difference expansion of triplets, *Proceedings of the International Conference on Image Processing*, Barcelona, Spain, Vol. 1, pp. 501–504, September 2003.

Alattar, A. M., Reversible watermark using difference expansion of quads, *Proceedings of the International Conference on Acoustics, Speech, and Signal Processing*, Montreal, Canada, pp. 377–380, 2004a.

Alattar, A. M., Reversible watermark using the difference expansion of a generalized integer transform, *IEEE Trans. Image Process.*, 13(8), 1147–1156, 2004b.

Caldelli, R., Filippini, F., and Becarelli, R., Reversible watermarking techniques—an over-view and a classification, *EURASIP J. Inf. Sec.*, 2010, 19, 2010.

Chan, Y. K., Chen, W. T., Yu, S. S., Ho, Y. A., Tsai, C. S., and Chu, Y. P., A HDWT-based reversible data hiding method, *J. Syst. Softw.*, 82(3), 411–421, 2009.

Chang, C. C., Hsieh, Y. P., and Lin, C. Y., Lossless data embedding with high embedding capacity based on declustering for VQ-compressed codes, *IEEE Trans. Inf. Forensics Sec.*, 2(3), 341–349, 2007a.

Chang, C. C. and Kieu, T. D., A reversible data hiding scheme using complementary embed-ding strategy, *Inf. Sci.*, 180(16), 3045–3058, 2010.

Chang, C. C., Kieu, T. D., and Chou, Y. C., Reversible information hiding for VQ indices based on locally adaptive coding, *J. Vis. Commun. Image Represent.*, 20(1), 57–64, 2009.

Chang, C. C. and Lin, C. Y., Reversible steganography for VQ-compressed images using side matching and relocation, *IEEE Trans. Inf. Forensics Sec.*, 1(4), 493–501, 2006.

Chang, C. C. and Lin, C. Y., Reversible steganographic method using SMVQ approach based on declustering, *Inf. Sci.*, 177(8), 1796–1805, 2007.

Chang, C. C. and Lu, T. C., Reversible index-domain information hiding scheme based on side-match vector quantization, *J. Syst. Softw.*, 79(8), 1120–1129, 2006.

Chang, C. C., Pai, P. Y., Yeh, C. M., and Chan, Y. K., A high payload frequency-based reversible image hiding method, *Inf. Sci.*, 180(11), 2286–2298, 2010.

Chang, C. C., Wu, W. C., and Hu, Y. C., Lossless recovery of a VQ index table with embedded
secret data, *J. Vis. Commun. Image Represent.*, 18(3), 207–216, 2007b.

Fallahpour, M. and Sedaaghi, M. H., High capacity lossless data hiding based on histogram
modification, *IEICE Electron. Express*, 4(7), 205–210, 2007.

Hong, W., Chen, T. S., Chang, Y. P., and Shiu, C. W., A high capacity reversible data hiding
scheme using orthogonal projection and prediction error modification, *Signal Process.*,
90(11), 2911–2922, 2010.

Jo, M. and Kim, K. D., A digital image watermarking scheme based on vector quantization,
IEICE Trans. Inf. Syst., E85-D(6), 1054–1056, 2002.

Kamstra, L. and Heijmans, H. J. A. M., Reversible data embedding into images using wavelet
techniques and sorting, *IEEE Trans. Image Process.*, 14(12), 2082–2090, 2005.

Kim, H. J., Sachnev, V., Shi, Y. Q., Nam, J., and Choo, H. G., A novel difference expansion
transform for reversible data embedding, *IEEE Trans. Inf. Forensics Sec.*, 3(3), 456–465,
2008.

Lee, C. C., Wu, H. C., Tsai, C. S., and Chu, Y. P., Adaptive lossless steganographic scheme
with centralized difference expansion, *Pattern Recognit.*, 41(6), 2097–2106, 2008.

Lee, J. D., Chiou, Y. H., and Guo, J. M., Reversible data hiding based on histogram modification of SMVQ indices, *IEEE Trans. Inf. Forensics Sec.*, 5(4), 638–648, 2010.

Lee, S., Yoo, C. D., and Kalker, T., Reversible image watermarking based on integer to integer
wavelet transform, *IEEE Trans. Inf. Forensics Sec.*, 2(3), 321–330, 2007.

Lin, C. C. and Hsueh, N. L., A lossless data hiding scheme based on three-pixel block differences, *Pattern Recognit.*, 41(4), 1415–1425, 2008a.

Lin, C. C., Tai, W. L., and Chang, C. C., Multilevel reversible data hiding based on histogram
modification of difference images, *Pattern Recognit.*, 41(12), 3582–3591, 2008b.

Lin, C. C., Yang, S. P., and Hsueh, N. L., Lossless data hiding based on difference expansion
without a location map, In D. Li and G. Deng, (eds.), *Proceedings of the Congress on
Image and Signal Processing*, Sanya, China, pp. 8–12, May 2008c.

Linde, Y., Buzo, A., and Gary, R. M., An algorithm for vector quantization design, *IEEE Trans.
Commun.*, 28(4), 84–95, 1980.

Lu, T. C. and Chang, C. C., Lossless nibbled data embedding scheme based on difference
expansion, *Image Vis. Comput.*, 26(5), 632–638, 2008.

Lu, Z. M. and Sun, S. H., Imaging coding using fast correlation based VQ, *Chin. J. Image
Graphics*, 5A(6), 489–492, 2000.

Lu, Z. M., Wang, J. X., and Liu, B. B., An improved lossless data hiding scheme based on
image VQ-index residual value coding, *J. Syst. Softw.*, 82(6), 1016–1024, 2009.

Luo, H., Yu, F. X., Chen, H., Huang, Z. L., Li, H., and Wang, P. H., Reversible data hiding
based on block median preservation, *Inf. Sci.*, 181(2), 308–328, 2011.

Ni, Z., Shi, Y. Q., Ansari, N., and Su, W., Reversible data hiding, *IEEE Trans. Circuits Syst.
Video Technol.*, 16(3), 354–362, 2006.

Tai, W. L., Yeh, C. M., and Chang, C. C., Reversible data hiding based on histogram modification of pixel differences, *IEEE Trans. Circuits Syst. Video Technol.*, 19(6), 906–910,
2009.

Thodi, D. M. and Rodriguez, J. J., Expansion embedding techniques for reversible watermarking, *IEEE Trans. Image Process.*, 16(3), 721–730, 2007.

Tian, J., Reversible data embedding using a difference expansion, *IEEE Trans. Circuits Syst.
Video Technol.* 13(8), 890–896, 2003.

Tsai, P. Y., Hu, Y. C., and Yeh, H. L., Reversible image hiding scheme using predictive coding
and histogram shifting, *Signal Process.*, 89(6), 1129–1143, 2009.

Wang, J. X. and Lu, Z. M., A path optional lossless data hiding scheme based on VQ index
table joint neighboring coding, *Inf. Sci.*, 179(19), 3332–3348, 2009.

Weng, S., Zhao, Y., and Pan, J. S., Lossless data hiding based on companding technique and difference expansion of triplets, *IEICE Trans. Fundam. Electron. Commun. Comput. Sci.*, E90-A(8), 1717–1718, 2008.

Xuan, G., Shi, Y. Q., Yang, C., Zheng, Y., Zou, D., and Chai. P., Lossless data hiding using integer wavelet transform, and threshold embedding technique, *Proceedings of the International. Conference on Multimedia and Expo*, Amsterdam, Netherlands, pp. 1520–1523, July 2005.

Xuan, G., Zhu, J., Chen, J., Shi, Y., Ni, Z., and Su, W., Distortionless data hiding based on integer wavelet transform, *Electron. Lett.*, 38(25), 1646–1648, 2002.

Yang, B., Lu, Z. M., and Sun, S. H., Reversible watermarking in the VQ compressed domain, *Proceedings of the Fifth IASTED International. Conference on Visualization*, Benidorm, Spain, pp. 298–303, September 2005.

Yang, B., Schmucker, M., Niu, X. M., Busch, C., and Sun, S. H., Reversible image watermarking by histogram modification for integer DCT coefficients, *IEEE 6th Workshop on Multmedia Signal Processing*, Siena, Italy, pp. 143–146, September 2004.

Yang, C. H. and Lin, Y. C., Reversible data hiding of a VQ index table based on referred counts, *J. Vis. Commun. Image Represent.*, 20(6), 399–407, 2009.

8 Watermarking Based on Local Binary Pattern Operators

Wenyin Zhang and Frank Y. Shih

CONTENTS

8.1 INTRODUCTION

The successes of the Internet and digital consumer devices have been profoundly changing our society and daily lives by making the capture, transmission, and storage of digital data extremely easy and convenient. However, this raises large concern on how to secure these data and prevent unauthorized modification. This issue has become problematic in many areas, such as copyright protection including (Schyndel et al., 1994; Pitas, 1996; Swanson et al., 1996; Xia et al., 1997), content authentication (Yeo and Kim, 2001; Cvejic and Tujkovic, 2004), hiding of information (Tanaka et al., 1990), and covered communications (Chen and Wornell, 2001). Digital watermarking technologies are considered to be an effective means to address this issue. Many researchers have developed various algorithms of digital watermarking methods which intend to embed some secret data (called watermark) in digital content to mark or seal the digital data content, such as (Cox et al., 1997; Podilchuk and Zeng, 1998; Kaewkamnerd and Rao, 2000; Piva et al., 2000; Reed and Hannigan, 2002;

Bas et al., 2003; Shih and Wu, 2003; Shih, 2008; Tsui et al., 2008; Sachnev et al., 2009; Wei and Ngan, 2009; Yang et al., 2009; Luo et al., 2010; Wu and Cheung, 2010). The watermark embedded into a host image is such that the embedding-induced distortion is too small to be noticed. At the same time, the embedded watermark must be robust enough to withstand common degradations or deliberate attacks.

During the last 20 years, digital watermarking techniques have achieved considerable progress, from the spatial domain to the transformed domain, from robustness to fragileness, and from irreversibility to reversibility. The earliest work of digital watermarking schemes can be traced back to the early 1990s (Tanaka et al., 1990), which presented the least significant bit (LSB) method to embed watermarks in the LSB of the pixels in spatial domain. Patchwork methods (Yeo and Kim, 2001; Cvejic and Tujkovic, 2004) process pairs of pixels of the image to embed or extract watermarks. Spread-spectrum modulation techniques (Chen and Wornell, 2001) embed information by linearly combining the host image with a small pseudo-noise signal that is modulated by the embedded watermark.

In the frequency domain, watermarks are inserted into the coefficients of a transformed image, for example, using the discrete Fourier transform (DFT), (Reed and Hannigan, 2002; Bas et al., 2003; Shih, 2008; Tsui et al., 2008), discrete cosine transform (DCT), (Cox et al., 1997; Shih and Wu, 2003; Wei and Ngan, 2009), and discrete wavelet transform (DWT) (Podilchuk and Zeng, 1998; Kaewkamnerd and Rao, 2000; Piva et al., 2000). There are more literatures on the frequency domain than on the spatial domain, mainly because watermarking in the frequency domain can be easily combined with the human visual system (HVS). Recently, more attention has been paid on the reversible watermarking and tamper detection and recovery (Sachnev et al., 2009; Yang et al., 2009; Luo et al., 2010; Wu and Cheung, 2010).

Recently, more attention has been paid on image content authenticity and integrity as well as image tamper and recovery. Based on the objective of image certification, the watermarking technologies can be classified into complete fragile watermarking and semi-fragile watermarking. Complete fragile watermarking does not require any modification on the image, which means the any tamper operation on the image will lead to the failure of the image verification. Semi-fragile watermarking emphasizes much on the watermarking information, and has more robustness than the complete fragile watermarking, but more fragility than robust watermarking. The semi-fragile watermarked image can suffer some common-used image processing such format change, additive noise, compression, luminance change, and contrast adjustment and so on, but it still passes the verification. In fact, most images are stored or transferred with the compressed format, so semi-fragile watermarking technologies have much wider applications.

In this chapter, we introduce the local binary pattern (LBP) operators into image watermarking fields and provide a new semi-fragile watermarking method (Zhang and Shih, 2011). The original LBP operator, which measures the local contrast of pixels, is widely used in the texture classification and face recognition (Maenpaa et al., 2000; Ojala et al., 2002; Ahonen et al., 2004; Maenpaa, 2008). By its extension, we define Boolean function operations on calculating LBP patterns, and adjust one or more of the pixels in the neighborhood to make the function results consistent with the bits of embedded watermarks to realize the watermark embedding in the spatial domain. Firstly, we explain the principle of watermark embedding and extraction processes by using the single-level watermarking technique. Furthermore, we discuss the technique of applying multilevel watermarking methods based on different scale or radii LBP operators or other enhanced or improved LBP operators, such as improved LBP of (Zhang and Jin, 2009; Zhang et al., 2010) and complete LBP of (Zhenhua et al., 2010).

The remainder of this chapter is organized as follows. In Section 8.2, we introduce the basic knowledge of LBP operators. In Section 8.3, we propose the spatial single-level watermarking technique based on LBP operators. The experimental results and analysis are presented in Section 8.4. Section 8.5 provides a multilevel watermarking scheme and its analysis. Finally, we conclude the chapter in Section 8.6.

8.2   LBP OPERATOR AND ITS EXTENSIONS

8.2.1   ORIGINAL LBP OPERATORS

The LBP operator was proposed to measure the local contrast in texture analysis (Maenpaa et al., 2000). It has been successfully applied to visual inspection and image retrieval (Ojala et al., 2002; Ahonen et al., 2004). The LBP operator is defined in a circular local neighborhood. Using the center pixel as the threshold, its circularly symmetric P neighbors within a certain radii R are individually labeled as 1 when the value is larger than the center, or labeled as 0 when the value is smaller than the center. Note that $P = (2R + 1)^2 - 1$. Then, the LBP code of the center pixel is produced by multiplying the thresholded values (i.e., 1 or 0) by weights given to the corresponding pixels, and summing up the result. Take a 3×3 neighborhood as an example, the LBP of a 3×3 window (where $R = 1$ and $P = 8$) uses the center pixel as a threshold value, and the values of the thresholded neighbors are multiplied by the binomial weight and summed to obtain the LBP pattern. In this way, the LBP can produce a pattern from 0 to 255. Figure 8.1 gives an example of this. The entire LBP patterns composite a texture spectrum of an image with 256 gray levels, which are often used to extract image features, such as histograms or statistics for classification or recognition.

Given parameters P and R, which control the quantization of the angular space and spatial resolution, respectively, the LBP pattern, denoted by LBP_p, which presents the local contrast in a pixel neighborhood, is defined as

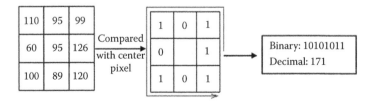

FIGURE 8.1   An example of LBP.

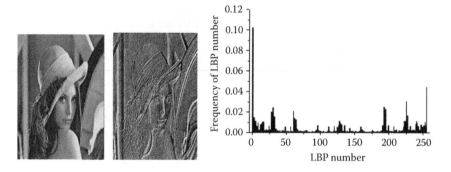

FIGURE 8.2 The texture spectrum and its histogram of image Lena processed by (8, 1) LBP operator.

$$\text{LBP}_P = \sum_{p=0}^{P-1} S(g_p - g_c) \times 2^p \tag{8.1}$$

where g_c denotes the gray level of the center pixel c in the P neighborhood, g_p denotes the gray level of the neighboring pixels p, and $S(x)$ refers to the sign function defined as

$$S(x) = \begin{cases} 1, & \text{if } x \geq 0 \\ 0, & \text{otherwise} \end{cases} \tag{8.2}$$

Let $I(M, N)$ be an image with a size of $M \times N$, $LBP_p(i, j)$ be the LBP pattern in the position of pixel $I(i, j)$, $h(k)$ be the histogram of texture spectrum, $k = 0,1,2, \ldots, 255$, then

$$h(k) = \frac{1}{M \times N} \sum_{i=0}^{M-1} \sum_{j=0}^{N-1} \delta(i, j) \tag{8.3}$$

where $\delta(i, j) = \begin{cases} 1, & \text{LBP}_p(i, j) = k \\ 0, & \text{otherwise} \end{cases}$.

Figure 8.2 shows the texture spectrum and its histogram of image Lena processed by an (8, 1) LBP operator.

8.2.2 EXTENSIONS OF ORIGINAL LBP OPERATORS

Similar to the original ($P = 8$, $R = 1$) LBP operator, we can extend the P and R to get different LBP operators of circle neighborhood by bilinear interpolation of Ahonen et al. (2004). Figure 8.3 presents some circle neighborhood. Compared with the original LBP operator, the extended LBP operators arrange their neighbor pixels along

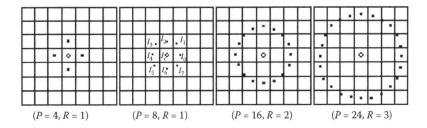

$(P = 4, R = 1)$ $(P = 8, R = 1)$ $(P = 16, R = 2)$ $(P = 24, R = 3)$

FIGURE 8.3   Some extended circle neighborhoods.

the circle, and describe the texture spectrums in different scales. The more the values of P and R, the more detailed the texture. We can choose different P and R according to our requirement.

In order to lower the computational cost, Maenpaa (2008) provided another extended LBP operator called uniform LBP operator. By linking the head to tail, we can make the binary LBP pattern into a ring. In the ring, if times of spatial transitions from 0 to 1 or 1 to 0 are not more than 2, the LBP pattern can be a uniform LBP pattern. Experiments show that the uniform LBP pattern has a lower rate of recurrence, but describe most of the texture features. In the $(P = 8, R = 1)$ LBP texture spectrum, uniform LBP pattern takes up 38%, but can present more than 90% texture pattern. Furthermore, it can reduce the quantity of texture features.

Let U be a uniform value of LBP pattern with (P, R), U can be computed as

$$U(\text{LBP}_{P,R}) = \sum_{i=0}^{P-1} \left| S(I_i - I_c) - S(I_{i+1} - I_c) \right| \tag{8.4}$$

where $I_P = I_0$. Any LBP pattern with $U \leq 2$ belongs to uniform pattern. In a circle neighborhood, there are $P(P - 1) + 2$ uniform patterns. Figure 8.4 shows some uniform LBP patterns and nonuniform LBP patterns of $(P = 8, R = 1)$ neighborhood.

From the nature of texture, symmetric texture patterns should be regarded as a same pattern (see the example Figure 8.5), but they have different pattern values. By rotation of circle neighborhood, we can get a series of LBP pattern values and choose the minimum as the LBP pattern value of the circle neighborhood, as shown by Equation 8.5. Figure 8.6 gives an example of pattern rotation.

$$\text{LBP}_{P,R}^{ri} = \min\left\{\text{ROL}(\text{LBP}_{P,R}, i)\right\} \tag{8.5}$$

Therefore, to achieve rotation invariance, a locally rotation invariant pattern can be defined as

$$\text{LBP}_{P,R}^{riu2} = \begin{cases} \sum_{i=0}^{P-1} S(I_i - I_c) & U \leq 2 \\ P + 1 & \text{otherwise} \end{cases} \tag{8.6}$$

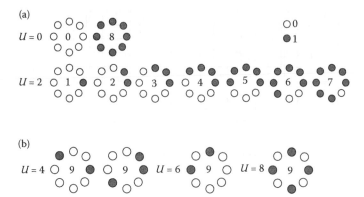

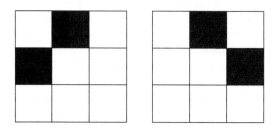

FIGURE 8.4 Some uniform LBP patterns and nonuniform LBP patterns of ($P = 8$, $R = 1$) neighborhoods. (a) Uniform LBP pattern and (b) nonuniform LBP pattern.

FIGURE 8.5 Examples of symmetric texture patterns.

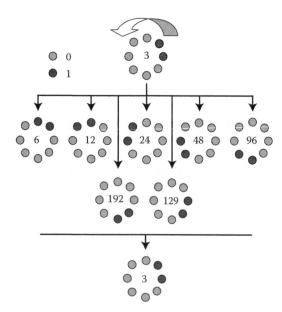

FIGURE 8.6 An example of pattern rotation.

where superscript "riu2" means rotation invariant uniform patterns with $U \leq 2$. The mapping from $LBP_{P,R}$ to $LBP_{P,R}^{riu2}$, which has $P + 2$ distinct output values, can be implemented with a lookup table. More detailed information about the LBP operators and their applications can be referred to Ahonen et al. (2004) and Maenpaa (2008).

8.2.3 Other LBP Operators

The LBP operators discussed above ignore the gray-level changes of pixels and illumination adaptability. In Zhang and Jin (2009), we improved the LBP operator by considering the magnitude of gray-level differences, which is important in texture description. It is noted that the improved LBP intends to concentrate on the visually most important texture pattern parts of images and disregard the unimportant details. The improved LBP operator introduced a parameter α to control the difference between neighboring pixels. If the difference between two pixels does not reach an extent controlled by α, it regards the two pixels as same. The improved LBP operator is defined as

$$LBP_p^\alpha = \sum_{p=0}^{p-1} S\left(\frac{g_p - g_c}{g_c} - \alpha\right) \times 2^p \qquad (8.7)$$

The improved LBP operator is based on the features of video frames captured by the two-phase flow monitoring system and ignores the nonsignificant details, so that it can extract the most important texture patterns. If we change the sign function $S(x)$ as:

$$S_\uparrow(x) = \begin{cases} 1, & x > 0 \\ 0, & x \leq 0 \end{cases} \qquad (8.8)$$

$$S_\downarrow(x) = \begin{cases} 0, & x > 0 \\ 1, & x \leq 0 \end{cases} \qquad (8.9)$$

we can obtain two LBP patterns, one of which (noted as $LBP_\uparrow$) describes the distribution of pixels with gray levels larger than the center pixel, and the other (noted as $LBP_\downarrow$) describes the distribution of pixels with gray levels lower than the center one. When reflecting on LBP texture spectrum, $LBP_\uparrow$ captures the bright part of texture map, and $LBP_\downarrow$ captures the dark part of texture map, they are compensatory with each other. The improved LBP operator has been applied in the gas–liquid two-phase flow regimes analysis and excellent results have been obtained. Similar to the improved LBP operator in thought, Tan and Triggs (2010) provided an enhanced LBP operator named LTP for face recognition and got satisfactory results.

Zhenhua et al. (2010) presented a completed modeling of the LBP operator (named by The Completed LBP) for texture classification. A local region or a neighborhood is represented by its center pixel and a local difference sign-magnitude transform. It decomposes the image local differences into two complementary components: the signs and the magnitudes, and obtains two LBP patterns. In fact, they performed another LBP operation on the magnitude and obtained another LBP pattern, besides sign LBP pattern. This completed LBP has been successfully used in the texture pattern recognition.

8.3 SPATIAL WATERMARKING BASED ON LBP OPERATORS

8.3.1 DEFINITIONS ON A (P, R) LOCAL REGION

Before presenting the proposed watermarking algorithms, we first provide some definitions. Let g_c denote the gray level of the center pixel c in the P neighborhood, and let g_p denote the gray level of the neighboring pixels p. For a (P, R) local region, we describe it as follows:

$$g_p = \{g_i \,|\, i = 0,\ldots,c,\ldots,P-1\} \tag{8.10}$$

$$m_p = \{m_i \,|\, m_i = |g_i - g_c|, i = 0,\ldots,P-1\} \tag{8.11}$$

$$s_p = \{s_i \,|\, s_i = \text{sign}(g_i - g_c), i = 0,\ldots,P-1\} \tag{8.12}$$

Note that Equation 8.12 uses the *sign* function, which is equivalent to Equation 8.2. In this way, we divide the local region into three parts (Zhenhua et al., 2010): g_p is a vector composed of P pixels in the R radius, m_p is a vector built by the magnitude obtained from the difference between the p pixels and the center pixel g_c, and s_p is a sign vector from the difference. Figure 8.7 shows an example of the three parts in a $(P = 8, R = 1)$ local region.

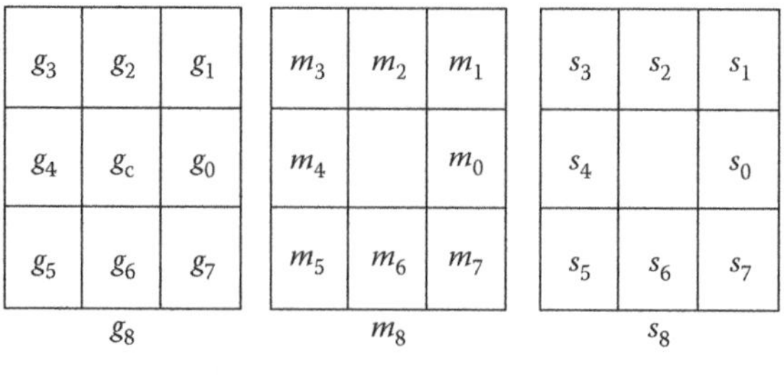

g_3	g_2	g_1
g_4	g_c	g_0
g_5	g_6	g_7

g_8

m_3	m_2	m_1
m_4		m_0
m_5	m_6	m_7

m_8

s_3	s_2	s_1
s_4		s_0
s_5	s_6	s_7

s_8

FIGURE 8.7   An example of a (8, 1) local region, as divided into three parts: g_8 is the pixel vector, m_8 is the magnitude vector, s_8 is the sign vector.

In order to embed watermarks, we define Boolean functions $f(s_p)$ to be applied to the binary sign vector part s_p. Two types of Boolean functions are chosen for illustration purposes, which are defined as follows:

$$f_\oplus(s_p) = s_0 \oplus s_1 \oplus \cdots \oplus s_{P-1} \tag{8.13}$$

$$f_\#(s_p) = \text{Bool}(\#1(s_p) - \#0(s_p) > N) \tag{8.14}$$

In Equation 8.13, $\oplus$ is the exclusive OR (XOR) operator. Obviously, $f_\oplus(s_p) \in \{0, 1\}$. It satisfies the associative and commutative properties, so any circular bit shifted on s_p clockwise or counterclockwise does not change the function value. However, any one bit change in s_p from 0 to 1 or from 1 to 0 will reverse the function value.

In Equation 8.14, $\#1(s_p)$ means the number of pixels with value "1" in s_p, $\#0(s_p)$ is the number of "0" in s_p, N is an integer, and $N \leq P - 1$. If $\#1(s_p) - \#0(s_p) > N$, then $f_\#(s_p)$ returns 1; otherwise, it returns 0. Obviously, $f_\#(s_p)$ is immune to bit shift and rotation.

8.3.2 WATERMARK-EMBEDDING ALGORITHM

We embed the watermarks by changing the value of $f(s_p)$ in a local region. The value of $f(s_p)$ is changed by altering the bits in s_p. These changes are reflected by modification of pixels in the spatial local region. Different Boolean functions correspond to different algorithms.

For instance, we use Boolean function $f_\oplus(s_p)$. In a (P, R) neighborhood, we select a pixel with the minimal magnitude in m_p to alter for embedding the watermark, so that we affect the quality of the original image block to its least degree. In other words, we keep the value of $f_\oplus(s_p)$ consistent with the corresponding bit of watermarks without reducing the quality of the image block much.

The watermark embedding procedure can be summarized in the following two steps:

1. The original image is divided into (P, R) non-overlapping local region blocks. The LBP pattern is computed to obtain m_p and s_p, and then obtain the value of $f_\oplus(s_p)$. Let w be one of bits in the watermarks and β be the watermarking intensity factor.
2. For each (P, R) local neighborhood, if the value of $f_\oplus(s_p)$ is equal to the value of w, we do nothing to the pixels in the neighborhood. Otherwise, we modify one of the pixels by making the value of $f_\oplus(s_p)$ consistent with the corresponding w. That is,

```
if (w = 1 and f⊕(sp) = 0) or (w = 0 and f⊕(sp) = 1),
then {select mi = min(mp);
    if si = 1 then gi = (gi − mi) × (1 − β);
        else gi = (gi + mi) × (1 + β)}
```

Note that min( ) is the minimal function. If there are more than one minimum, we select any one of the minimums to determine the pixel to be changed. If a block's pixels are all "0" or "1," we will modify the center pixel based on the corresponding watermarking bit before embedding it to the block.

8.3.3 WATERMARK EXTRACTION ALGORITHM

The watermark extraction procedure in the proposed method becomes straightforward. We judge the value of $f_\oplus (s_p)$ in the watermarked image to extract the watermark w. That is

```
if f⊕(sₚ) =1 then w = 1 else w = 0
```

8.4 EXPERIMENTAL RESULTS AND ANALYSIS

We use the Lena image of size 256×256 to test the performance of the algorithms shown in Section 8.3. The watermark is a binary image of size 84×84. The neighborhood is (8, 1), which is a 3×3 local region. One local region embeds one bit of watermark. Therefore, the watermarking capacity is 1/9 of the original image size.

The notations are given below. $W(i, j)$ denotes the original watermark binary image of size $M \times M$, $W^*(i, j)$ denotes the extracted watermarked binary image of size $M \times M$, $F(i, j)$ denotes the original image of size $N \times N$ to be watermarked, and $F^*(i, j)$ denotes the watermarked image. We use PSNR (peak signal-to-noise ratio), EBR (error bit rate), and NC (normalized correlation), as shown in Equations 8.8, (8.9), and (8.10), respectively, to evaluate the performance.

The EBR is used to compute the rate of error bits on the whole watermark accurate bits. The NC is used to locate a pattern on the extracted watermark image that best matches the specified reference pattern from the original image base, see Kung et al. (2009). Evidently, NC measures the amount of altered information which is originally "1", and we name it as white NC (WNC). In order to accurately calculate the effect of the attack, the amount of altered information which is originally "0" is also considered, and we name it as black NC (BNC). Note that the formula of BNC is the same as Equation 8.9 with all 1's being changed to 0's and vice versa. The PSNR is often used in engineering to measure the signal ratio between the maximum power and the power of corrupting noise. We use it to compare the original and the embedded images in the spatial domain.

$$EBR = \frac{\sum_{i=0}^{M-1} \sum_{j=0}^{M-1} (W(i,j) \oplus W^*(j,j))}{M \times M} \tag{8.15}$$

$$NC = \frac{\sum_{i=1}^{M} \sum_{j=1}^{M} W(i,j)W^*(i,j)}{\sum_{i=1}^{M} \sum_{j=1}^{M} [W(i,j)]^2} \tag{8.16}$$

$$PSNR = 10\log_{10}\left[\frac{255^2}{\sum_{i=1}^{N} \sum_{j=1}^{N} [F(i,j) - F^*(i,j)]^2 / N^2}\right] \tag{8.17}$$

By experiments, the proposed (8, 1) LBP-based watermarking algorithm shows better transparency and robustness against some commonly used image-processing operations, such as additive noise, luminance variation, contrast adjustment, and color balance. Some examples of applying various operations on the watermarked image are shown in Figure 8.8, where Figure 8.8a is the original Lena image, Figure 8.8b is the original watermark, Figure 8.8c is the watermarked Lena by the proposed algorithm with PSNR 42.67 and intensity factor $\beta = 0.08$, and Figure 8.8d is the extracted watermark with WNC = 1 and BNC = 1.

From Figure 8.8e–z, all processes are carried out in Figure 8.8c. Figure 8.8e is the resulting image after adding 10% noise, and Figure 8.8f is the extracted watermark with EBR = 3.85%, WNC = 0.959, and BNC = 0.962. Figure 8.8g is the resulting image after adding 30% noise, and Figure 8.8h is the extracted watermark with EBR = 10.01%, WNC = 0.887, and BNC = 0.905. Figure 8.8i is the resulting image after logarithm transform of darkening, and Figure 8.8j is the extracted watermark with EBR = 5.33%, WNC = 0.948, and BNC = 0.946. Figure 8.8k is the resulting image after logarithm transform of brightening, and Figure 8.8l is the extracted watermark with EBR = 2.98%, WNC = 0.979, and BNC = 0.966. Figure 8.8m is the resulting image after contrast enhancement of +10%, and Figure 8.8n is the extracted watermark with EBR = 0.47%, WNC = 0.995, and BNC = 0.995. Figure 8.8o is the resulting image after contrast reduction of −50%, and Figure 8.8p is the extracted watermark with EBR = 14.24%, WNC = 0.875, and BNC = 0.851.

Figure 8.8q is the resulting image after coloring by Photoshop 7.0, and Figure 8.8r is the extracted watermark with EBR = 1.03%, WNC = 0.990, and BNC = 0.989. Figure 8.8s is the resulting image after color saturation adjustment, and Figure 8.8t is the extracted watermark with EBR = 5.75%, WNC = 0.946, and BNC = 0.940. Figure 8.8u is the resulting image after destroying some parts, and Figure 8.8v is the extracted watermark with EBR = 7.17%. Figure 8.8w is the resulting image after cut from the original image, and Figure 8.8x is the extracted watermark. Figure 8.8y is the resulting image after JPEG compression by Photoshop 7.0 with quality 12, and Figure 8.8z is the extracted watermark with EBR = 19.20%, WNC = 0.80, and BNC = 0.81. Figure 8.8A is the resulting image after JPEG compression with quality 11, and Figure 8.8B is the extracted watermark with EBR = 35.12%, WNC = 0.648, and BNC = 0.656.

By experiments, the proposed method shows better image tamper detection ability. Figure 8.9 provides two examples. In Figure 8.9a, the enclosed face area of the watermarked Lena image (see Figure 8.8c) is replaced by the original (unwatermarked) face area in Figure 8.8a. The extracted watermark in Figure 8.9b reveals the modification, and Figure 8.9c shows the corresponding location of the modification.

Another example is the automobile license plate number forgery. Figure 8.9d shows an original license plate image, and Figure 8.9e is the watermarked image. Figure 8.9f is the tampered image by using the digit "2" to replace the character "M" in the license plate of Figure 8.9e. Figure 8.9g shows the result of tamper detection and location.

Figures 8.10 through 8.14 show some validation results. From Figure 8.10, as the intensity factor increases, the PSNR declines slowly, but maintains satisfactory

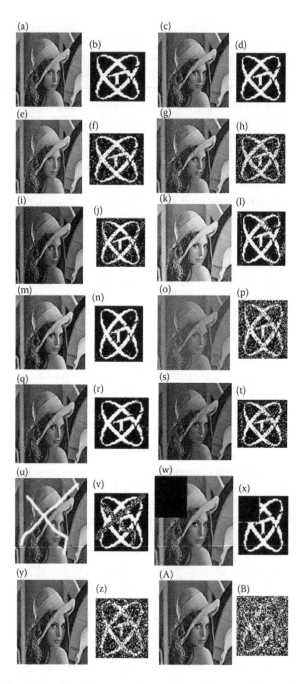

FIGURE 8.8 Examples of applying some image-processing operations on the watermarked image. (a) The original Lena image, (b) the original watermark, (c) the watermarked Lena by the proposed algorithm with β = 0.08, (e)–(z): results by image-processing operations carried out on (c). See context for more explanation. (A): is the resulting image after JPEG compression and (B): is the extracted watermark.

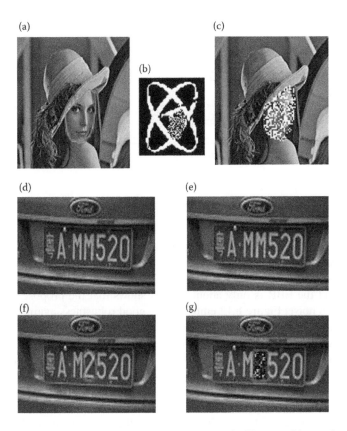

FIGURE 8.9 Tamper detection and location examples. (a) Tampered image by replacing face area, (b) the extracted watermark showing the tampering area, (c) tamper location, (d) original license plate, (e) watermarked license plate, (f) tampered license plate, and (g) tamper detection and location.

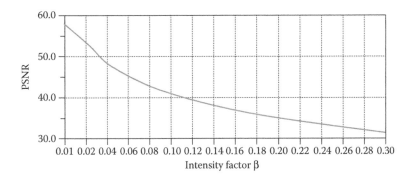

FIGURE 8.10 The relationship between PSNR and intensity factor.

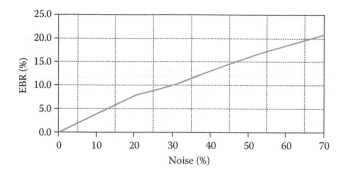

FIGURE 8.11 The relationship between EBR and noise.

values. When β reaches 0.3, the PSNR is still above 30, which demonstrates that the proposed method keeps good image quality. Figure 8.11 shows its better power against noise. When adding noise is 50%, the watermarked image is nearly destroyed, but the EBR is only about 16%. It shows that the proposed method is very robust to noise. Figure 8.12 embodies its good robustness against luminance modification. Luminance does not have much effect on LBP operators, which is shown in Figure 8.12. The best characteristic of the proposed method is its anticontrast adjustment shown in Figure 8.13, where the EBR keeps very low values especially when contrast adjustment increases from 0 to 10. When contrast increases to 50% or decreases to −50%, the EBR are below 15%. Figure 8.14 demonstrates that it is only robust against slight JPEG compression. When compression keeps better quality, the method has EBR of <20%. However, the proposed method is fragile to medium filter, image blurring, pixel interpolation, and other operations on a window neighborhood.

Although the function $f_\oplus (s_p)$ is invariant to rotation, the method achieves better results when the rotations are close to the multiples of 90°. It is because if any one of the bits in s_p changes from 0 to 1 or from 1 to 0, the value of $f_\oplus (s_p)$ will change to its inverse. To improve the robustness against rotation and compression, we use the

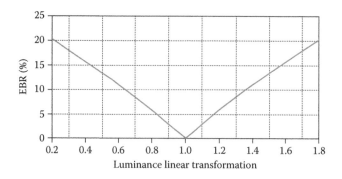

FIGURE 8.12 The relationship between EBR and luminance linear transformation.

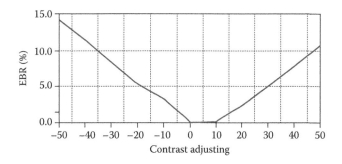

FIGURE 8.13   The relationship between EBR and contrast adjustment.

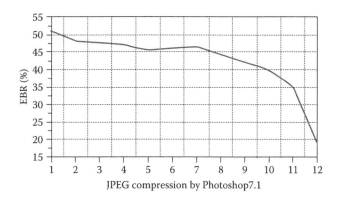

FIGURE 8.14   The relationship between EBR and JPEG compression.

function $f_\#$ (s_p) with $N = 1$ and watermark intensity factor $\beta = 0.02$. In the experiment, we modify the center pixel to satisfy the consistence between $f_\#$ (s_p) and the watermark bits. The watermark embedding algorithm is described as follows:

```
if (w = 1 and f_#(s_p) = 0) then do {g_c = g_c × (1 + β);
   Compute f_#(s_p);
   } while not f_#(s_p)
if (w = 0 and f_⊕(s_p) = 1) then do {g_c = g_c × (1 - β);
   Compute f_#(s_p);
   } while f_#(s_p)
```

By experiments, we observe that the function $f_\#$ (s_p) is more robust against additive noise, luminance change, contrast adjustment, JPEG compression, and rotation than the function $f_⊕$ (s_p) is. Figures 8.15 and 8.16 show the results with respect to rotation and JPEG compression. From Figure 8.15, we see that the watermarked image rotates from 5°, 15°, 30°, 45°, 60°, 75°, 90°, the EBR is, respectively, 15.1, 11.8, 17.2, 25.9, 18.7, 16.8, and 0%. When the rotation angle is 45°, the result is the worst. Except for 90°, the angle 15° corresponds to the best result. From Figure 8.16, we see that the JPEG compression quality factors change from 12 to 6, the EBR is respectively 2.3, 5.9, 8.6, 12.4, 16.5,

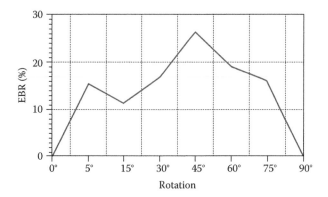

FIGURE 8.15   The relationship between EBR and rotation.

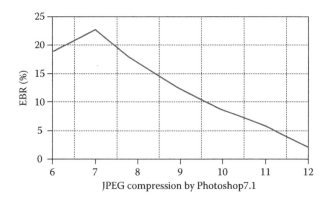

FIGURE 8.16   The relationship between EBR and JPEG compression.

25.1, and 18.2%. When the factor is 7, the EBR is the worst. With the decrease of the compression factors from 12 to 7, the EBR keeps an approximate lineal increase.

8.5   MULTILEVEL WATERMARKING BASED ON LBP OPERATORS

We can extend the aforementioned watermarking algorithm to multilevel watermarking techniques to achieve higher-embedding capacity and better robustness. We firstly present a double-level watermarking algorithm and conduct analysis on its experimental results. Then, we extend it to a general framework for multilevel watermarking schemes.

8.5.1   DOUBLE-LEVEL WATERMARKING

We divide the neighborhood s_p into two parts: even and odd neighbors, denoted as s_p^e and s_p^o. We perform $f_\oplus (s_p)$ on them and realize the embedding of two bits in the (P, R) neighborhood. In this way, the watermarking capacity is doubled. Figure 8.17

S_3^o	S_2^e	S_1^o
S_4^e		S_0^e
S_5^o	S_6^e	S_7^o

FIGURE 8.17 The s_p^e and s_p^e of (8, 1) LBP pattern. s_p^e denotes even neighbors and s_p^e denotes odd neighbors, $p = 0 \dots 7$.

shows an example of the (8, 1) LBP pattern, which in fact is equivalent to two (4,1) neighborhoods.

An example of embedding two watermark images into the Lena image is shown in Figure 8.18, where Figure 8.18a is the original Lena image, Figure 8.18b and c are two watermark images denoted by W1 and W2, and Figure 8.18d is the water-marked image with PSNR = 36.5 and β = 0.08. Figures 8.18e and f are the two extracted watermarks from Figure 8.18d. Figure 8.18g is the resulting image after adding 10% noise, and Figure 8.18h and i are the extracted two watermarks with EBR = 1.96% and 2.47%, WNC = 0.980 and 0.966, BNC = 0.980 and 0.977, respectively. Figure 8.18j is the resulting image after adding noise 120%, and Figure 8.18k and l are the extracted two watermarks with EBR = 8.73% and 9.04%, WNC = 0.894 and 0.885, BNC = 0.919 and 0.916, respectively.

Figure 8.18m is the resulting image after luminance reduction of −50%, and Figure 8.18n and o are the extracted two watermarks with EBR = 10.23% and 7.74%, WNC = 0.916 and 0.949, BNC = 0.890 and 0.914, respectively. Figure 8.18p is the resulting image after luminance enhancement of +50%, and Figure 8.18q and r are the extracted two watermarks with EBR 10.08% and 7.55%, WNC = 0.923 and 0.946, BNC = 0.916 and 0.949, respectively. Figure 8.18s is the resulting image after contrast reduction of −50%, and Figure 8.18t and u are the extracted two watermarks with EBR = 10.07% and 8.86%, WNC = 0.918 and 0.942, BNC = 0.892 and 0.908, respectively. Figure 8.18v is the resulting image after JPEG compression with quality 12 by Photoshop 7.0, and Figure 8.18w and x are the extracted two watermarks with EBR = 12.47% and 8.42%, WNC = 0.882 and 0.925, BNC = 0.872 and 0.912, respectively.

Note that the embedding and extraction of two watermarks do not interfere with each other. Figures 8.19 through 8.22 show the performance curves after applying some image-processing operations. We observe that the double-level watermarking technique performs better robustness than the single-level one. In Figure 8.19, when the double-level watermarked image is added by 50% noise, the extracted two watermarks are EBR 8.73 and 9.04%, but for single-level watermarking, EBR is 16.02%. In Figure 8.20, when the double-level watermarked image is compressed by JPEG with quality factor 12, the extracted two watermarks are EBR 12.47 and 8.42%, but for single-level watermark, EBR is 19.02%. In Figures 8.21 and 8.22, when the double-level watermarked image is applied by luminance or

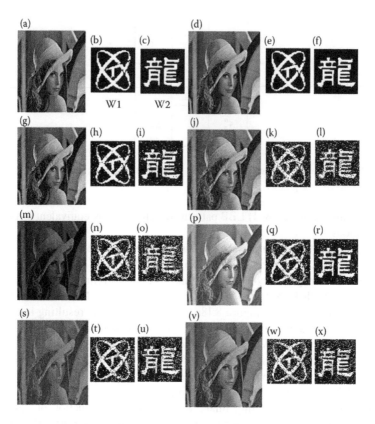

FIGURE 8.18 Examples of multilevel watermarking based on (8, 1) LBP pattern. (a) The original Lena image, (b,c) two watermark images W1 and W2, (d) the watermarked image, (e,f) the two extracted watermarks. (g,j,m,p,s,v) The resulting images by image-processing operations carried on (d). (h,i,k,l,n,o,q,r,t,u,w,x) The extracted watermarks from (g,j,m,p,s,v), respectively. See text for more explanation.

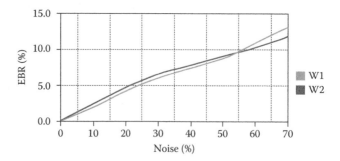

FIGURE 8.19 The relationship between EBR and Noise by double-level watermarking.

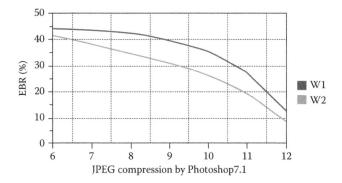

FIGURE 8.20 The relationship between EBR and JPEG compression by double-level watermarking.

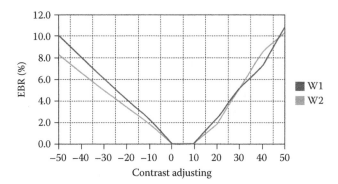

FIGURE 8.21 The relationship between EBR and contrast adjustment by double-level watermarking.

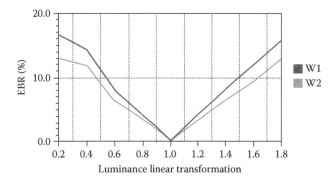

FIGURE 8.22 The relationship between EBR and luminance linear transformation by double-level watermarking.

contrast adjustment, the extracted two watermarks are EBR 3–5% lower than the single-level one.

8.5.2 EXTENSION TO MULTILEVEL WATERMARKING

Based on double-level watermarking, we can extend it to multilevel watermarking using variant (P, R) blocks to embed multiple watermarks. For example, four-level watermarking on the 5×5 neighborhood block is shown in Figure 8.23, which is divided into four parts: s_i^1, s_i^2, s_j^3, s_j^4, $i = 0 \cdots 3$, $j = 0 \cdots 7$. For s_i^1 and s_i^2, we use $f_\oplus(s_p)$ to embed watermarks, and for s_j^3 and s_j^4, we use $f_\#(s_p)$ on any one to embed watermarks and use $f_\oplus(s_p)$ on the other to embed watermarks. Therefore, we can embed four watermarks individually without mutual interference.

Let W_i, $i = 0 \cdots 3$ be the four watermarks. In experiment, we firstly embed W_2 and W_3, one of which is embedded by modifying the value of the center pixel (watermark factor $\beta = 0.02$), and the other by changing one of noncenter pixels (watermark factor $\beta = 0.08$). Then, we embed W_0 and W_1 based on Section 8.5.1.

Figure 8.24 shows some examples of multilevel watermarking. Figures 8.24a–d are the original images of size 256×256, and Figures 8.24e–h are the four watermark images of size 51×51. Figures 8.24i–l are the watermarked images with PSNR 36.11, 35.01, 38.24, and 36.7, respectively, and the four watermark images can be extracted exactly. Because the embedding procedures of the four watermarks do not affect each other, their performances are basically consistent with the results provided previously in Sections 8.4 and 8.5.

Although the watermarked images achieve better PSNR, we can observe from Figure 8.24 that some pixels in the smooth white or black region of these images are changed obviously, just like additive noises. In Figure 8.24l, we see that several points are protruding in smooth regions, while in Figure 8.24j, it is difficult to see those points. Therefore, the proposed multilevel watermarking technique is very suited for the images with more complicated textures.

s_3^3	s_2^4	s_2^3	s_1^4	s_1^3
s_3^4	s_1^2	s_1^1	s_0^2	s_0^4
s_4^3	s_2^1		s_0^1	s_0^3
s_4^4	s_2^2	s_3^1	s_3^2	s_6^4
s_5^3	s_5^4	s_6^3	s_6^4	s_7^3

FIGURE 8.23 The four parts of s_p in a 5×5 block. s_i^1, s_i^2, s_j^3, s_j^4, $i = 0 \cdots 3$, $j = 0 \cdots 7$ are used to embed the four watermarks, respectively.

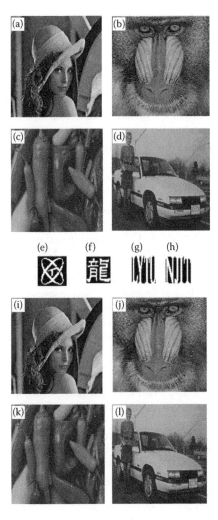

FIGURE 8.24 Multilevel watermarking examples. (a)–(d) are the four original images, (e)–(h) are the four watermark images, and (i)–(l) are the watermarked images extracted from (e)–(h), respectively. See text for further explanation.

The proposed method can be similarly extended to other LBP operators with different (P, R). We can design many multilevel watermarking schemes by jointly using $f_{\oplus}(s_p)$ and $f_{\#}(s_p)$ or using other different functions. Furthermore, the proposed method can be applied to the improved and complete LBP operators (Zhang and Jin, 2009; Zhenhua et al., 2010) to embed multilevel watermarks.

8.6 CONCLUSIONS

In this chapter, a new semi-fragile spatial watermarking scheme based on the LBP operator is proposed, whose single-level and multilevel watermarking methods are described and analyzed. The proposed methods are robust against some commonly used image-processing operations, such as additive noise, luminance change, and contrast adjustment. At the same time, they maintain good fragility to some window operations, such as filtering and blurring, and have better sensitivity to image tampering. It can also achieve tamper detection and location.

For future research, we will focus on the comprehensive comparison of different watermarking schemes based on different LBP operators, their reversibility, and security. Also, we will conduct research on steganalysis based on LBP operators, as enlightened by Lafferty and Ahmed (2004).

REFERENCES

Ahonen, T., Hadid, A., and Pietikainen, M., Face recognition with local binary patterns, in *Proc European Conf Computer Vision*, Prague, Czech, pp. 469–481, 2004.

Bas, P., Bihan, N. L., and Chassery, J., Color watermarking using quaternion Fourier transform, in *Proc ICASSP*, Hong Kong, China, pp. 521–524, Jun. 2003.

Chen, B. and Wornell, G., Quantization index modulation methods for digital watermarking and information embedding of multimedia, *J VLSI Signal Process*, 27, 7–33, 2001.

Cox, I., Kilian, J., Leighton, F., and Shamoon, T., Secure spread spectrum watermarking for multimedia, *IEEE Trans Image Process*, 6(12), 1673–1687, 1997.

Cvejic, N. and Tujkovic, I., Increasing robustness of patchwork audio watermarking algorithm using attack characterization, in *Proc. IEEE Int Symp Consumer Electronics*, U.K., pp. 3–6, 2004.

Kaewkamnerd, N. and Rao, K. R., Wavelet based image adaptive watermarking scheme, *Electron Lett*, 36, 518–526, 2000.

Kung, C.-M., Chao, S.-T., Tu, Y.-C., Yan, Y.-H., and Kung, C.-H., A robust watermarking and image authentication scheme used for digital content application, *J Multimed*, 4(3), 112–19, 2009.

Lafferty, P. and Ahmed, F., Texture based steganalysis: Results for color images, mathematics of data/image coding, compression, and encryption VII, with applications, in *Proc. of SPIE*, 5561, 145–151, 2004.

Luo, L., Chen, Z., Chen, M., Zeng, X., and Xiong, Z, Reversible image watermarking using interpolation technique, *IEEE Trans. Forensics Sec*, 5(1), 187–196, 2010.

Maenpaa, T., The local binary pattern approach to texture analysis extensions and applications, 2003, website http://herkules.oulu.fi/isbn9514270762/. Last accessed Dec. 20, 2008.

Maenpaa, T., Pietikainen, M., and Ojala, T., Texture classification by multi-predicate local binary pattern operators, in *Proc 15th Int Conf Pattern Recognition*, Barcelona, Spain, pp. 951–954, 2000.

Ojala, T., Pietikainen, M., and Maenpaa, T., Multiresolution gray scale and rotation invariant texture analysis with local binary pattern, *IEEE Trans Pattern Anal Mach Intell*, 24(7), 971–987, 2002.

Pitas, I., A method for signature casting on digital images, in *Proc IEEE Int Conf Image Process*, III, 215–218, 1996.

Piva, A., Bartolini, F., Boccardi, L., Cappellin, V. De, Rosa A., and Barni, M., Watermarking through color image bands decorrelation, in *Proc IEEE Int Conf Multimedia Expo.*, New York, pp. 1283–1286, Jul. 30–Aug. 2, 2000.

Podilchuk, C. and Zeng, W., Image-adaptive watermarking using visual models, *IEEE J Sel Areas Commun*, 16(4), 525–539, 1998.

Reed, A. and Hannigan, B., Adaptive color watermarking, in *Proc SPIE*, 4675, 222, 2002.

Sachnev, V., Kim, H. J., Suresh, J. N. S., and Shi, Y. Q., Reversible watermarking algorithm using sorting and prediction, *IEEE Trans Circuits Syst Video Technol*, 19(7), 989–999, 2009.

Schyndel, R., Tirkel, A., and Osborne, C., A digital watermark, in *Proc IEEE Int Conf Image Process*, II, 86–90, 1994.

Shih, F. Y., *Digital Watermarking and Steganography: Fundamentals and Techniques*, CRC Press, Boca Raton, FL, 2008.

Shih, F. Y. and Wu, S., Combinational image watermarking in the spatial and frequency domains, *Pattern Recognit*, 36, 969–975, 2003.

Swanson, M., Zhu, B., and Tewfik, A., Transparent robust image watermarking, in *Proc IEEE Int Conf Image Process*, III, 211–214, 1996.

Tanaka, K., Nakamura, Y., and Matsui, K., Embedding secret information into a dithered multi-level image, in *Proc IEEE ILCOM Int Conf*, pp. 216–220, 1990.

Tan, X. and Triggs, B., Enhanced local texture feature sets for face recognition under difficult lighting conditions, *IEEE Trans Image Process*, 19(6), 1635–1650, 2010.

Tsui, T. K., Zhang, P., and Androutsos, D., Color image watermarking using multidimensional Fourier transforms, *IEEE Trans Forensics Sec*, 3(1), 16–28, 2008.

Wei, Z. and Ngan, K. N., Spatio-temporal just noticeable distortion profile for grey scale image/video in DCT domain, *IEEE Trans Circuits Syst Video Technol*, 19(3), 337–346, 2009.

Wu, H.-T. and Cheung, Y.-M., Reversible watermarking by modulation and security enhancement, *IEEE Trans Instrum Meas*, 59(1), 221–228, 2010.

Xia, X.-G., Boncelet, C. G., and Arce, G. R., A multiresolution watermark for digital images, in *Proc IEEE Int Conf Image Process*, I, 548–551, 1997.

Yang, Y., Sun, M., Yang, H., Li, C. T., and Xiao, R., A contrast-sensitive reversible visible image watermarking technique, *IEEE Trans Circuits Syst Video Technol*, 19(5), 656–667, 2009.

Yeo, I.-K. and Kim, H. J., Modified patchwork algorithm: A novel audio watermarking scheme, in *Proc Int Conf Inf Technol: Coding Comput*, pp. 237, 2001.

Zhang, W. Y. and Jin, N. D., Improved local binary pattern based gas-liquid two-phase flow regimes analysis, in *Proc 6th Int Conf Fuzzy Syst Knowl Discov*, August 2009.

Zhang, W. Y. and Shih, F. Y., Semi-fragile spatial watermarking based on local binary pattern operators, *J Opt Commun*, 284(16), 3904–3912, 2011.

Zhang, W. Y., Shih, F. Y., Jin, N., and Liu, Y., Recognition of gas-liquid two-phase flow patterns based on improved local binary pattern operator, *Int J Multiph Flow*, 36, 793–797, 2010.

Zhenhua, G., Zhang, L., and Zhang, D., A completed modeling of local binary pattern operator for texture classification, *IEEE Trans Image Process*, 19(6), 1657–1663, 2010.

9 Authentication of JPEG Images Based on Genetic Algorithms

Venkata Gopal Edupuganti and Frank Y. Shih

CONTENTS

9.1 INTRODUCTION

Digital images are transmitted more often over the Internet now than ever before. Unfortunately, free-access digital multimedia communication provides virtually unprecedented opportunities to pirate copyrighted material. Due to wide availability of image manipulation software, digital images suffer from different kinds of attack, such as modification and removal of the content. Therefore, the tasks of detecting and tracing copyright violations have stimulated significant interest among engineers, scientists, lawyers, artists, and publishers, to name a few. As a result, the research in watermark authentication has become very active in recent years, and the developed techniques have grown and been improved a great deal.

Authentication of digital images can be carried out in two ways: *digital signature* and *digital watermarking* (Celik et al., 2002; Cox et al., 2002; Shih, 2007). A digital signature, such as hash value, which is a cryptographic technique, has two drawbacks in authenticating images. First, in order to check the authenticity, the verifier must know the digital signature beforehand, so the sender has to send it through a separate secure communication channel. This generates an overhead on bandwidth.

Second, although digital signatures authenticate the images, they fail to locate which parts of the images were altered. One way of achieving tamper localization is dividing the image into blocks and sending the digital signature associated with each block to the receiver. However, this increases the overhead on bandwidth.

Digital watermarking is categorized into three types: *robust, fragile,* and *semi-fragile*. Robust watermarking (Cox et al., 2002; Shih, 2007) is used to protect the copyright of digital media. Such a watermark stands with the image even after signal-processing operations, such as compression, translation, and rotation, are applied on the image. Fragile watermarking (Cox et al., 2002; Lin et al., 2004, 2005; Shih, 2007; Lee and Lin, 2008) on the other hand becomes invalid even if the image is slightly modified. This feature allows us to verify the authenticity of digital images. Semi-fragile watermarking (Fridrich, 1998; Lin and Chang, 2000; Cox et al., 2002; Ho and Li, 2004; Li, 2004; Shih, 2007), which comes in between robust and fragile, can distinguish between malicious (such as cropping, modification, etc.) and nonmalicious attacks (such as compression, smoothing, etc.).

In general, there are two ways to perform digital watermarking: one is in spatial domain and the other is in the frequency domain (Cox et al., 2002; Shih, 2007). Spatial-domain techniques are simple as they embed the watermark directly into the pixels, such as least-significant bit (LSB) modification. The disadvantage is that the attacker can easily detect the watermark. On the other hand, frequency-domain approaches involve complex calculations, but they enhance the data security as compared with spatial-domain techniques. Such approaches first apply discrete Fourier transform (DFT), discrete cosine transform (DCT), or discrete wavelet transform (DWT) to transform the image data from spatial domain to frequency domain, and then embed the watermark into the frequency-domain coefficients.

Most of fragile watermarking methods for image authentication have been designed for authenticating uncompressed images (Fridrich, 1998; Lin et al., 2004; Lee and Lin, 2008). Even though they have the advantage of recovering the tampered blocks in addition to tamper localization, their storage space and bandwidth required to transfer an uncompressed image are very high. As JPEG compression is a widely-used image format, several authentication methods for JPEG images (Lin and Chang, 2000; Ho and Li, 2004; Li, 2004; Wang et al., 2008) have been proposed.

Lin and Chang (2000) designed a semi-fragile watermarking system for authenticating JPEG visual content using two invariant properties of DCT coefficients. Ho and Li (2004) proposed a semi-fragile watermarking scheme for authentication of JPEG images using nine-neighborhood block consideration to provide block dependency. This scheme is tolerant to JPEG compression to a predetermined lowest-quality factor. Li (2004) proposed a fragile watermarking scheme for authentication of JPEG images by considering all the DCT coefficients of each block. Wang et al. (2008) proposed a fragile watermarking method for authentication of JPEG images that generates the watermark using the initial values of the quantized DCT coefficients as an input to the chaotic system and embeds the watermark into the LSB of the quantized DCT coefficients. As each block is watermarked individually (i.e., using only the information present in that block), the method suffers from copy-paste (CP) (Lee and Lin, 2008) and vector quantization (VQ) (Holliman and

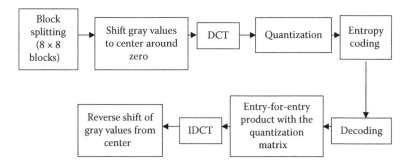

FIGURE 9.1   The JPEG procedure.

Memon, 2000) attacks. Note that all the above methods require the receivers to know the quantization tables beforehand for image authentication.

Figure 9.1 shows the often-used JPEG procedure to compress an image. It involves splitting an input image into blocks of size 8×8, DCT, quantization, entropy coding, and so on (Wang et al., 2008). As aforementioned, the existing fragile and semi-fragile watermarking methods need the receiver to know the quantization table beforehand to authenticate the image. As different JPEG quality factors use different quantization tables, the receiver has to maintain a lot of quantization tables. We intend to overcome this problem to allow the receiver to authenticate the image irrespective of the quality factor and the quantization table used in creating the watermarked image. In this chapter, we propose a genetic algorithm (GA)-based method that can adjust the image such that the modified image after JPEG compression contains the authentication information in the DCT coefficients of each block to authenticate the image (Edupuganti and Shih, 2010). Furthermore, the generation of authentication information guarantees the uniqueness with respect to each block (depending on block position) and each image for thwarting the CP and VQ attacks.

The rest of this chapter is organized as follows. Section 9.2 presents the GA-based watermark-embedding method. Section 9.3 describes the authentication procedure. Section 9.4 shows experimental results. Finally, conclusions are drawn in Section 9.5.

9.2   GA-BASED WATERMARK-EMBEDDING METHOD

9.2.1   Overall Watermark-Embedding Procedure

The block diagram of the GA-based watermark-embedding procedure is shown in Figure 9.2. Each step is briefly described below and will be explained in more detail in later sections. Let A be an original grayscale image of size $N \times N$, where N is a multiple of 8. Let the quality factor of JPEG compression be denoted as QF.

> **Step$_e$ 1.** As JPEG uses blocks of size 8×8, we divide the image A into a set of 8×8 blocks, named $\{b(i, j)|\ i, j = 1,2, \ldots, N/8\}$, where i and j, respectively, denote the row and column numbers of the block. Mark each block $b(i, j)$

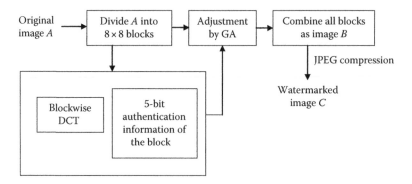

FIGURE 9.2 The overall watermark embedding procedure.

with a number X by traversing the blocks in the image from top-to-bottom and left-to-right as

$$X(i, j) = (i - 1) \times (N/8) + (j - 1). \qquad (9.1)$$

Step$_e$ 2. Apply a chaotic map (Lee and Lin, 2008) to transform the block number X into a new mapping block number X' as

$$X'(i, j) = \{(k \times X(i, j)) \bmod M\} + 1, \qquad (9.2)$$

where $M = N/8 \times N/8$ as the total number of blocks in the image, X, $X' \in [0, M-1]$, and $k \in [2, M-1]$ (a secret key) is a prime number. Note that if k is not a prime number, the one-to-one mapping cannot be achieved.

Step$_e$ 3. For each block $b(i, j)$, we generate the authentication information which will be described in Section 9.2.2.

Step$_e$ 4. For each block $b(i, j)$, we adjust its values using GA, so the modified block $b'(i, j)$ after JPEG compression contains the authentication information generated in step 3 in the LSB of integer part of five upper-left DCT coefficients as shown in Figure 9.3.

Step$_e$ 5. Combine all modified blocks $b'(i, j)$ to generate a modified image B.

Step$_e$ 6. Compress the modified image B using JPEG compression with QF to obtain the watermarked image C.

9.2.2 AUTHENTICATION INFORMATION GENERATION

In order to thwart the CP and VQ attacks, the authentication information needs to be unique per block and per image. Figure 9.4 illustrates the authentication information generation. The chaotic-map block number $X'(i, j)$, computed in Section 9.2.1, is unique to each block. The 11-bit clock information is obtained by using the bit-XOR " $\oplus$ " (i.e., bitwise exclusive-or) operation of six elements of image capture identity (which was recorded in the image header): year, month, day, hour, minute, and second. We attach

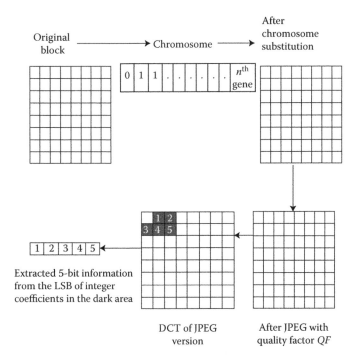

FIGURE 9.3 GA procedure to adjust the block.

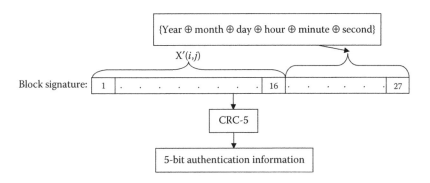

FIGURE 9.4 The generation of the authentication information.

the 11-bit information to the end of 16-bit stream of $X'(i, j)$ to generate the 27-bit block signature. Finally, the 5-bit authentication information is computed by subjecting the 27-bit block signature to CRC-5 checksum. The secret key k of the chaotic map is shared with the receiver using the public key cryptographic algorithm (Lin et al., 2004).

9.2.2.1 Cyclic Redundancy Checksum

In the computation of cyclic redundancy checksum (CRC) (Tanebaum, 2003; Lin et al., 2004), a k-bit string is represented by the coefficients of a $k-1$ degree polynomial, that is, the coefficients of x^0 to x^{k-1}. The most significant bit (MSB) of the k-bit string becomes the coefficient of x^{k-1}, the second MSB becomes the coefficient of x^{k-2}, and continuously the LSB becomes the coefficient of x^0. For example, the polynomial representation of the bit string 1101 is $x^3 + x^2 + 1$. To compute the CRC-r authentication information of a bit string, we pad "r" zeros to the end of the bit string. Then we divide the resulting bit string with an r-degree polynomial. We subtract the remainder from the bit string padded with zeros. The resultant bit string after subtraction is the authentication information. The subtraction is done in modulo-2 arithmetic. To check if a CRC-r generated bit string is modified or not, we divide the bit string with the r-degree polynomial used in the generation stage. If there is a nonzero remainder after division, it means that the bit string is modified.

9.2.3 ADJUSTMENT BY GA

GA is a randomized, parallel, and global search approach based on the mechanics of natural selection and natural genetics to find solutions of problems (Holland, 1975). Generally, GA starts with some randomly selected genes in the first generation, called *population*. Each individual, called *chromosome*, in the population corresponds to a solution in the problem domain. An objective (or fitness function) is used to evaluate the quality of each chromosome. The chromosomes with high quality will survive and form the population of the next generation. By using the reproduction, crossover and mutation operations, a new generation is recombined to find the best solution. This process will repeat until a prespecified condition is satisfied, or a constant number of iterations is reached.

We apply GA to adjust the pixel values in each block $b(i, j)$, such that the adjusted block $b'(i, j)$ after compression with quality factor QF contains the authentication information (generated as in Section 9.2.2) in the LSB of the integer part of the five DCT coefficients as indicated in Figure 9.3. This algorithm is divided into three steps: substitution of chromosome in the original block, chromosome evaluation, and GA. Section 9.2.3.3 presents the GA adjustment process in two different approaches. The two approaches use the details in Sections 9.2.3.1 and 9.2.3.2 in the evaluation of each chromosome to find a suitable adjusted block.

9.2.3.1 Substitution of Chromosome in the Original Block

The idea is to modify the p LSBs of each pixel in the block, where p is equal to the chromosome length divided by 64. For example, if the chromosome length is 64, then only one LSB is modified in each pixel. We traverse the pixels from left-to-right and top-to-bottom, and substitute the first gene in the LSB of the first pixel, the second gene in the LSB of the second pixel, and so on. For another example, if the chromosome length is $64p$, then the first p genes are used to replace the p LSBs of the first pixel, the next p genes are used to replace the p LSBs of the second pixel, and so on.

9.2.3.2 Objective Function for Chromosome Evaluation

The objective function is the bit difference between the five extracted LSBs from the integer part of the DCT coefficients (as shown in Figure 9.3) and the 5-bit authentication information (as described in Section 9.2.2). We intend to minimize the difference, with "0" being treated as the best solution. The objective function can be expressed as

$$Objective\_function = \sum_{i=1}^{5} |a(i) - r(i)|,$$ (9.3)

where a is the 5-bit authentic information, and r is the retrieved LSBs from the integer part of the five DCT coefficients.

9.2.3.3 GA to Adjust the Block

We present two types of block adjustment: *fixed-length chromosome* and *variable-length chromosome*. It is a trade-off between speed and watermarked image quality. The fixed-length chromosome algorithm runs faster, but produces a lower-quality watermarked image as compared with the variable-length chromosome algorithm.

9.2.3.3.1 Fixed-Length Chromosome

In the fixed-length chromosome algorithm, the chromosome length is fixed and determined by the quality factor QF of JPEG compression. The chromosome length designed for the corresponding QF range is listed in Table 9.1. It aims at adjusting the block with acceptable image quality under reduced time consumption of the GA adjustment process. The GA adjustment process by the fixed-length chromosome algorithm is presented as follows:

Step_f 1. Define the chromosome length, population size, crossover rate, replacement factor, and mutation rate. The initial population is randomly assigned with 0's and 1's.

Step_f 2. Substitute each chromosome in the original block as explained in Section 9.2.3.1 and evaluate the objective function for each corresponding chromosome by Equation 9.3.

Step_f 3. Apply reproduction, crossover, and mutation operators to generate the next generation of chromosomes.

TABLE 9.1

The Chromosome Lengths Associated with Different QF Values

Quality Factor (QF)	Chromosome Length
$QF > 80$	192
$65 \leq QF \leq 80$	256
$55 \leq QF \leq 65$	320
$45 \leq QF \leq 55$	384
$QF < 45$	448

Step$_f$ 4. Repeat steps 2 and 3 until the objective function value is equal to zero.

Step$_f$ 5. Substitute the final chromosome in the original block to obtain the modified block.

9.2.3.3.2 Variable-Length Chromosome

In the variable-length chromosome algorithm, for improving the watermarked image quality, we start the adjustment process with a chromosome size of 64. If a solution cannot be obtained before the termination condition is satisfied, we increase the chromosome size by 64 and execute the adjustment process again. This process is repeated until the termination condition is satisfied. The final chromosome size is selected based on the QF of the JPEG compression in Table 9.1. The termination condition for the final chromosome size is to obtain an objective function value of zero and that of the intermediate chromosome sizes is a generation limit of 100. The GA adjustment process by the variable-length chromosome algorithm with $QF = 65$ is presented as follows and its flowchart is shown in Figure 9.5.

Step$_v$ 1. Define the objective function, number of genes $n = 64$, population size, crossover rate, replacement factor, and mutation rate. The initial population is randomly assigned with 0's and 1's.

Step$_v$ 2. If $n < 256$, the termination condition is the maximum of 100 generations; otherwise, the termination condition is the objective function value being zero.

Step$_v$ 3. Substitute each chromosome in the original block as explained in Section 9.2.3.1 and evaluate the objective function for each corresponding chromosome by Equation 9.3.

Step$_v$ 4. While the termination condition is not satisfied, apply reproduction, crossover, and mutation operators to generate the next generation of chromosomes, and compute the objective function for each corresponding chromosome.

Step$_v$ 5. If the objective function value of population is zero, substitute the chromosome in the original block to obtain the modified block; otherwise, $n = n + 64$ and go to step 2.

9.3 AUTHENTICATION

Authentication is the procedure that validates the authenticity of the received image. The main objective of authentication is to detect the tampered blocks with low false acceptance ratio (FAR). Our authentication algorithm is presented as follows:

Step$_a$ 1. Let D be a testing image, which is divided into blocks of size 8×8. Let the group of blocks be denoted as *attack_blocks*.

Step$_a$ 2. Retrieve the secret key k, the polynomial used in CRC checksum, and the clock values.

Step$_a$ 3. Compute the block numbers and the corresponding mapping block number in the same way as in Section 9.2.1.

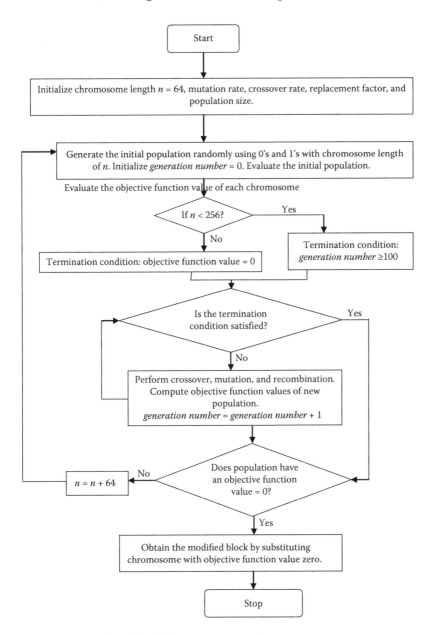

FIGURE 9.5   The variable-length chromosome procedure.

Stepa **4**. (Level-1 detection) For each block *attack_blocks(i, j)* ∈ *D*, where
$1 \le i, j \le N/8$, compute the authentication information as in Section 9.2.2
and denote it as *auth*. Compute the DCT of *attack_blocks(i, j)*. Retrieve
the 5-bit information from the LSB of integer part of the five DCT coefficients as shown in Figure 9.3, and denote it as *retv*. Follow the same order

of DCT coefficients used in the embedding stage while retrieving. If $\sum_{k=1}^{5} | \text{auth}(k) - \text{retv}(k) | \neq 0$, mark *attack_blocks(i,j)* as invalid; otherwise, mark *attack_blocks(i,j)* as valid.

Step$_a$ 5. (Level-2 detection) This level is used to reduce the FAR of invalid blocks. For each valid block after level-1 detection, if it has five or more invalid blocks in its 3×3 neighborhood, mark the block as invalid; otherwise, mark the block as valid.

The proposed authentication algorithm not only can detect the tampered blocks, but also can reduce the burden on the verifier as it does not require quantization of the DCT coefficients in the authentication process. Furthermore, the two-level detection strategy can reduce the FAR of invalid blocks.

9.4   EXPERIMENTAL RESULTS

To evaluate experimental results, we use the peak signal-to-noise ratio (PSNR) to measure the quality of the watermarked image. Let the image size be $m \times n$. The PSNR in decibels (db) of an image A with respect to an image B is defined by

$$\text{PSNR} = 20 \log_{10} \left(\frac{255}{\sqrt{\text{MSE}}} \right), \tag{9.4}$$

where MSE is the mean square error defined by

$$\text{MSE} = \frac{1}{mn} \sum_{i=0}^{m-1} \sum_{j=0}^{n-1} [A(i, j) - B(i, j)]^2. \tag{9.5}$$

Note that the higher the PSNR is, the less distortion there is to the host image and the retrieved one.

We use the FAR as a metric for performance evaluation of the authentication system. The FAR is defined as the ratio of total number of undetected blocks to the total number of attacked blocks. Its value varies in between 0 and 1. The lesser the FAR value, the better the performance of the authentication system.

$$\text{FAR} = \frac{\text{total number of undetected blocks}}{\text{total number of attacked blocks}}. \tag{9.6}$$

Figure 9.6 shows the results of Lena images before and after the watermarking algorithms by fixed-length chromosome and variable-length chromosome with $QF = 75$, secret key $k = 13$, and clock = 2008/11/27 12:45:40. As expected, the variable-length chromosome algorithm provides better watermarked image quality than the fixed-length one. More results of the PSNR values by variable-length chromosome on Barbara, Baboon, and Cameraman images are listed in Table 9.2.

FIGURE 9.6   (a) The original Lena image, (b) the watermarked image by fixed-length chromosome algorithm with PSNR = 36.67, and (c) the watermarked image by variable-length chromosome algorithm with PSNR = 45.12.

We conduct CP, cropping, and VQ attacks on the watermarked image. In the CP attack (Lee and Lin, 2008), a portion of a given image is copied and then used to cover some other objects in the given image. If the splicing is perfectly performed, human perception would not be able to notice that identical (or virtually identical) regions indeed exist in an image. The ideal regions for the CP attack are textured areas with irregular patterns, such as grass. Because the copied areas will likely blend with the background, and it will be very difficult for the human eye to detect any suspicious artifact. Another fact which complicates the detection is that the copied regions come from the same image. They therefore have similar properties, such as the noise component or color palette. It makes the use of statistical measures to find irregularities in different parts of the image impossible.

TABLE 9.2
PSNR Values of Four Watermarked Images

	Lena	Barbara	Baboon	Cameraman
PSNR	45.12	35.34	45.12	38.13

In the case of blockwise authentication systems, if the system has no blockwise dependency or unique signature that differentiates each block, then this kind of attack is not detected.

Holliman and Memon (2000) presented a counterfeiting attack on blockwise independent watermarking schemes, referred to as VQ or collage attack. In such a scheme, the attacker generates a forgery image by the collage of authenticated blocks from different watermarked images. Because of the blockwise nature of embedding and authentication processes, the forgery image is authenticated as valid. Additionally, if the database of watermarked images is huge, the attacker can easily generate a collage image that is identical to the unwatermarked image.

Figure 9.7 shows the results of a watermarked car image along with the CP-attacked version and the authentication images after level-1 and level-2 detections. The white portion in the authentication image indicates the modified part. Figure 9.8 shows the results of a watermarked donkey image. We also test the proposed algorithm under the cropping attack using 11 different single-chunk cropping criteria, as shown in Figure 9.9, and their resulting FAR values are listed in Table 9.3. Figure 9.10 shows the resulting images under different cropping patterns.

FIGURE 9.7   (a) The watermarked image by the variable-length chromosome algorithm with PSNR = 43.36, (b) the CP-attacked watermarked image with plate number removed, (c) the authentication image after level-1 detection, and (d) the authentication image after level-2 detection.

FIGURE 9.8 (a) The original donkey image, (b) the watermarked image using variable-length chromosome algorithm with PSNR = 41.14, (c) the CP-attacked image, and (d) the authentication image.

FIGURE 9.9 Different cropping criteria: (a) top-64, (b) bottom-64, (c) left-64, (d) right-64, (e) top-128, (f) bottom-128, (g) left-128, (h) right-128, (i) outer, (j) center-25%, and (k) center-63.7%.

TABLE 9.3

The *FAR* Values under Attacks Using Different Cropping Criteria in Figure 9.9

Attack Type	A	b	c	d	e	f	g	h	i	j	k
FAR	0	0	0	0.0039	0	0	0	0	0.038	0	0

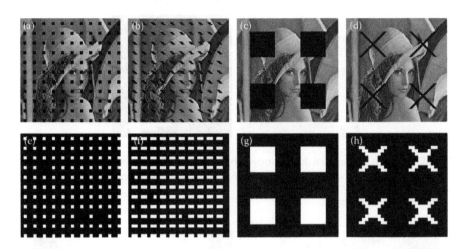

FIGURE 9.10   (a), (b), (c), and (d) Different cropping patterns; (e), (f), (g), and (h) corresponding detection images.

9.5   CONCLUSIONS

The described new efficient authentication method for JPEG images is based on GA, which adjusts the image so that the modified image contains authentication information in the DCT coefficients after JPEG compression. In the watermark extraction stage, the receiver obtains the authentication information without quantizing the DCT coefficients; this reduces the burden on the receiver in terms of computation and space required for storing quantization tables. In addition, with the generation of authentication information that is unique to each block and each image, the presented GA-based method is superior to the existing methods as it thwarts the VQ and CP attacks.

This chapter also presents the variation of image adjustment by fixed-length and variable-length chromosome to meet different application requirements, such as time and watermarked image quality. Furthermore, the two levels of detection in the authentication procedure reduce the FAR of invalid blocks.

REFERENCES

Celik, M. U., Sharma, G., Saber, E., and Tekalp, A. M., Hierarchical watermarking for secure image authentication with localization, *IEEE Trans Image Process*, 11(6), 585–594, 2002.

Cox, I. J., Miller, M. L., and Bloom, J. A., *Digital Watermarking Principles & Practice*, Morgan Kaufmann, San Francisco, 2002.

Edupuganti, V. G. and Shih, F. Y., Authentication of JPEG images based on genetic algorithms, *Open Artifi Intell J*, 4, 30–36, 2010.

Fridrich, J., Image watermarking for tamper detection, *Proc. IEEE Intl. Conf. Image Processing*, Chicago, IL, vol. 2, pp. 404–408, Oct. 1998.

Ho, C. K. and Li, C. T., Semi-fragile watermarking scheme for authentication of JPEG images, *Proc. Intl. Conf. Information Technology: Coding and Computing*, Las Vegas, NV, vol. 1, pp. 7–11, 2004.

Holland, J. H., *Adaptation in Natural and Artificial Systems*, University of Michigan Press, Ann Arbor, 1975.

Holliman, M. and Memon, N., Counterfeiting attacks on oblivious block wise independent invisible watermarking schemes, *IEEE Trans Image Process*, 9(3), 432–441, 2000.

Lee, T. Y. and Lin, S. D., Dual watermark for image tamper detection and recovery, *Pattern Recogn*, 41(11), 3497–3506, 2008.

Li, C. T., Digital fragile watermarking scheme for authentication of JPEG images, *Proc. IEE Proc. Vision, Image and Signal Processing*, 151(6), 460–466, 2004.

Lin, C. Y. and Chang, S.-F., Semi-fragile watermarking for authenticating JPEG visual content, *Proc. SPIE Security and Watermarking of Multimedia Content II*, San Jose, CA, vol. 3971, pp. 140–151, Jan. 2000.

Lin, P. L., Hsieh, C. K., and Huang, P. W., A hierarchical digital watermarking method for image tamper detection and recovery, *Pattern Recognit*, 38(12), 2519–2529, 2005.

Lin, P. L., Huang, P. W., and Peng, A. W., A fragile watermarking scheme for image authentication with localization and recovery, *Proc. IEEE Sixth Intl. Symp. Multimedia Software Engineering*, Miami, FL, pp. 146–153, Dec. 2004.

Shih, F. Y., *Digital Watermarking and Steganography: Fundamentals and Techniques*, CRC Press, Boca Raton, FL, 2007.

Tanebaum, A. S., *Computer Networks*, Fourth Edition, The Netherlands Pearson Education International, Upper Saddle River, New Jersey, 2003.

Wang, H., Ding, K., and Liao, C., Chaotic watermarking scheme for authentication of JPEG images, *Proc. Intl. Symp. Biometric and Security Technologies*, Islamabad, Pakistan, 1–4, 2008.

10 An Efficient Block-Based Fragile Watermarking System for Tamper Localization and Recovery

Venkata Gopal Edupuganti, Frank Y. Shih, and I-Cheng Chang

CONTENTS

10.1 INTRODUCTION

Digital information and data are transformed more often over the Internet now than ever before. Creative approaches to storing, accessing, and distributing data have generated many benefits for the digital multimedia field, mainly because of properties such as distortion-free transmission, compact storage, and easy editing. But due to the availability of multimedia manipulation software in the market, a malicious attacker can easily modify the image content. Therefore, the authentication and recovery of the multimedia content has been becoming increasingly important in recent years. Digital watermarking (Cox et al., 2001; Shih, 2007) is the leading technique for tamper localization and recovery of the modified content.

Digital watermarking (Cox et al., 2001; Shih, 2007) can be categorized into three types based on its purposes. The first type is *robust* watermarking, which provides the content protection of multimedia content by resisting the watermark against signal-processing operations. The second type is *fragile* watermarking, which authenticates the multimedia content. The watermark becomes invalid even if a single bit is modified in the content. The third type is *semi-fragile* watermarking, which acts in between robust and fragile watermarking. It resists the watermark against legitimate attacks and becomes invalid against malicious attacks. Although fragile (Lin et al., 2004, 2005; Lee and Lin, 2008) and semi-fragile (Fridrich, 1998; Kundur and Hatzinakos, 1999) techniques provide image authentication, the fragile watermarking technique has the advantage of tamper localization with recovery of the modified content.

The fragile watermarking uses two ways to authenticate an image—blockwise independent and blockwise dependent. Blockwise-independent (Celik et al., 2002) techniques use image ids, block indices, and hierarchical signatures to authenticate images and resist the vector quantization (VQ) (Holliman and Memon, 2000) attack. They need more authentication information to store in order to counteract the VQ attack. Blockwise-dependent techniques (Tanebaum, 2003; Lin et al., 2004, 2005; Lee and Lin, 2008), on the other hand, map the blocks using a secret key. The mapping is one-to-one. The feature of each block together with the authentication information of the corresponding mapping block is embedded into the mapping block. Each block is authenticated based on the information stored in that block together with the feature stored in the corresponding mapping block. It has the advantage of recovering the tampered block information from the feature stored in the corresponding mapping block along with the localization of tampered area. In this chapter, we focus on blockwise-dependent fragile watermarking for the purpose of image authentication and recovery.

Lin et al. (2004) designed a fragile watermarking scheme for image authentication with localization and recovery with a block size of 8×8. The scheme suffers from poor localization accuracy; that is, if a single pixel is modified, the whole 8×8 block is detected as tampered and poor recovery rate although it counteracts the VQ attack. Lin et al. (2005) came up with a different technique with a block size of 4×4, that is, better localization accuracy, but it still suffers from lack of recovery information if the corresponding mapping block is also marked as tampered. Wang and Tsai (2008) proposed a technique based on fractal code embedding and image inpainting. Their technique embeds the recovery information of blocks of ROI (region of interest) into the blocks other than ROI. This method also suffers from the lack of recovery information. Wang and Chen (2007) proposed a tamper detection and recovery scheme for color images based on a majority voting scheme with a block size of 2×2. The method still suffers from the lack of recovery information.

Recently, Lee and Lin (2008) proposed a dual-watermarking scheme for tamper detection and recovery with a block size of 2×2, which embeds the recovery information in two blocks instead of one block. Although it achieves impressive tamper localization and recovery information, it suffers from two main drawbacks. The first drawback is that it fails to detect the bit changes in the five most significant bits

(MSBs) of each pixel. The second drawback is the lack of support against the VQ attack. Because each block is authenticated individually, that is, considering only the information stored in that block, we can substitute different watermarked image blocks from vector quantization code book to form an authenticated image. This method also uses parity bit checksum (Tanebaum, 2003) for authentication. The disadvantage of parity sum is that it only detects even number of changes. This drawback is discussed in experimental results. In this chapter, we propose a watermarking system that not only thwarts the VQ attack, but also has good tamper localization accuracy and recovery rate (Edupuganti et al., 2011).

The rest of the chapter is organized as follows. Section 10.2 presents the related work to our method. Our proposed method is presented in Section 10.3. Experimental results and discussions are presented in Section 10.4. Finally, conclusions are drawn in Section 10.5.

10.2 RELATED WORK

10.2.1 VECTOR QUANTIZATION ATTACK

Holliman and Memon (2000) presented a counterfeiting attack on blockwise-independent watermarking schemes. In such a scheme, the attacker generates a forgery image by the collage of authenticated blocks from different watermarked images. Because of the blockwise nature of embedding and authentication processes, the forgery image is authenticated as valid. Additionally, if the database of watermarked images is huge, the attacker can easily generate a collage image that is identical to the unwatermarked image.

10.2.2 CYCLIC REDUNDANCY CHECKSUM

In the computation of cyclic redundancy checksum (CRC) (Tanebaum, 2003; Lin et al., 2004), a k-bit string is represented by the coefficients of a k-1-degree polynomial, that is, coefficients of x^0 to x^{k-1}. The MSB of the k-bit string becomes the coefficient of x^{k-1}, second MSB becomes the coefficient of x^{k-2}, and continuously the least-significant bit (LSB) becomes the coefficient of x^0. For example, the polynomial representation of the bit string 1011 is $x^3 + x + 1$. To compute the CRC-r authentication information of a bit string, we pad "r" zeros to the end of the bit string. Then we divide the resulting bit string with an "r" degree polynomial. Then, we subtract the remainder from the bit string padded with zeros. The resultant bit string after subtraction is the authentication information. The subtraction is done in modulo-2 arithmetic. To check if a CRC-r-generated bit string is modified or not, we divide the bit string with the r-degree polynomial used in the generation stage. If there is a remainder after division, it means that the bit string is modified; otherwise, the bit string is not manipulated.

10.2.3 COMPARISON OF EXISTING METHODOLOGIES

In Table 10.1, a comparison of three existing blockwise-dependent fragile watermarking schemes is made using different characteristics. The three schemes

TABLE 10.1

Comparison of Three Existing Schemes

Characteristic	Lin et al. (2004)	Lin et al. (2005)	Lee and Lin (2008)
Minimum block size of tamper localization	8×8	4×4	2×2
Copies of recovery information	1	1	2
Detection of VQ attack	Yes	Yes	No
Number of bits modified in each pixel of the original image	2 (LSB)	2 (LSB)	3 (LSB)
Blockwise dependent detection	Yes	Yes	No
Detection of changes in the higher order bits of each pixel	Yes, if the corresponding mapping block is valid	Yes, if the corresponding mapping block is valid	No
Authentication method used	MD5 with CRC	Parity bit	Parity bit

considered are Lin et al. (2004, 2005) and Lee and Lin (2008). Our proposed algorithm is designed to achieve the best characteristics of these three schemes, that is, tamper localization size of 2×2, two copies of recovery information, support against the VQ attack, and detection of the modification in higher-order bits. Our proposed scheme can be applied to grayscale and color images. In the proposed algorithm, the probability to thwart the VQ attack and higher-order bit modification is higher than that of Lin et al. (2004, 2005) because of the two copies of recovery information. In order to achieve the dual recovery, we adopt the generation of lookup table, which is similar to Lee and Lin (2008).

10.2.4 Lookup Table Generation

We outline the procedure of the lookup table generation as follows:

1. Let A be a host image of size $M \times M$, where M is a multiple of 2.
2. Divide the image A into blocks of size 2×2. Calculate the block number for each block as

$$X(i, j) = (i - 1) \times (M/2) + (j - 1) \tag{10.1}$$

where $i, j \in [1, M/2 \times M/2]$
3. Calculate the corresponding mapping block for each block number as

$$X'(i, j) = \{(k \times X(i, j)) \bmod N\} + 1 \tag{10.2}$$

(a)

0	1	2	3	4	5	6	7
8	9	10	11	12	13	14	15
16	17	18	19	20	21	22	23
24	25	26	27	28	29	30	31
32	33	34	35	36	37	38	39
40	41	42	43	44	45	46	47
48	49	50	51	52	53	54	55
56	57	58	59	60	61	62	63

(b)

1	4	7	10	13	16	19	22
25	28	31	34	37	40	43	46
49	52	55	58	61	0	3	6
9	12	15	18	21	24	27	30
33	36	39	42	45	48	51	54
57	60	63	2	5	8	11	14
17	20	23	26	29	32	35	38
41	44	47	50	53	56	59	62

(c)

4	7	13	22	1	10	16	19
28	31	37	46	25	34	40	43
52	55	61	6	49	58	0	3
12	15	21	30	9	18	24	27
36	39	45	54	33	42	48	51
60	63	5	14	57	2	8	11
20	23	29	38	17	26	32	35
44	47	53	62	41	50	56	59

FIGURE 10.1 (a) Original table, (b) the mapping table obtained by Equation 10.2, and (c) the lookup table after column transformation.

where $i, j \in [1, M/2 \times M/2]$; $X, X' \in [0, N-1]$ are the corresponding one-to-one mapping block numbers, $k \in [2, N-1]$ (secret key) is a prime number, and $N = M/2 \times M/2$ is the total number of blocks in the image. Note that if k is not a prime number, we will not achieve one-to-one mapping.

4. Shift the columns in X', so the blocks that belong to the left half of the image are moved to the right half and the blocks that belong to the right half of the image are moved to the left half. Denote the new lookup table as X''. Figure 10.1 shows the generation of a lookup table for a 16×16 image with $k = 3$.

10.3 PROPOSED METHOD

The design of our proposed scheme is divided into three modules as watermark embedding, tamper detection with localization, and recovery of tampered blocks.

10.3.1 WATERMARK EMBEDDING

The watermark-embedding process consists of the following six steps:

1. Let A be a host image of size $M \times M$, where M is a multiple of 2. Compute the block numbers (X) and lookup table (X'') as described in Section 10.2.4.
2. Divide the host image A horizontally into two halves as shown in Figure 10.2. The corresponding blocks in the two halves form a partner-block, that is, blocks 0 and 32, 1 and 33, 16 and 48, and so on.

(a)

0	1	2	3	4	5	6	7
8	9	10	11	12	13	14	15
16	17	18	19	20	21	22	23
24	25	26	27	28	29	30	31

32	33	34	35	36	37	38	39
40	41	42	43	44	45	46	47
48	49	50	51	52	53	54	55
56	57	58	59	60	61	62	63

(b)

4	7	13	22	1	10	16	19
28	31	37	46	25	34	40	43
52	55	61	6	49	58	0	3
12	15	21	30	9	18	24	27

36	39	45	54	33	42	48	51
60	63	5	14	57	2	8	11
20	23	29	38	17	26	32	35
44	47	53	62	41	50	56	59

FIGURE 10.2   (a) Two-halves of a 16×16 image with block numbers and (b) two-halves of the corresponding lookup table generated with $k = 3$.

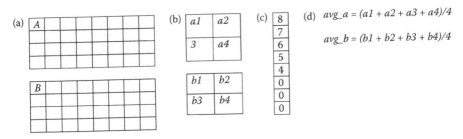

FIGURE 10.3   (a) The partner block a,b of image A, (b) 2×2 view of a, b, (c) replacing three LSBs of each pixel with zero, and (d) computation of the average intensity.

3. For each partner-block a,b in the host image A, we set the three LSBs of each pixel in the blocks a and b to zero. Compute the average intensities avg_a and avg_b as shown in Figure 10.3.

4. Determine x,y corresponding mapping partner block of the partner block a,b from the look up table X''. The 10-bit feature of partner block a,b is shown in Figure 10.4. Pad two zeros to the end of the feature computed. We use the CRC (Tanebaum, 2003; Lin et al., 2004) to compute authentication information. Obtain the CRC-2 by dividing the 12-bit feature with a secret polynomial of degree 2 and subtracting the remainder from the feature.

5. The 12-bit watermark of the partner block a,b is shown in (Figure 10.5b). Embed the computed 12-bit watermark into the three LSBs of each pixel of the corresponding mapping partner block x,y as shown in (Figure 10.5c). Note that the same watermark is embedded into two blocks, that is, x and y.

6. After completing the embedding process for each partner block in the host image A, we will obtain the watermarked image B.

10.3.2   TAMPER DETECTION

The tamper detection is done in four levels. In each level, we try to reduce the false alarms along with the localization of the tampered area. We denote the image under detection as B'.

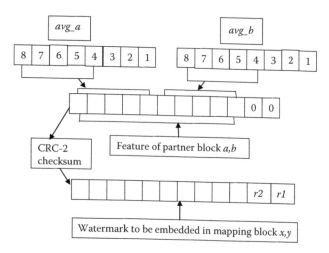

FIGURE 10.4 Feature of partner block a,b.

1. The image under the detection process is divided into blocks of size 2×2, and the mapping sequence is generated as explained in Section 10.2.4.
2. Level-1 detection: For each block b, the 12-bit watermark is extracted from the three LSBs of each pixel. Divide the extracted watermark with the secret polynomial used in the embedding stage. If there is a remainder, it means that the block is tampered in the least significant three bits and mark b as invalid; otherwise, mark b as valid. Level-2 and Level-3 are used to reduce the false alarms.

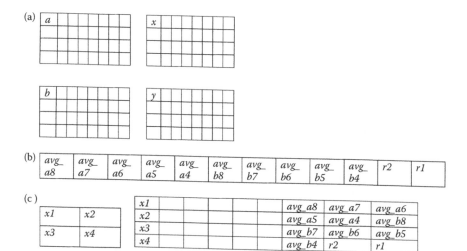

FIGURE 10.5 (a) Partner block a,b and its mapping partner block x,y, (b) 12-bit watermark, and (c) embedding sequence of 12-bit watermark in each block of the mapping partner-block.

3. Level-2 Detection: For each valid block *b* after level-1 detection, we conduct the following:

```
If b is a block in the border of the image, then
    If number of invalid blocks in 3×3 neighborhood of b≥3,
    mark b as invalid;
Else
    If number of invalid blocks in 3×3 neighborhood of b≥5,
    mark b as invalid;
```

4. Level-3 detection: For each valid block *b* after level-2 detection, we conduct the following:

```
If b is a block in the border of the image, then
    If number of invalid blocks in 5×5 neighborhood of
    b≥8,mark b as invalid;
Else
    If number of invalid blocks in 5×5 neighborhood of b≥13,
    mark b as invalid;
```

In general, if a block is tampered, then there are more chances that the neighboring blocks are also tampered. Thus, Level-2 and Level-3 reduce the false acceptance ratio of invalid blocks and help to reduce the false alarms.

5. Level-4 detection: This level is to detect any changes in the five MSBs of each pixel of the valid blocks and to thwart the VQ attack. In this level, we mark the block invalid only if there is a mismatch of the feature with both of the mapping blocks, so the false alarms can be reduced. For each valid block *b* after Level-3 detection, we set the three LSBs of each pixel in block *b* to zero. Compute the average intensity of block *b*, which is denoted as *avg_b*. Determine the corresponding mapping partner block *x,y* from the lookup table, and perform the following task:

```
If x is valid, then
    extract the 12-bit watermark from x;
    If b ∈{upper half of the image}, then
        compare 5 MSBs of avg_b against bits 1 to 5 of the
        extracted watermark;
    Else
        compare 5 MSBs of avg_b against bits 6 to 10 of the
        extracted watermark;
    If there is any mismatch in bit comparison, then mark b
    as invalid;
    If b is invalid, then
        If y is valid, then
            extract the 12-bit watermark from y;
            If b ∈{upper half of the image}, then
```

```
                compare 5 MSBs of avg_b against bits 1 to 5 of
                the extracted watermark;
            Else
                compare 5 MSBs of avg_b against bits 6 to 10
                of the extracted watermark;
            If there is any mismatch in bit comparison, then
            mark b as invalid;
            Else mark b as valid;
        Else if y is valid, then
            extract the 12 - bit watermark from y;
            If b ∈{upper half of the image}, then
                compare 5 MSBs of avg_b against bits 1 to 5 of
                the extracted watermark;
            Else
                compare 5 MSBs of avg_b against bits 6 to 10
                of the extracted watermark;
        If there is any mismatch in bit comparison, then
        mark b as invalid;
```

10.3.3 RECOVERY OF TAMPERED BLOCKS

As in Lee and Lin (2008), the recovery of tampered blocks is done in two stages. Stage-1 is responsible for recovery from the valid mapping blocks, which is the recovery from the embedded watermark. In stage-2, we recover the image by averaging the 3×3 neighborhood.

10.3.3.1 Stage-1 Recovery

1. For each invalid block b, find the mapping block x from the lookup table.
2. If x is valid, then x is the candidate block and go to step 5.
3. Determine the partner block of x and denote it as y.
4. If y is valid, then y is the candidate block; otherwise, stop and leave b as invalid.
5. Retrieve the 12-bit watermark mark from the candidate block.
6. If block b is in the upper half of the image, then the 5-bit representative information of block b starts from the first bit; otherwise, it starts from the sixth bit.
7. Pad three zeros to the end of 5-bit representative information.
8. Replace each pixel in block b with this new 8-bit intensity and mark b as valid.

10.3.3.2 Stage-2 Recovery

For each invalid block b, we recover the block by averaging the surrounding pixels of block b in the 3×3 neighborhood as shown in Figure 10.6.

10.3.4 EXTENSION TO COLOR IMAGES

To extend the proposed scheme to color images, we consider red, green, and blue components of the color image. The embedding process is done in each of the three

FIGURE 10.6   The 3×3 neighborhood of block b.

components individually, and then each component is assembled to obtain a water-marked image as shown in Figure 10.7a. Like embedding process, the detection is also done on individual components. Then each block in the three components is validated, such that if the block is invalid in any of the red, green, and blue compo-nents, then mark the block invalid in all the three components. Finally, each compo-nent is assembled back to obtain the watermarked image as shown in Figure 10.7b.

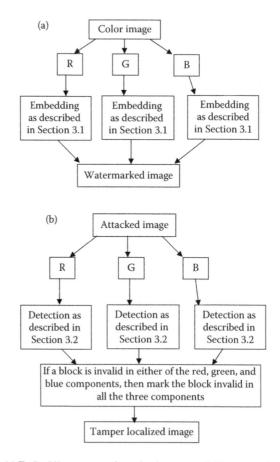

FIGURE 10.7   (a) Embedding process for color images and (b) tamper detection process for color images.

The recovery process is conducted in each of the components individually as described in Section 10.3.3 by considering the localization image in Figure 10.7b. Finally, combining the red, green, and blue components gives the recovered color image.

10.4 RESULTS AND DISCUSSIONS

In this section, we use two different measures: one is peak signal-to-noise ratio (PSNR) to measure the quality of the watermarked image, and the other is percentage of recovery (PR) to compare the recovery rate of our technique with the existing techniques. The PSNR of an image A with respect to an image B is defined by

$$PSNR = 20 \times \log_{10}\left(\frac{MAX_I}{\sqrt{MSE}}\right) \tag{10.3}$$

where MAX_I is the maximum gray value of the images A and B, and MSE is the mean square error defined by

$$MSE = \frac{1}{m \times n}\sum_{i=0}^{m-1}\sum_{j=0}^{n-1}\left\|A(i,j) - B(i,j)\right\|^2 \tag{10.4}$$

where $m \times n$ is the size of the images A and B. The more the PSNR ratio, the less is the distortion, that is, the more is the image quality. The PR of an image I is defined by

$$PR(I) = \left(\frac{SIZE_R(I)}{SIZE_M(I)}\right) \times 100 \tag{10.5}$$

where $SIZE_M(I)$ is the size of the modified portion in the image I, and $SIZE_R(I)$ is the size of recovered portion of the modified portion after stage-1 recovery, that is, recovery from features stored in mapping block. The more the PR, the more is the recovery and vice versa. Experiments are conducted on grayscale images of size 256×256 with $k = 13$ and 101 as the secret polynomial. Table 10.2 lists the PSNR values of different watermarked images by applying the proposed method.

TABLE 10.2
PSNR Values of Different Watermarked Images by the Proposed Method

	Lena	Baboon	Barbara	Cameraman	Peppers
PSNR	41.14	41.14	41.14	40.35	41.14

We conduct various types of cropping attacks in the experiments. There are 12 distinct cropping patterns used as shown in Figure 10.8. Table 10.3 lists the resulting *PR* values of applying Lin et al. (2004, 2005), and the proposed schemes under these 12-pattern cropping attacks. We observe that the proposed algorithm outperforms the other two algorithms.

We also measure the performance of the proposed technique in terms of *PR* by considering different percentages of cropping. The cropping is carried out from the top-left corner of the image. The graph in Figure 10.9 gives the performance of the proposed scheme against Lin et al. (2004, 2005). It is observed that the proposed

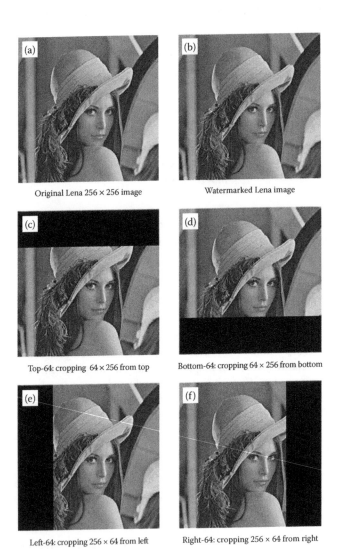

Original Lena 256 × 256 image   Watermarked Lena image

Top-64: cropping 64 × 256 from top   Bottom-64: cropping 64 × 256 from bottom

Left-64: cropping 256 × 64 from left   Right-64: cropping 256 × 64 from right

FIGURE 10.8   Different cropping styles.

Top-128: cropping 128 × 256 from top

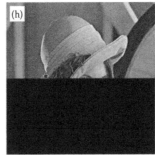

Bottom-128: cropping 128 × 256 from bottom

Left-128: cropping 256 × 128 from left

Right-128: cropping 256 × 128 from right

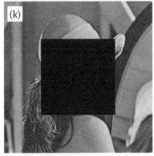

Center cropping: 128 × 128 in the center

Outer: keep the center 128 × 128

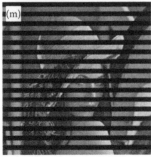

Horizontal: crop the alternate 8 × 256 from top

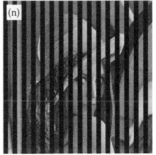

Vertical: crop the alternate 256 × 8 from left

FIGURE 10.8 Continued.

TABLE 10.3

Comparisons of the *PR* Values under the 12-Pattern Cropping Attacks

Cropping style	Lin et al. (2004) PR	Lin et al. (2005) PR	Proposed PR
Top -64	75.00	69.34	100.00
Bottom-64	75.00	69.34	100.00
Left-64	75.00	62.50	100.00
Right-64	75.00	62.50	100.00
Top-128	50.00	46.24	100.00
Bottom-128	50.00	46.24	100.00
Left-128	50.00	40.63	100.00
Right-128	50.00	40.63	100.00
Outer-64	27.86	23.57	50.80
Center-64	83.59	90.33	100.00
Horizontal-8	50.00	46.09	45.90
Vertical-8	50.00	50.00	48.44

scheme produces a much higher number of recovery pixels than the other two schemes. Figure 10.10 shows the application of our technique on a color image, where the image of the license plate number of the car was removed and can still be recovered. Figure 10.11 demonstrates that the use of random shape cropping cannot affect the accurate recovery of the Barbara image.

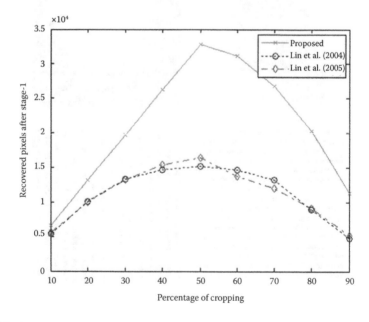

FIGURE 10.9 Comparisons of the proposed scheme with different percentages of cropping.

FIGURE 10.10 (a) A watermarked image containing a car, (b) a modified image with the license plate number removed, (c) the result of tamper detection, and (d) the result of the recovered image.

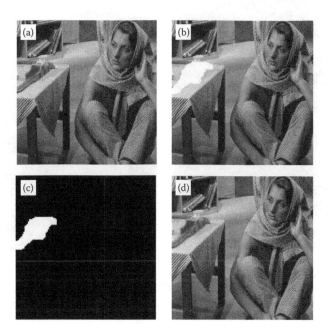

FIGURE 10.11 (a) A watermarked Barbara image, (b) a modified image with random shape cropping, (c) the result of the tamper localization image, and (d) the result of the recovered image.

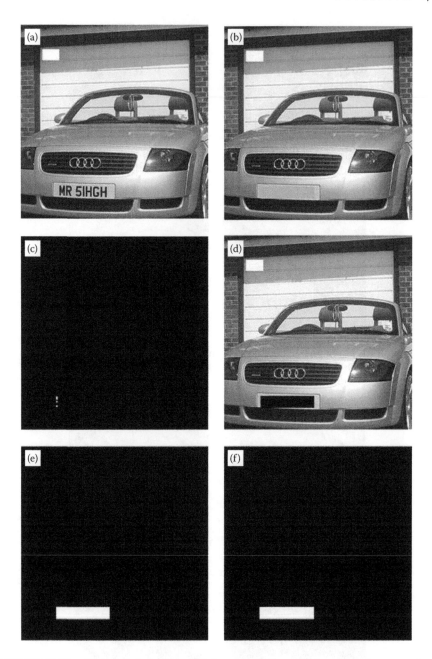

FIGURE 10.12 (a) An original image, (b) the modification of license plate by gray value 193, (c) the detection of (b) by Lee and Lin (2008), (d) the modification of license plate by black value 0, (e) the detection of (d) by Lee and Lin (2008), and (f) the detection of (b) and (d) by our proposed scheme.

Lee and Lin (2008) used the parity sum for authenticating the individual blocks. This approach has a problem of detecting only even number of changes in the bit string. In their authentication method, the fist authentication bit is the XOR of first 10 bits of the feature, and the second authentication bit is the opposite of first authentication bit. Therefore, if each pixel is modified with a gray value whose two LSBs are opposite to each other, then it is less likely for their method to detect the modification. Figure 10.12 shows the results of the modification and its corresponding localization with a different gray value other than black. Their method detects the modification in the case of black pixel replacement, but fails to detect the modification by replacing some other gray values. In Figure 10.12(f), we show that the proposed method can detect both black pixel and other gray value modifications.

10.5 CONCLUSIONS

The described new efficient fragile watermarking technique is proposed for authentication and recovery. Our method is designed to meet the desired characteristics of a fragile watermarking system for tamper localization and recovery. The characteristics include tamper localization size of 2×2, support against the VQ attack, dual recovery, and reduced false alarms. We apply the CRC-2 checksum for authentication of the feature stored in each block. The proposed method can be applied to both grayscale and color images. Experimental results show that our method has better recovery rate than Lin et al. (2004, 2005) and has the advantages of support against the VQ attack, changes in higher-order bits, and the detection of modification with any gray value over Lee and Lin (2008).

REFERENCES

Celik, M. U., Sharma, G., Saber, E., and Tekalp, A. M., Hierarchical watermarking for secure image authentication with localization, *IEEE Trans Image Process*, 11(6), 585–594, 2002.

Cox, I. J., Bloom, J., Miller, M., and Cox, I., *Digital Watermarking Principles & Practice*, Morgan Kaufmann, San Francisco, 2001.

Edupuganti, V. G., Shih, F. Y., and Chang, I., An efficient block-based fragile watermarking system for tamper localization and recovery, *Autosoft J Intell Autom Soft Comput*, 17(2), 257–267, 2011.

Fridrich, J., Image watermarking for tamper detection, *Proc IEEE Intl Conf Image Process*, Chicago, IL, vol. 2, pp. 404–408, Oct. 1998.

Holliman, M. and Memon, N., Counterfeiting attacks on oblivious blockwise independent invisible watermarking schemes, *IEEE Trans Image Process*, 9(3), 432–441, 2000.

Kundur, D. and Hatzinakos, D., Digital watermarking for telltale tamper proofing and authentication, *Proc IEEE*, 87(7), 1167–1180, 1999.

Lee, T. Y. and Lin, S. D., Dual watermark for image tamper detection and recovery, *Pattern Recognit*, 41(11), 3497–3506, 2008.

Lin, P. L., Hsieh, C. K., and Huang, P. W., A hierarchical digital watermarking method for image tamper detection and recovery, *Pattern Recognit*, 38(12), 2519–2529, 2005.

Lin, P. L., Huang, P. W., and Peng, A. W., A fragile watermarking scheme for image authentication with localization and recovery, *Proc IEEE Sixth Intl Symp Multimedia Software Eng*, Miami, FL, pp. 146–153, Dec. 2004.

Shih, F. Y., *Digital Watermarking and Steganography: Fundamentals and Techniques*, CRC Publisher, Boca Raton, FL, 2007.

Tanebaum, A. S., *Computer Networks*, Fourth Edition, The Netherlands Pearson Education International, Upper Saddle River, New Jersey, 2003.

Wang, M. S. and Chen, W. C., A majority voting based watermarking scheme for color image tamper detection and recovery, *Comput Stand Interfaces*, 29(5), 561–570, 2007.

Wang, S. S. and Tsai, S. L., Automatic image authentication and recovery using fractal code embedding and image painting, *Pattern Recognit*, 41(2), 701–712, 2008.

Part III

Steganography

11 Survey of Image Steganography and Steganalysis

Mayra Bachrach and Frank Y. Shih

CONTENTS

11.1 INTRODUCTION

The term, steganography, comes from Greek and means "covered writing" (Johnson and Jajodia, 1998a; Petitcolas et al., 1999). The intent of steganography is to hide a message in a medium in such a way that no one but the intended recipient knows that the message is even there.

Steganography is not a new discipline. Throughout history, there has been the need to send hidden messages through covert means. The literature provides many examples of the rustic uses of steganography. In ancient times, slaves were used as messengers. The head of a slave was shaved and a message was tattooed on it. The hair was allowed to grow back in order to hide the message. The slave was then sent to the intended recipient of the message who would have the hair shaved to reveal the message (Johnson and Jajodia, 1998a; Petitcolas et al., 1999; Provos and Honeyman, 2003; Kessler, 2004). Music scores were used in the 1600s to hide messages. Each note in the music score would correspond to a letter of the message. Acrostic writings and invisible inks used during the world wars were also early examples of steganography. With the development of photography, tiny images containing secrets were sent out by pigeon post, or hidden under ears, nostrils, and finger nails of human messengers (Petitcolas et al., 1999).

Today, computer technology is used to embed hidden messages in images using steganographic algorithms and techniques. The images are then transmitted electronically. There are several steganographic software systems available that can be used to easily embed messages in images. Some of this software is available license free. The military and other government agencies are interested in steganography in order to be able to safely and secretly transmit information. They are also interested in being able to detect secret messages transmitted by criminals, terrorists, and other unfriendly forces. It is suspected that Al-Qaeda made use of steganography during the planning of the World Trade Center attack. Experts in the field have conducted studies to determine whether images modified by steganography are widely present on the Internet (Provos and Honeyman, 2001).

Businesses and corporations are also interested in steganography and steganalysis. Disgruntled or unethical employees can use steganography to transmit private corporate information through digital communication. Companies want effective steganalytic tools that would help them detect unauthorized distribution of their private information and trade secrets (Shih, 2007; Shih and Edupuganti, 2009).

With the advancement of technology and the advent of the Internet and personal computers, digital steganography and steganalysis have become topics of interest (Bachrach and Shih, 2011). Research in the area of steganography and steganalysis is driven by the need for protection of copyrights on digital works, detection of copyright infringement, the need for unobtrusive communication by military and intelligence agencies, and detection by law enforcement agencies of steganographic methods used in criminal activities (Provos and Honeyman, 2001).

11.2 STEGANOGRAPHY

Using steganography hidden information in the form of files, text, or other images can be embedded in digital images. Digital watermarking is used to embed information (watermark) in an image that can be used to identify and verify the owner and authorized users of an image. Steganography and digital watermarking are related disciplines. Although many of the concepts and techniques used in watermarking and steganography are similar, the two disciplines differ in purpose. The existence of the watermark is usually known to users of the image. The watermark may or may not be visible although it usually does not alter the appearance of the image. Both watermarking and steganography are used to embed information that cannot be easily detected or altered in an image. However, watermarking is used mainly to prevent the violation of copyrights by unauthorized users of an image whereas steganography is used to exchange hidden information. In the case of watermarking, hidden information is embedded so that the image can be transmitted digitally while preventing copyright infringement and illegal distribution. In the case of steganography, the intent is to embed hidden information in order to transmit the information without detection.

Steganalysis is the art and science used to determine whether undetectable messages have been embedded in images using steganography. The steganographer uses steganalysis to detect, extract, disable, or modify the message before it reaches the recipient.

A steganographer can take an active or passive approach to steganalysis. This is explained in the literature through the classic example of the prisoner's problem (Simmons, 1998). Alice and Bob are prisoners planning an escape. Wendy is the warden. Alice and Bob are allowed to exchange messages freely through channels which allow Wendy to inspect the messages. Wendy can take two approaches. The "passive" approach is to examine all messages exchanged to see if any secrets are being exchanged. The "active" approach is for Wendy to alter all the messages exchanged between Bob and Alice, so that it is very difficult or impossible for Bob and Alice to extract the secret messages. In this classic example, Alice and Bob are using steganography to hide the secret information in their messages. Wendy is using steganalysis to detect and distort or remove the hidden information (Provos and Honeyman, 2001; Wu and Shih, 2006).

Many techniques, algorithms, and software have been developed to both embed hidden information in images (steganography) and to detect whether an image has an embedded hidden message (steganalysis).

11.3   IMAGE REPRESENTATION

Image files that are common on the Internet today are a very good medium for digital steganography. They are easily and frequently shared on the Internet through email, posting on sites, and other digital means. They are large in size which makes it easier to embed information in such a way that it is not noticed by the uninformed user. Steganographic techniques exploit the properties of images and are used to manipulate bits in the image and embed information.

Computer images are represented as arrays of pixel values. A pixel is a point of an image. The numeric value of each pixel is stored in 3 bytes (24 bits) and represents a digital color. Each of the 3 bytes defines the amount of light intensity in each of the primary colors: red, green, and blue (RGB) and can hold values from 0 to 255 (values which can be stored in 8 bits, 2^8). For example, 255 for the red value and 0 for the blue and green will render the color red. Other values close to 255 for the red component and zeroes for the blue and green will also render a shade of red which may appear to be the same color to the human eye (Johnson and Jajodia, 1998a; Manoj, 2009).

The large size of image files can be attributed to how pixels are represented. For example, a 24-bit image (3 bytes per pixel) 600 pixels wide and 600 pixels in height would consist of $600 \times 600 \times 24$ bits (8,640,000 bits). Some of the pixels could be adjusted in value to correspond to letters without any noticeable effect on the appearance of the image.

Large image files are often compressed for faster transmission. There are two kinds of compression, lossless and lossy. Lossless compression maintains all the information in the image when compressed. It is used for GIF files among others. Lossy compression which is used for JPEG files may result in some loss of information. Steganography software employs different techniques for processing images based on their compression algorithm (Johnson and Jajodia, 1998a; Kessler, 2004).

When selecting an image for steganography, both the image and color palette are considered. Images with large areas of solid color are more likely to show variations that can occur as a result of embedding messages. Embedded messages will be less likely to be noticed in gray scale images and images with subtle color variations.

11.4 SPATIAL DOMAIN AND FREQUENCY DOMAIN

There are two classifications of image steganography methods: spatial domain and frequency domain. The techniques in the spatial domain embed the secret messages directly into the intensity values of the image pixels. In the frequency domain, images are first manipulated with algorithms and transforms and then the messages are embedded in the image. The methods in the spatial domain are considered not only simplest but also more susceptible to steganalytic attacks, less robust. The spatial domain is sometimes called the image domain. The frequency domain is also known as the transform domain (Wu and Shih, 2006).

11.4.1 COMMON APPROACHES

There are two common approaches for image steganography: Least-significant bit substitution (LSB) and algorithms and transformations (Shih, 2007). LSB substitution is simpler but less robust than steganographic methods which use algorithms and transformations.

11.4.1.1 Least-Significant Bit (LSB) Substitution

LSB substitution involves embedding information in the LSB of a number of bytes of a carrier image. By overwriting the LSB of any pixel, information is embedded in the LSB of a byte representing an RGB color value. The change resulting from modifying the LSB of scattered bytes results in such slight changes that it is not detected by the human eye. Overwriting any other bit, especially the most significant bit, would result in a much more noticeable change and distortion of the image. LSB substitution is more effective on 24-bit images than on 8-bit images. LSB substitution is more effective with images that use lossless compression such as BMP and GIF files (Johnson and Jajodia, 1998b; Morkel et al., 2005). LSB substitution is in the Spatial (image) domain.

A clear example of the result of embedding the letter "A" in a 24-bit image is provided by Johnson and Jajodia (1998a, pp. 28–29). In their example, three pixels (9 bytes) of an image contain the binary values:

```
(00100111  11101001  11001000)
(00100111  11001000  11101001)
(11001000  00100111  11101001)
```

The binary value for the letter "A" is 10000011. Inserting the binary value for the letter "A" in the three pixels results in:

```
(00100111 1110100 0 11001000)
(0010011 0 11001000 1110100 0)
(11001000 00100111 1110100 1)
```

The letter "A" was embedded into the LSB of each byte. But only the 3 bits underlined in the 9 bytes were modified as a result. All the other bits remained the same. LSB substitution is also used in watermarking. However, LSB substitution is more easily detected or made unusable by steganalytic tools than other methods of steganography.

There are many variations of the LSB algorithm some of which are less susceptible to detection than others. The LSB algorithm may be easily used with grayscale images. Each pixel in a grayscale image consists of 8 bits. The first bit to the left is the most significant digit and the first bit on the right is the least significant digit. The pseudo-code below can be used to explain the processing to embed a text message in a grayscale image by replacing the LSB of each pixel:

```
pic=cover image
msg=secret message
n = number of chars in msg
for i=1 to n
  get char from msg
  for each bit in char
    get a pixel from pic
    if the bit=1
      insert a 1 in the least significant bit of the pixel
    else
      insert a 0 in the least significant bit of the pixel
      replace the pixel in pic
  end for
end for
```

For a 24-bit image, the algorithm can embed more information per pixel. A 24-bit image utilizes 3 bytes, 24 bits to store the value for each pixel. The first 8 bits in each pixel represent the color red, the second 8 bits represent the color green and the last 8 bits represent the color blue. The pseudo-code below can be used to explain a simple LSB insertion algorithm used with 24-bit images:

```
pic=cover image
msg=secret message
n=number of chars in msg
for i=1 to n
  get char from msg
  for each 3 bits in char
    get a pixel from pic
    get the red value of the pixel
```

```
  if the first bit=1
    insert a 1 in the least significant bit of the red value
  else
    insert a 0 in the least significant bit of the red value
    get the green value of the pixel
  if the second bit=1
    insert a 1 in the least significant bit of the green
    value
  else
    insert a 0 in the least significant bit of the green
    value
    get the blue value of the pixel
  if the third bit=1
    insert a 1 in the least significant bit of the blue
    value
  else
    insert a 0 in the least significant bit of the blue
    value
    replace the value in pic
  end for
end for
```

The changes resulting from the LSB insertion algorithm applied to the LSB are not noticeable to the human eye. Even changes made to the second and third LSBs of each pixel are not visible to the human eye. In order to embed a larger message, information is sometimes hidden in the second and third bits of each pixel. Generally speaking, the LSB algorithm has a higher capacity than other embedding techniques. This means that a greater amount of information can be embedded per image (Shih, 2007). The down side is that message embedded using LSB insertion can be easily destroyed by compressing, filtering, or cropping the image (Venkatraman et al., 2004).

Steganographic techniques based on the LSB algorithm vary in complexity and robustness. The simple algorithm described above inserts the bits of the hidden message sequentially into the cover image. As a result, it is easy to detect and extract the message. Some variations of the LSB insertion algorithm insert the bits randomly into the cover image based on a stego-key. For example, one variation of LSB insertion uses the random pixel manipulation technique to insert the message into random pixels in the cover image. The random pixel manipulation technique utilizes a stego-key. The stego-key provides a seed value for a random number generator. Using the seed value, random pixels in the image are selected for embedding the message. The stego-key is then used to extract the message by using the same seed number to generate the random pixels where the data have been inserted (Venkatraman et al., 2004). Although inserting the message in random pixels makes detection and extraction of the hidden message less likely, the hidden message can still be destroyed by compression and other image manipulation such as filtering or cropping (Johnson and Jajodia, 1998a). Some of the steganographic tools available modify the color palettes of the image in order to make detection and extraction of the message less likely. Examples of tools using these techniques, such as S-Tools and EZStego, will be described later in this chapter.

11.4.1.2 Algorithms and Transformations

Steganographic methods utilizing algorithms and transformations are more complex but also more robust. They are in the frequency (transform) domain. Some of these methods utilize the discrete Fourier transform (DFT), the discrete cosine transform (DCT), the discrete wavelet transform (DWT), and the genetic algorithm (GA) (Johnson and Jajodia, 1998a,b; Kessler, 2004; Shih, 2007). Steganographic methods in the frequency domain apply transformations such as the DFT or DCT to an image and then embed information in the bits of the coefficients obtained from the transforms (Kharrazi et al., 2004).

The DFT is based on the Fourier series used to represent the continuous time periodic signal (Shih, 2007). The Fourier series which is the basis for the DFT states that any periodic function can be expressed as the sum of sines and/or cosines of different frequencies multiplied by a different coefficient (Gonzalez and Woods, 2008). The DFT decomposes an image into its sine and cosine function. Using the inverse DFT, an image that has been transformed using DFT can be transformed back into its spatial-domain equivalent image. Steganographic methods which use the DFT embed information by modifying the bits of the resulting DFT coefficients. As the calculation of the DFT for an image is processing intensive, fast Fourier transform (FFT) is used to derive the results. Languages such as MATLAB® which are used for image processing provide functions that can be used to derive the DFT of an image.

The two-dimensional Fourier transform (McAndrew, 2004) takes a two-dimensional image matrix, $f(x,y)$ as input and outputs another two-dimensional matrix, $F(u,v)$. The calculations using Fourier transform are resource intensive. Languages such as MATLAB, rich in image processing tools, are used to derive the transforms. In particular, MATLAB has the FFT (fft, fft2) and inverse FFT (ifft, ifft2) functions that can be used in processing images.

The DCT is used to compress JPEG image files. Using steganographic techniques information can be hidden in JPEG images during the compression process. Using the DCT, blocks of 8×8 pixels of a JPEG image are transformed into 64 DCT coefficients. The DCT coefficients are quantized using a 64-element quantization table. The LSBs of the quantized DCT coefficients are used to embed the hidden information (Johnson and Jajodia, 1998b; Provos and Honeyman, 2003; Shih, 2007). The image can be transformed back into its spatial-domain equivalent using the inverse DCT. During JPEG image compression, the DCT is used to convert 8×8 blocks of the image into 64 DCT coefficients.

When using the DCT coefficients in steganographic algorithms, a quantization table $Q(u,v)$ is used to quantize the coefficients according to the formula

$$F^Q(u,v) = \frac{F(u,v)}{Q(u,v)}.$$

The bits of a hidden message can then be embedded in the least significant digits of the quantized DCT coefficients.

The calculations using these formulas are resource intensive. Languages such as MATLAB, rich in image processing tools, are used to derive the transforms. In particular, MATLAB has the DCT and inverse DCT (IDCT) functions which can be used in processing images.

There are variations of steganography algorithms that use the DCT. Some of the algorithms embed bits of a hidden message sequentially in the quantized DCT coefficients. One such algorithm is used in the JPeg–JSteg tool explained in a later section of this chapter. Other steganography algorithms, for example, the F5 algorithm, decrement the values of the DCT coefficients using a process called matrix encoding in the embedding of a message into a stego-image. This use of the DCT results in a more robust algorithm; the hidden message is less likely to be detected and/or extracted.

The DWT is used in the JPEG 2000 compression algorithm. A wavelet is a function which integrates to zero above and below the z-axis. Using the DWT, an image can be decomposed into wavelet coefficients. The Haar DWT is the simplest of the wavelet transforms. After performing the DWT on an image, secret information can be stored in selected DWT coefficients (Chen and Lin, 2006; Shih, 2007). Wavelets are relatively new in image processing. They are less resource intensive and result in less image distortion than the DFT and DCT. Wavelets are used in image processing for noise reduction, edge detection, and compression (McAndrew, 2004). There is a wavelet toolbox in MATLAB that provides functions for processing wavelets. There are also many wavelet toolboxes available as freeware.

11.5   GENETIC ALGORITHM

The GA-based steganographic system proposed by Wu and Shih (2006) is also in the frequency domain. The GA generates the stego-image by artificially counterfeiting statistical features in such a way as to break the steganalytic system. The GA has its origin in natural genetics. It consists of the key concepts: reproduction, crossover, mutation, and the fitness function. As applied to steganography, the GA is used to correct rounding errors that occur in processing an image transformed by DCT. After applying the DCT, embedding the secret information and then applying the IDCT to transform the image back into the spatial domain, the GA is used to translate the real numbers into integers (Shih, 2007).

In summary, each of the approaches described have strengths and weaknesses that make them susceptible to steganalytic techniques and applicable to specific applications and image types.

11.6   STEGANALYSIS

Steganalysis is the process used to detect secret information embedded in images through steganography. Most techniques used in steganography alter the characteristics and statistics of the cover image in some way (Provos and Honeyman, 2002; Kessler, 2004). Statistical analysis of images can detect if an image has been modified with steganography. In correspondence with the steganographic techniques, steganalysis systems fall into the same two broad categories: spatial-domain steganalytic systems (SDSS) and frequency-domain steganalytic systems (FDSS). SDSS is used to analyze characteristics in the spatial-domain image statistics. FDSS is used to analyze characteristics in the frequency-domain image statistics (Wu and Shih, 2006).

Detecting hidden information can be quite complex without knowing which steganalytic technique was used or if a stego-key was used. But most of the steganographic techniques and tools alter the images in unique ways which serve as their "signature." This helps in the detection of altered images.

There are two main methods for detecting images modified with steganography: The first is visual analysis. Visual analysis compares the original image with the stego-image either visually or using a computer to detect hidden information. Some of the simpler steganographic tools in the image domain embed information in bits without regard to the content of the carrier image. The information may be inserted in bits that make the change in the image's appearance more easily detected through visual inspection. However, this method is often not feasible since the original image is not available. The other method is through statistical analysis. This method looks for anomalies in the structure or statistical measures of the image. Other steganalytic methods analyze statistical measures of an image that are considered the norm. A variation from the norm in the statistics points to an image having been altered with steganography (Kessler, 2004).

In addition to detecting the use of steganography, a more active approach to steganalysis involves extracting and destroying the embedded hidden information. It is easier to destroy messages embedded with LSB insertion than those embedded using the transforms.

Steganalytic techniques can vary greatly depending on what information is known about the carrier image, the stego-image, the message, and the algorithm used to embed the hidden message. For example, if an image is suspected of carrying a hidden message, it may be visually inspected for irregularities and then statistical analysis through analyzing means, variances, and chi-square tests may be conducted. However, many of the steganalytic techniques used currently depend on detecting the signature of the steganographic tool used to create the stego-image (Kessler, 2004). Such is the case with the Steg-Detect software explored in this article. It can be used only on JPEG images and it detects messages embedded with the known tools. For example, when Steg-Detect was used to analyze the image used to experiment with JSteg–JPeg for this article, it did not detect the embedded hidden message.

11.6.1 STEGANOGRAPHY SOFTWARE

Software systems have been developed to embed stego-messages in images. These systems utilize some of the steganographic techniques explained. Many of the systems available are freeware and can be readily downloaded, JSteg, JPHide, S-Tools, and others being among them. These systems have strengths and weaknesses that have been explored in the literature (Johnson and Jajodia, 1998a,b; Provos and

Honeyman, 2002, 2003; Kessler, 2004). In this section, we will explore S-Tools, EzStego, and JSteg–JPeg.

S-Tools is used to hide secret information in BMP, GIF, and WAV files. The messages embedded in the cover objects are encrypted using various encryption algorithms. This tool uses LSB substitution for lossless file formats. It also uses a pseudorandom number for the LSB substitution which makes extracting the message more difficult. S-Tools can be used to embed and extract the hidden information from stego-images (Johnson and Jajodia, 1998b; Shih, 2007).

EzStego is used to embed a secret message in a GIF file by modulating LSBs. EzStego compares the bits that it wants to hide with the LSB value for each pixel only changing the bit if needed. It is freely available as a Java application on the Web.

JSteg–JPeg uses the DCT transform for embedding stego messages in multiple image formats and saves the stego-image as a JPEG image file. Its embedding algorithm sequentially replaces the least significant digits of the DCT coefficients of the image.

11.6.1.1 S-Tools

S-Tools is available as freeware and can be readily downloaded from steganography sites on the Internet. The S-Tools software was developed by Andrew Brown. As an experiment for this article, messages were embedded in lena_256.bmp using S-Tools version 4. S-Tools is easy to use. It consists of a simple graphical user interface which is integrated with Windows Internet Explorer. The cover image and the file to be embedded can be dragged into the S-Tools work area from a windows explorer window. The cover image can be a WAV, GIF, or BMP file. The message to be embedded can be stored in another file such as a txt file. The hidden information is encrypted before it is embedded in the stego-image. The encryption algorithm to be used is selected from a drop down list. A password is also required to embed the information. The password is required to extract the hidden message using S-Tools.

For this experiment, text files of different sizes were embedded in Lena. One file, hide.txt, was very small and contained the phrase, "Happy Birthday." Another file, starspangled.txt contained the lyrics to the Star Spangled Banner. Before embedding text, Lena consisted of 66,614 bytes. The size of Lena did not change regardless of the amount of text stored in the image. A status message was displayed in the S-Tools work area, showing the maximum number of bytes that could be hidden in Lena (see Figure 11.1).

Extracting hidden text file was also very straightforward but required knowledge of the password and the encryption algorithm used when creating the stego-image. There appeared to be a lightening in the appearance of Lena when a small text file containing the text "Happy Birthday" was embedded in Lena. The difference in the image before and after Lena was not noticeable after embedding the text file containing the lyrics of Star Spangled Banner (see Figure 11.2). Without having both the cover and stego-image side by side, there would not be any indication to the human eye that the Lena image was altered in either case.

FIGURE 11.1   S-Tools work area.

FIGURE 11.2   Before and after Lena using S-Tools.

11.6.1.2   JPeg–JSteg

JPeg–JSteg is also available as freeware and can be readily downloaded from steganography sites on the Internet. The JPeg–JSteg (JSteg) software was developed by Derek Upham. JSteg runs at the DOS prompt using the commands cjpeg and djpeg. Although JSteg outputs stego-images as JPEG files, it does not read the JPEG file format. Cover images are converted using the DOS command djpeg to the Targa image file (.tga file extension). The cover image converted to the Targa image format and the hidden text file are input to JSteg using the DOS command cjpeg which produces the stego-image.

The algorithm used in JPeg–JSteg sequentially replaces the LSB of the DCT coefficients of the stego-images with bits from the hidden message to be embedded.

According to Niels Provos and Peter Honeyman, the algorithm used by JSteg can be described by the following pseudo-code (Provos and Honeyman, 2003, p. 34):

```
Msg=hidden message
Pic=cover image
Output=stego-image
While more characters in msg do:
  Get next DCT coefficient
  If DCT coefficient!= 0 and DCT coefficient !=1
    Get next LSB from message
    Replace DCT LSB with message LSB
  End if
  Insert DCT into stego-image
End While
```

The algorithm used in JPeg–JSteg inserts the steganographic secret data in the cover image during the JPEG compression algorithm after the DCT and quantization of the DCT coefficients and before the Huffmann coding step. The algorithm also embeds a length field in the stego-image which is used to extract the hidden message.

As an experiment for this chapter, messages of different sizes were embedded in a JPEG image, boys.jpg. Before embedding the message, the JPEG file was converted to a tga image file using the djpeg command in DOS mode. Using the cjpeg command in DOS mode, the lyrics of the Star Spangled Banner, starspangled.text, were embedded in the tga image file and then extracted. A password was not required for the embedding or extraction process. The hidden message was extracted without any problems. There appeared to be a slight lightening in the appearance of the cover image. Without having the images side by side, there was no indication that the cover image had been altered to embed a secret message (See Figures 11.3 and 11.4). The size of the image remained the same.

FIGURE 11.3   Boys.jpg—cover image used for JPeg–JSteg experiment.

FIGURE 11.4   Boyschanged.jpg—stego-image produced by embedding starspangled.txt.

11.6.1.3   Steganalysis Software

Software systems have also been developed to detect stego-messages embedded in images using steganography. Many of the steganography tools available are specific to the software used to embed the stego message. An example of such software is StegDetect, which was developed by Niels Provos.

StegDetect can be used to detect JPEG images that have been altered using Steg, JPhide, Invisible Secrets, Outguess, F5, and others. The tools which StegDetect targets all use some variation of modifying bits after applying the DCT transform (Provos and Honeyman, 2003). StegDetect uses knowledge of the technique used by the stenography software to embed the message in its detection mechanism. StegDetect can be downloaded in DOS form as freeware from the Internet. StegDetect was used to analyze the stego-image created for this chapter (Figure 11.4). It gave a negative reading and was not able to detect that the image had been tampered with.

11.7   CONCLUSIONS

Steganography and steganalysis are relatively new disciplines with many relevant applications in today's digital society. The use of steganalysis is likely to increase in computer forensics in the near future. There is significant research being conducted in academic circles on steganographic and steganalytic techniques. A number of steganography tools are available on the Internet as freeware. It is not known with certainty if the use of steganography for illegal activities is widespread.

There are a number of algorithms written and posted out on the Internet for anyone to download. This makes the use of steganography much easier and available for anyone who chooses to abuse the technology for illegal activities. A controlled repository of steganography and steganalysis software can be used by students of steganography for continued research.

Many of the articles on steganography and steganalysis are aimed at the scientific and academic community and there is not a great deal of literature available as a starting point for the novice student of steganography and steganalysis.

In conclusion, steganography and steganalysis are a growing discipline. The need for better techniques, algorithm, and software for steganalysis will continue to increase in our digital society.

REFERENCES

Bachrach, M. and Shih, F. Y., Image steganography and steganalysis, *WIREs Comput Stat*, 3(3), 251–259, 2011.

Chen, P.-Y. and Lin, H.-J., A DWT based approach for image steganography, *Intl J Appl Sci Eng*, 4(3), 275–290, 2006.

Gonzalez, R. C. and Woods, R., *Digital Image Processing*, 3rd edition, Pearson Prentice Hall, Upper Saddle River, NJ, 2008.

Johnson, N. F. and Jajodia, S., Exploring steganography: Seeing the unseen, *IEEE Computer*, 31(2), 26–34, 1998a.

Johnson, N. F. and Jajodia, S., Steganalysis of images created using current steganography software, *Lect Notes Comput Sci*, 1525, 273–289, 1998b.

Kessler, G. C., An overview of steganography for the computer forensics examiner, *Forensic Sci Commun*, 6(3), 1–29, 2004.

Kharrazi, M., Sencar, H., and Memon, N., Image steganography: Concepts and practices, *WSPC/Lecture Notes Series*, 11, 22–28, 2004.

Manoj, R., Understanding digital steganography, *InfoSecurity Mag*, 2009. Available at: http://fanaticmedia.com/infosecurity/archive/Sep09/Digital%20Steganography.htm

McAndrew, A., *Introduction to Digital Image Processing with MATLAB*, Boston, MA, Thomson Course Technology, 2004.

Morkel, T., Eloff, J. H. P., and Olivier, M. S., An overview of image steganography, *Proc Fifth Annual Information Security South Africa conference*, Sandton, South Africa, June/July 2005.

Petitcolas, F. A. P, Anderson, R. J., and Kuhn, M. G., Information hiding—A survey, *Proc IEEE*, special issue on protection of multimedia content, 87(7), 1062–1078, 1999.

Provos, N. and Honeyman, P., *Detecting Steganographic Content on the Internet*, CITI Technical Report 01–11, August 31, 2001.

Provos, N. and Honeyman, P., *Detecting Steganographic Content on the Internet*, Center for Information Technology Integration, University of Michigan, ISOC NDSS, 2002.

Provos, N. and Honeyman, P., Hide and seek: An introduction to steganography, *IEEE Security and Privacy Magazine*, 1(3), 32–44, 2003.

Shih, F. Y., *Digital Watermarking and Steganography: Fundamentals and Techniques*, Boca Raton, FL: CRC Press Inc, 2007.

Shih, F. Y. and Edupuganti, V. G., A differential evolution based algorithm for breaking the visual steganalytic system, *Soft Comput*, 13, 345–353, 2009.

Simmons, G. J., Prisoners problem and the subliminal problem, *Advances in Cryptology: Proc Crypto*, vol. 83, Berlin, Heidelberg: Springer-Verlag, 1998.

Venkatraman, S., Abraham, A., and Paprzycki, M., Significance of steganography on data security, *Proc Intl Conf Inform Technol: Coding Comput*, 2, 347–351, 2004.

Wu, Y.-Ta and Shih, F. Y., Genetic algorithm based methodology for breaking the steganalytic systems, *IEEE Trans Syst, Man,Cyber—Part, B Cybern*, 36(1), 24–31, 2006.

12 Digital Steganographic Schemes Based on Image Vector Quantization

Shinfeng D. Lin and Shih-Chieh Shie

CONTENTS

12.1 INTRODUCTION

Along with the progress relating to computer hardware and software, the Internet has become the most popular channel for transmitting various forms of digital media. Since the environment of the Internet is open, the protection of digital data transmitted on the network has become an important research topic in recent years. Information hiding is a common technique to achieve the purpose of data protection. It involves embedding significant data into various forms of digital media such as text, audio, image, and video secretly. Information hiding can be further divided into two main branches: steganography and digital watermarking.

The popularity of the Internet and the increasing bandwidth of communications provide great convenience to the transmission of multimedia data via networks. However, data transmissions over the Internet still have to face some security problems, such as illegal access, data security, and copyright protection. To safely transmit data through the Internet, some mechanisms have been proposed to protect important data from illegal interception. One of the most important mechanisms is data encryption (Bourbakis and Alexopoulos, 1992; Rhee, 1994; Jan and Tseng, 1996; Highland, 1997), which refers to the process of encoding secret data in such a way that only the receiver with the right key can successfully decode it. However, as an encrypted data usually flags the importance of the data, it may also attract eavesdroppers' attention. Another approach to increase the security of transmitted data, especially for digital images, is by hiding the secret data in images (Bender et al., 1996; Marvel and Retter, 1998; Petitcolas et al., 1999). The main purpose of image steganography is to embed a piece of secret information into a noncritical host image to distract opponents' attention (Kohn, 1996; Stallings, 1999), whereas the major goal of image watermarking is to protect the copyright of the marked image itself.

Digital steganography is a feasible means for covert communications. It has been implemented successfully for many forms of digital content, especially for digital image. In this chapter, we introduce different types of image steganographic schemes based on vector quantization (VQ). The concept of VQ will be introduced first as the background of the steganographic schemes discussed in this chapter. The main components in the VQ compression system, such as the input image to be compressed, the VQ codebook, the VQ indices of input image, and the reconstructed image, can be used for applying digital steganography. Based on the components of VQ, various steganographic schemes have been designed and proposed in the literature.

12.2   BACKGROUND FOR VQ-BASED IMAGE STEGANOGRAPHIC SCHEMES

We first describe the detailed processes of VQ for image compression in Section 12.2.1. Then, an efficient lossless coding scheme that further compresses the VQ indexes of image based on the search order of index is reviewed in Section 12.2.2. Furthermore, the simplest data-embedding technique (the greedy least-significant-bit substitution method) is presented in Section 12.2.3. Finally, some of the VQ-based image steganographic schemes are briefly reviewed in Section 12.2.4.

12.2.1   VECTOR QUANTIZATION

Image compression has become very essential for multimedia storage and transmission applications. Vector quantization technique has been extensively and successfully applied for image compression. One feature of VQ is that high compression ratios are possible with relatively small block sizes. The other attractiveness of VQ as source coding scheme derives from its optimality and the simplicity of the hardware implementation of the decoder. In a VQ coding system, the image to be coded is first divided into nonoverlapping blocks. Each image block is individually mapped to the closest codeword in the codebook based upon the minimum distortion rule.

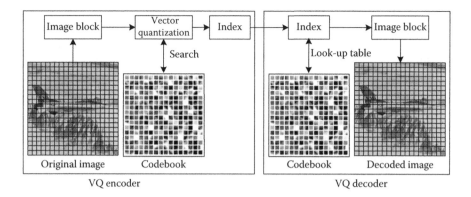

FIGURE 12.1 The encoding and decoding modules of VQ for image compression.

Compression is achieved by replacing these codewords with the corresponding indexes in transmission or storage applications. Reconstruction of image is then simply performed by table lookup using the index as an entry to the codebook.

A VQ coding system can be simply defined as a mapping from a k-dimensional Euclidean space R^k to a finite subset of space R^k. The finite subset $C = \{y_i: y_i \in R^k$ and $i = 0, 1, \ldots, N_c - 1\}$ is called a codebook, where N_c is the size of this codebook. Each $y_i = (y_{i,0}, y_{i,1}, \ldots, y_{i,k-1})$ in the codebook C is called a codeword or a codevector and is composed of k scalar elements. These codewords are generally generated, based on the Linde–Buzo–Gray (LBG) iterative clustering algorithm (Linde et al., 1980), from the training set. The compression performance is significantly dependent upon these codewords. An ordinary VQ system, as shown in Figure 12.1, has two parts: the encoder and the decoder, each of which is equipped with the same codebook. In the encoder, the closest codeword y_i to the input block x can be found by searching the codebook, and the index i will be sent to the decoder. Consequently, in the decoder, the codeword y_i can be found via the received index i, and the block x will be replaced by codeword y_i. In this case, a quantization error is unavoidable. Therefore, VQ belongs to lossy compression scheme. The distortion between the input block x and its quantized block y can be obtained by the squared Euclidean distance measure criterion given as Equation 12.1. Moreover, the overall distortion between the original image and the reconstructed image can be obtained by summing up all the distortion caused by quantizing image blocks.

$$d(x,y) = \left\| x - y \right\|^2 = \sum_{j=0}^{k-1} (x_j - y_j)^2 \qquad (12.1)$$

12.2.2 Search-Order-Coding Algorithm

In a traditional VQ system for images, each block of image is quantized independently and its corresponding index is sent to the decoder directly. To further improve the performance of compression, variable-length coding (VLC) techniques such as Huffman coding (Huffman, 1952) or arithmetic coding (Witten et al., 1987) can be applied to the VQ indices of image. However, this will introduce extra hardware cost in image compression. Therefore, Hsieh and Tsai (1996) proposed a lossless compression scheme for VQ index with search-order coding (SOC). The SOC algorithm exploits the interblock correlation in the index domain rather than in the pixel domain. The idea behind the SOC algorithm is that it searches the previous indices on the basis of a predefined search order and sends the corresponding search result to decoder to notify if any previous index is matched with the current index. Hsieh and Tsai pointed out that because the VQ index is a scalar, not a high-dimensional vector, the SOC algorithm is much simpler than conventional memory VQ and, therefore, it may achieve better compression performance. The SOC algorithm can be summarized as follows:

Step 1. Determine the number of bits n for encoding the search order.

Step 2. Input a VQ index and use it as a search center.

Step 3. Try to find a search point (SP) with the same VQ index value as the search center in the predefined search path on the VQ index table until the currently searched index is not a repetition SP and cannot be encoded with any of the SOC codes, $(0)_2 \sim (2^n - 1)_2$.

Step 4. If a matched SP is found, the currently searched index is encoded with a 1-bit indicator followed by the corresponding SOC code; otherwise, it is encoded with the indicator followed by its original index value (OIV).

Step 5. If there is another index to be processed, go to *Step 2*; otherwise, output the compressed index table of image to decoder.

12.2.3 DATA EMBEDDING BY LSB SUBSTITUTION

Many techniques concerning embedding significant data into the least significant bits (LSB) of image pixels have been proposed in earlier literatures. Wang et al. (2001) proposed an optimal LSB substitution method for embedding secret data into the selected LSBs of pixels in the cover image. The proof for the effectiveness of Wang et al.'s scheme in the worst-case condition is also described in Wang et al. (2001). In addition, a genetic algorithm that greatly reduces the huge computational time for finding the optimal selection of LSBs is developed as well. In order to find the optimal LSB substitution for pixels within an image, Chang et al. (2003) proposed another more efficient approach in which the dynamic programming strategy was applied. However, the most efficient and easiest substitution method for image data embedding in spatial domain is the greedy LSB substitution technique.

The general processes of data embedding based on the greedy LSB substitution method are given as follows. Let M be the binary representation of the secret data, and X be the image data where M will be embedded. Assume the length of M is l; therefore, M can be represented as Equation 12.2. Let X be an image with resolution $u \times v$ pixels and d bits per pixel. X can be represented as Equation 12.3.

$$M = \{m_i | 0 \le i < l,\ m_i \in \{0,\ 1\}\} \tag{12.2}$$

$$X = \{x_{ij} | 0 \le i < u,\ 0 \le j < v,\ x_{ij} \in \{0,1,\ldots,2^d - 1\}\} \tag{12.3}$$

Assume that M is to be embedded within the r rightmost LSBs of x_{ij} in X. Consequently, the following formula must be satisfied:

$$l \leq r \times u \times v \qquad (12.4)$$

The data-embedding procedure of the greedy LSB substitution method is processed by first replacing the first LSBs of all x_{ij} in X with the first $u \times v$ bits of M (i.e., m_0 to $m_{u \times v-1}$). Second, the second LSBs of all x_{ij} in X are replaced with the next $u \times v$ bits of M (i.e., $m_{u \times v}$ to $m_{2 \times u \times v-1}$). This procedure continues until all the binary bits of M are embedded into X.

12.2.4 VQ-BASED IMAGE STEGANOGRAPHIC SCHEMES

The purpose of image steganography is different from traditional cryptography (Davis, 1978; Rivest et al., 1978) and watermarking techniques (Cox et al., 1997; Barni et al., 2001). Cryptography encrypts messages into meaningless data while watermarking is used to protect the copyright. Image steganography covers the secret information with the host media as camouflage and is considered as an extension of traditional cryptography. Image steganography is mainly used for covert communication. It attempts to conceal the existence of secret data, whereas digital watermarking techniques try to maintain the existence of embedded data for future usage. Nevertheless, there is still some similarity between them. Both the embedded data may be images with certain purpose. Speaking of the requirements, digital watermarking techniques usually concentrate on the robustness, while image steganography may lay greater emphasis on the available capacity for hiding. In the prior researches, raw images without any compression are considered as cover media (Wu and Tsai, 2000). To solve the problems of inefficient capacity, Chen et al. (1998) proposed a solution that secret images should be compressed by VQ (Linde et al., 1980) and then encrypted before the embedding process, which is called virtual image cryptosystem. Hu (2003) proposed a revised algorithm of virtual image cryptosystem. Both the above-mentioned schemes have been simulated and proven that multiple images can be hidden concurrently while the quality of extracted images at the receiver is acceptable.

The algorithm of Hu's scheme can be summarized as follows. The goal of Hu's scheme is to hide multiple secret images into another meaningful cover image of the same size. To compress the secret images and reduce the volume of hidden information, secret images are first encoded using VQ. In this scheme, each pixel value of the cover image was split into two parts. The significant one is used for VQ codebook generation, and the insignificant one is used for information hiding based on greedy LSB substitution. Namely, the VQ codebook for secret image encoding is generated from the low-, middle-, and middle-high-frequency parts of the cover image. After the procedure of image encoding, the VQ indexes of secret images are encrypted and embedded into the high-frequency part of the cover image by greedy LSB modification. In the receiver, the same codebook can be generated from the stego-image since the low, middle, and middle-high-frequency parts of the received cover image is totally the same as the ones in the transmitter. And the VQ indexes of secret images can be directly extracted from the LSBs of the received cover image. Consequently, several secret images can be embedded into a meaningful cover image by Hu's scheme. Hu's scheme makes a good improvement when considering the proportion of the hiding capacity (the total file size of secret images) to the file size of the cover medium.

Image steganography can be implemented on the compressed domain of images as well. Under this circumstance, the hiding capacity for secret information is more restricted and the visual quality of cover images may be decreased as compared with those of the techniques applied on the noncompressed domain of images. Moreover, when secret information is hidden in the compressed domain of the host image, the bit rate of the compressed cover image becomes another major consideration. It should not cause apparent increase of the bit rate of compressed cover images after the secret information are hidden or embedded. In the following paragraphs, two recently proposed image steganography schemes that implemented on the compressed domain of images are briefly reviewed.

Chang et al. (2004) proposed an image steganographic scheme in which a reasonable amount of binary data can be embedded into the compressed codes of the host image. They pointed out that as compared with various image steganographic schemes applied on the spatial domain of image, VQ-based steganographic schemes have not been paid much attention. Therefore, they proposed an image steganographic scheme based on the SOC (Hsieh and Tsai, 1996) compression technique of VQ indices. In this scheme, a cover image is first compressed based on traditional VQ and an index table for the cover image is generated. After that, the SOC algorithm is applied on the index table and a more compact index table is obtained. The compact index table consists of two kinds of compression codes: the SOC and the OIV codes. Therefore, additional one bit has to be added in front of each SOC code and OIV code for distinguishing purpose. The receiver can distinguish these two different codes according to the one-bit indicator. Chang et al. found that based on this characteristic, secret data can be embedded into the compression codes without inducing additional coding distortion. Specifically, the receiver determines that each bit of secret data is "0" or "1" based on whether the received compression code is SOC or OIV. In the hiding process of Chang et al.'s scheme, there are four categories taken into consideration. Two of the four categories induce more additional bits for the compression codes that have to be translated according to the bits of secret data. Chang et al.'s scheme is the first one that technically and directly embeds secret information into the VQ-compressed codes of the host image. In addition, they also announced that a good and acceptable compression ratio of the image after the hiding procedure of secret information has been confirmed by the experiments. The details of the description are available in Chang et al. (2004).

Du and Hsu (2003) proposed an adaptive data-hiding method to embed secret data into VQ-compressed images. This method adaptively varies the embedding process according to the amount of hidden data. More specifically, the secret data were embedded into the VQ-compressed images by first rearranging all the codewords in the codebook into exclusive groups based on codeword's similarity. And then, according to the secret bits to be hidden, a codeword was chosen from the group that contains the nearest codeword of the currently encoded block to replace the nearest one in the VQ encoding procedure. Accordingly, Du and Hsu's method achieved good improvements both on the hiding capacity and on the visual quality of cover images as compared with ordinary fixed embedding methods such as the mean gray-level embedding (MGLE) method and the pair-wise nearest-neighbor embedding (PNNE) method. More details on this method can be found in Du and Hsu (2003).

12.3 IMAGE STEGANOGRAPHIC SCHEME BASED ON VQ CODEBOOK MODIFICATION

Secret image transmission or communication has attracted much attention in these years, and many researches about image steganography have been proposed in the literature (Du and Hsu, 2003; Hu, 2003; Thien and Lin, 2003; Chan and Cheng, 2004; Chang et al., 2004). Image steganography involves embedding a large amount of image data into another cover medium with minimal perceptible degradation. However, the embedding capacity and the distortion of the cover medium are a trade-off since more embedded images always result in more degradation of the cover medium. Besides, the successful delivery for a large set of secret images becomes a challenge when the bandwidth of the communication channel is narrow. An image steganographic scheme suitable for a narrow communication channel is introduced in this section. A set of secret images can be simultaneously and efficiently delivered to the receiver via a small and meaningless data stream. To reduce the volume of secret images to be transmitted, a codebook is generated and these secret images are encoded into binary indexes based on the VQ technique. The compressed message of secret images is then embedded into this VQ codebook by an adaptive LSB modification technique. For the purpose of security, the slightly modified codebook is further encrypted into a meaningless data stream by the DES cryptosystem (National Bureau of Standards, 1997). Simulation results show that this image steganographic scheme provides an impressive improvement both in the visual quality of the extracted secret images at the receiver and in the hiding capacity of the cover medium.

In this section, a novel image steganographic scheme that focuses both on the hiding capacity of the cover medium and on the quality of extracted secret images at the receiver is introduced. To design a high-capacity and high-quality image steganographic scheme, we incorporate the VQ technique into our scheme to compact the volume of secret images. Moreover, to guarantee the visual quality of extracted secret images at the receiver, the VQ codebook utilized in the encoding procedure is adopted as the cover medium. This scheme provides a new approach to secretly transmit a set of images, especially for limited-bandwidth communication channel.

12.3.1 IMAGE STEGANOGRAPHIC SCHEME BY VQ AND LSB MODIFICATION

In this image steganographic scheme, the input of the transmitter is a set of secret images, and the output of the transmitter is a series of encrypted data stream. The goal of this scheme is to transmit a set of secret images via an encrypted meaningless data stream. The size of this encrypted data stream is relatively small as compared with the file size of all secret images.

Assume that there are t secret images to be transmitted and these images are gray-level images of $w \times h$ pixels. To compress the secret images with VQ, a codebook should be generated before the encoding procedure. Let the size of VQ codebook be N_c and the codeword be composed of $m \times n$ elements. These N_c codewords of codebook are generated based on the iterative LBG algorithm using the secret images as the training set. After the codebook C is generated, each of these secret images is partitioned into blocks of $m \times n$ pixels. Each image block is then encoded into a binary index of codeword. The compressed message of all secret images is obtained by merging these binary indexes. Therefore, the overall size l of the compressed message for secret images can be defined by

$$l = t \times \lceil w/m \rceil \times \lceil h/n \rceil \times \log_2 N_c \qquad (12.5)$$

In order to transmit a set of secret images via a small and encrypted data stream, we propose a novel idea to embed the compressed message of secret images into the codebook associated with these images. In addition, the parameters used in this scheme, t, w, h, m, n, and N_c, have to be preserved well for future use at the receiver. These parameters are also embedded into the codebook. The embedding procedure of this image steganographic scheme at the transmitter is accomplished based on the following steps:

Step 1. Sort the codewords of codebook into ascending order based on the referred frequency in the procedure of secret images' encoding.

Step 2. Select the first $p + q$ codewords of the sorted codebook based on the size of the compressed message for secret images. The first p codewords are taken for embedding the information of related parameters, while the rest q codewords are used for embedding the compressed message of secret images.

Step 3. Embed the parameters and the compressed message of secret images into the selected codewords by directly modifying the LSBs of codeword elements.

A modified codebook C' is obtained after all the compressed message of secret images and the information of related parameters are embedded into the original codebook C. For security, the modified codebook C' is then encrypted by the DES cryptosystem (National Bureau of Standards, 1997) to create an encrypted message. Finally, the encrypted data stream covering a set of secret images is generated. Figure 12.2 illustrates the flowchart for transmitting a set of secret images by this scheme. Note that only the parameter p and the arguments used in the procedure of DES encryption have to be sent to the receiver via another safe channel.

The procedure of secret images' extraction is quite simple at the receiver. To reconstruct the secret images, the received data stream is first decrypted based on the DES decryption procedure. After the decryption process, the modified codebook C' is directly obtained. Therefore, all the parameters, including the number of secret images t, the image size w and h, the codeword size m and n, and the codebook size N_c, can be easily extracted from the first p codewords of C'. According to the extracted

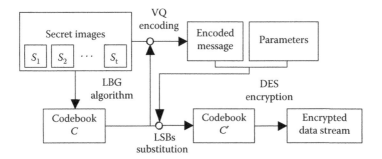

FIGURE 12.2   Flowcharts of the image steganographic scheme at the transmitter.

parameters, the compressed message of secret images can be directly obtained from the next q codewords of C'. The compressed message is then decoded by performing table look-up operation on the modified codebook C'. Finally, all the secret images are reconstructed.

12.3.2 EXPERIMENTAL RESULTS

The image steganographic scheme introduced in this section has been performed on a set of five images (Lena, Pepper, Boats, Goldhill, and Toys) and a set of 10 images (Lena, Airplane, Boats, Girl, Goldhill, Lenna, Pepper, Tiffany, Toys, and Zelda), respectively. These test images are all with size 512×512 pixels and 256 levels per pixel. In order to apply the VQ compression on the test images, each image was divided into image blocks with 8×8 pixels. In addition, the size of the VQ codebook is 4096 and each codeword consists of 8×8 elements. Consequently, there are 262,144 ($4096 \times 8 \times 8$) elements in this codebook and the length of each codeword index is 12 bits. Therefore, the compressed message for each secret image is 49,152 ($12 \times (512 \times 512)/(8 \times 8)$) bits.

In the first experiment, a set of five secret images has to be simultaneously embedded into the cover medium. Therefore, 3840 codewords were selected from the VQ codebook to embed the compressed message of five secret images. The least significant bits of the codeword elements in each selected codeword were directly replaced by the compressed message of secret images. Table 12.1 lists the

TABLE 12.1

Quality (in dB) of Secret Images Before and After Secret Message Embedding

Secret Images	Lena	Pepper	Boats	Goldhill	Toys
VQed image (at transmitter)	32.574	33.071	31.554	30.806	34.226
Extracted image (at receiver)	32.513	33.001	31.506	30.767	34.136

quality of extracted secret images at the receiver, together with the original quality of VQ-compressed images at the transmitter. The extracted secret images at the receiver are shown in Figure 12.3. Note that the peak signal-to-noise ratio (PSNR) criterion is adopted to evaluate the image quality in our experiments. The PSNR criterion is defined as

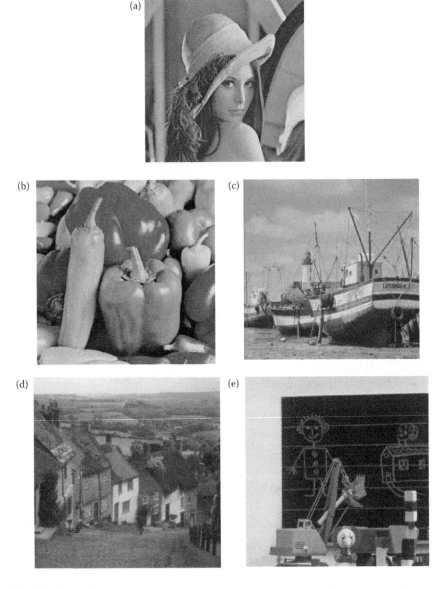

FIGURE 12.3 Five secret images simultaneously extracted at the receiver. (a) Lena, 32.513 dB (b) Pepper, 33.001 dB (c) Boats, 31.506 dB (d) Goldhill, 30.767 dB, and (e) Toys 34.136 dB.

$$\text{PSNR} = 10\log_{10}\frac{E_{\max}^2 \times W_I \times H_I}{\sum\left(I_{m,n} - I'_{m,n}\right)^2} \tag{12.6}$$

where W_I and H_I are the width and height of the cover image. $I_{m,n}$ is the original pixel value of the coordinate (m, n) and $I'_{m,n}$ is the altered pixel value of the coordinate (m, n). $E_{\max}$ is the largest energy of the image pixels (e.g., $E_{\max} = 255$ for 256 grayscale images). As shown in Table 12.1 and Figure 12.3, the experimental result shows that the degradation of image quality before and after secret image transmission is very small. It also reveals that the visual quality of these five secret images at the receiver is quite good.

In the second experiment, 10 secret images were simultaneously embedded into the cover medium and transmitted to the receiver. To embed the compressed message of 10 secret images, the first LSBs of all codeword elements and the second LSBs of some codeword elements in the codebook have to be modified. For evaluation by the human visual system, Figure 12.4 illustrates these 10 secret images simultaneously extracted at the receiver. The PSNR values of the secret images are also

FIGURE 12.4 Ten secret images simultaneously extracted at the receiver. (a) Lena, 31.24 dB, (b) Airplane, 30.77 dB, (c) Boats, 29.63 dB, (d) Girl, 31.28 dB, (e) Goldhill, 29.89 dB, (f) Lenna, 29.65 dB, (g) Pepper, 31.72 dB, (h) Tiffany, 32.47 dB, (i) Toys, 32.24 dB, and (j) Zelda, 32.71 dB.

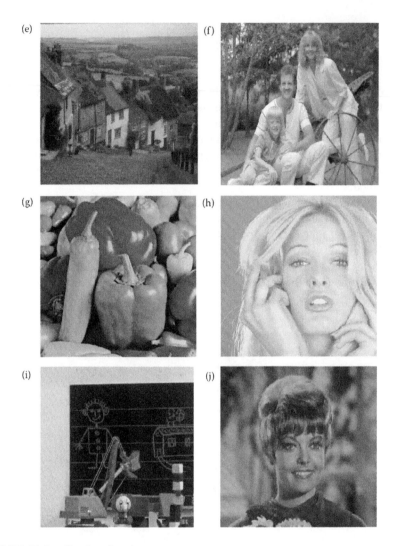

FIGURE 12.4 Continued.

provided. Note that the final cover medium of this scheme is a meaningless data stream as shown in Figure 12.5.

To verify the performance of the image steganographic scheme, the embedding capacity (the maximum number of secret images) with respect to the size of data stream (the file size of the cover medium) is listed in Table 12.2. Here, each secret image file is of size 256 kB. This table also illustrates that more embedding capacity can be provided by enlarging the codebook size or modifying more LSBs within a single codeword element in this scheme. However, to provide better visual quality for secret images at the receiver, it is suggested that the number of modified LSBs within each codeword element should not exceed three. Under these circumstances, in terms of mean-squared error (MSE), the degradation of image quality between the VQ-compressed secret

FIGURE 12.5   The final encrypted cover medium of this scheme.

TABLE 12.2
Embedding Capacity (Maximum Number of Secret Images)

Codebook Size	Size of Data Stream	Modified LSBs		
		1	2	3
2048	128 kB	2	5	8
4096	256 kB	5	10	16
8192	512 kB	9	19	29

image at the transmitter and the extracted secret image at the receiver is bounded by $(2^r - 1)^2$. Here, r is the number of modified LSBs within each codeword element.

A novel image steganographic scheme capable of delivering a set of secret images to the receiver at the same time has been introduced in this section. The proposed scheme provides impressive improvement both in the visual quality of extracted secret images and in the hiding capacity of the cover medium. The achievement of this scheme is based on the following reasons: (1) The secret images are included in the training set in the codebook generation procedure. (2) The volume of secret images to be delivered is greatly reduced by applying the VQ technique on these images. (3) The cover medium is composed of all the VQ codewords associated with these secret images. (4) The compressed message of secret images is embedded into the cover medium based on an adaptive greedy LSBs substitution technique. In addition, the modified cover medium is further encrypted into a meaningless data stream by the DES cryptosystem for more security. Moreover, the file size of the encrypted data stream that has to be transmitted to the receiver is quite small. That is, the image steganographic scheme introduced in this section is feasible for narrow communication channel.

12.4   IMAGE STEGANOGRAPHIC SCHEME IN VQ-COMPRESSED DOMAIN OF IMAGE

The VQ-based steganographic technique has not received much attention compared to various spatial domain-based steganographic techniques previously developed for digital image. In this section, we would like to introduce a new image steganographic scheme which was applied on the VQ-compressed domain of the host image. Image steganography involves embedding a large amount of secret data into a cover image with minimal perceptible degradation of the image quality. However, there is always a trade-off between the hiding capacity for secret data and the distortion of the cover image since more hidden data always result in more degradation of visual quality in the cover image. Moreover, when the image steganographic technique is implemented on compressed images, the hiding capacity and the visual quality of the cover images are even more restricted. Furthermore, the bit rate of the compressed cover image becomes another major consideration in image steganographic applications. Hiding secret data should not cause an apparent increase in the bit rate of the compressed cover images.

The image steganographic scheme introduced in this section focuses on the hiding capacity of the host image in the VQ-compressed domain. The major goal of this scheme is to embed secret data into the VQ-compressed codes of the host image such that the interceptors will not notice the existence of secret information. In addition, this scheme attempts to keep an acceptable bit rate for the compressed cover image after a large amount of secret data are embedded into the host image. To design a low-bit-rate image steganographic scheme, we incorporate the VQ technique (Linde et al., 1980) into our scheme to compress the host image. Moreover, to provide more hiding capacity for secret data and keep an acceptable bit rate for the compressed host images, the SOC algorithm (Hsieh and Tsai, 1996) is implemented to compress the VQ indices of the host image. During the process of data hiding, this scheme adaptively embeds secret data into the compressed VQ indices of image according to the amount of hidden data. To prevent the interceptors from being aware of the existence of secret data, the data is embedded into the compression codes of the host image directly. The embedding process induces no extra coding distortion and adjusts the bit rate according to the size of secret data for the compressed host image. The receiver can efficiently receive both the secret data and the compressed image with an acceptable bit rate at the same time.

12.4.1   IMAGE STEGANOGRAPHIC SCHEME IN VQ-COMPRESSED DOMAIN

The goal of this scheme is to transmit a set of binary information secretly via a VQ-compressed host image at an acceptable bit rate. Assume that the host image X is a gray-level image with $w \times h$ pixels. To compress the host image with VQ, a codebook should be generated before image compression. Let the size of VQ codebook be N_c and a codeword be composed of $m \times n$ elements. These N_c codewords of codebook are generated based on the iterative LBG algorithm (Linde et al., 1980). After a codebook is generated, X is partitioned into nonoverlapping blocks of $m \times n$ pixels. Each image block of X is then encoded into an index of codeword. Consequently, an index table T with $\lceil w/m \rceil \times \lceil h/n \rceil$ elements is constructed after all the image blocks of X are encoded by VQ. T is then ready to be transmitted to the receiver for the purpose of image compression. The bit rate BR_{VQ} of ordinary VQ compression can be obtained by

$$BR_{VQ} = \log_2 N_c /(m \times n) \tag{12.7}$$

In order to simultaneously hide secret information into the index table T and reduce the bit rate of the VQ-compressed host image, we further incorporate the SOC algorithm into this image steganographic scheme. The SOC algorithm takes advantages of the high correlation among adjacent indices in T to encode the traditional VQ indices with fewer bits. Here, the high correlation means that there may be many image blocks of X encoded with the same VQ indices in the neighborhood. The SOC algorithm encodes each index of T one by one in raster scan order. It tries to find the same index around the current processed index in a predefined search path. If the same index is found within the search path, the current processed index will be denoted as an SOC code and replaced with a d-bits code (d is much smaller than $\log_2 N_c$). Otherwise, the current processed index will remain unchanged and be denoted as an OIV code. Note that a one-bit indicator is needed for each processed index in T to distinguish SOC code from OIV code. The performance of SOC algorithm depends on the amount of SOC codes it can determine in the index table. Let the amount of SOC codes in T be s; therefore, the bit rate BR_{VQSOC} of the compressed host image after applying the SOC algorithm can be computed by

$$\text{BR}_{\text{VQSOC}} = ((1 + \log_2 N_c) \times \left(\lceil w/m \rceil \times \lceil h/n \rceil - s\right) + (1 + d) \times s)/(w \times h) \quad (12.8)$$

To hide binary information secretly in the VQ-compressed domain of the host image, we introduce a novel idea to hide information while applying the SOC algorithm on T. In this scheme, an unfixed or fixed amount of secret bits can be hidden into each SOC code in T. In order to hide a randomly unfixed amount of secret bits into an SOC code, a seed key k and an integer i are needed to randomly generate the amount ranging from 1 to i. However, if a fixed amount p of secret bits is hidden in each SOC code, the bit rate BR_{VQSOCDH} of the compressed host image after information embedding can be calculated by

$$\text{BR}_{\text{VQSOCDH}} = ((1 + \log_2 N_c) \times \left(\lceil w/m \rceil \times \lceil h/n \rceil - s\right) + (1 + d + p) \times s)/(w \times h)$$

$$(12.9)$$

The following steps summarize the embedding procedure of this image steganographic scheme:

Step 1. Encode the host image X into its corresponding index table T by applying the VQ algorithm.

Step 2. Apply the SOC algorithm on the index table T. If the current processed index is encoded as an SOC code, then extract a number of bits from the secret information and embed these secret bits into the SOC code. Otherwise, the current processed index is encoded as an OIV code and it remains unchanged to preserve an acceptable bit rate for the compressed host image.

Step 3. Encrypt and transmit the modified index table and the associated parameters to the receiver.

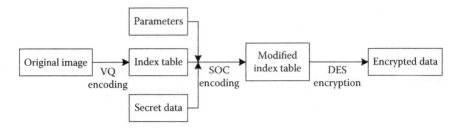

FIGURE 12.6   Flowchart of the data-embedding procedure at the transmitter.

A modified index table T' for the compressed host image X is obtained after the secret information has been embedded. To transmit the index table T' to the receiver, all the SOC codes and OIV codes in T' are merged into a binary bit stream based on the raster scan order. For more security, the bit stream can be further encrypted by the DES cryptosystem (National Bureau of Standards, 1997) to create an encrypted message. Finally, the encrypted data stream covering the secret information is generated. Note that the associated parameter set used in this scheme, d, k, and i for variable-length embedding or d and p for fixed-length embedding, has to be preserved well for future use at the receiver. These parameters can be embedded into T' or transmitted to the receiver via a secure channel. Figure 12.6 illustrates the flowchart of the data-embedding procedure at the transmitter.

The procedure of secret information extraction is quite simple at the receiver. To reconstruct the compressed codes of the host image, the received data stream is first decrypted based on the DES decryption procedure. After the decryption process, the modified index table T' of the compressed host image X is directly obtained. The parameters used in the information-embedding procedure can be extracted from T' or obtained via a predefined secure channel. Moreover, the SOC codes in T' can be easily specified with these parameters and the one-bit indicator that distinguishes the SOC code from OIV code. Finally, all the secret bits are extracted and the secret information is then reconstructed. Figure 12.7 illustrates the flowchart of the data extraction procedure at the receiver.

12.4.2   EXPERIMENTAL RESULTS

In the experiments, the image steganographic scheme introduced in this section has been performed on six host images (Airplane, Boat, Girl, Lena, Peppers, and Toys) with size 256×256 pixels and 256 levels per pixel, respectively. To simulate various

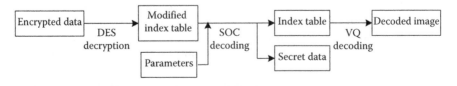

FIGURE 12.7   Flowchart of the data extraction procedure at the receiver.

types of secret information, the secret data used in this experiment are composed of randomly generated bit streams. The VQ codebook utilized in the simulation is generated by using the LBG algorithm and five standard images (Airplane, Boat, Lena, Sailboat, and Toys) with size 512×512 pixels and 256 gray levels. In addition, the codeword size is 4×4 pixels and the codebook size is 256. Consequently, each host image is partitioned into 4096 $((256 \times 256)/(4 \times 4))$ image blocks with 4×4 pixels. And the bit rate of the primitive VQ-compressed image is 0.5 bits per pixel. In order to hide secret information while keeping an acceptable bit rate for the compressed host images, the SOC algorithm is applied on the VQ indices of the host image and the parameter d for the predefined search path is set to be 2 in the experiment.

To verify the performance of the image steganographic scheme introduced here, Table 12.3 lists the bit rates for embedding different-sized secret information into the host images. As shown in Table 12.3, the secret information with size 4–16 K bits can be embedded into the compression codes of the host image by the introduced image steganographic scheme. In fact, there is no limitation on the hiding capacity of this scheme. The reason is that the secret bits are directly embedded into the compression codes of the host image. Nevertheless, this image steganographic scheme decreases the compression rate of host image after secret information embedding. As for the bit rate of the compressed host image, the increase of the bit rate is steady with that of information size. In addition, the increase in the bit rate of the compressed host image is independent of the property of the binary information.

Recently, Shie et al. (2006) proposed another VQ-based data-hiding technique that takes advantages of the prediction property of the side-match VQ (SMVQ) state codebook. The major idea of this technique is to hide secret data in VQ-compressed codes of cover images based on a modified SMVQ-encoding process, taking advantage of the concept of prediction. To compare the image steganographic scheme introduced in this section with Shie et al.'s scheme, the benchmark images "Lena" and "F16" are utilized as cover images. The VQ-codebook utilized in this experiment is generated from the cover images themselves. In addition, the codeword size is 4×4 pixels and the codebook size is 256. Consequently, each cover image is partitioned into 16,384 $((512 \times 512)/(4 \times 4))$ image blocks with 4×4 pixels. Note that this experimental environment is almost the same as that of Shie et al.'s scheme, except for the codebook size. The bit rate of the original VQ-compressed image is 0.5 bpp. In addition, the quality of the cover images is evaluated by the PSNR criterion defined in Equation 12.6.

TABLE 12.3

Bit Rates of the Compressed Cover Images Hidden with Different Amounts of Secret Data

Size of Secret Data (in Bits)	Airplane	Boat	Girl	Lena	Pepper	Toys
4 K	0.4227	0.4439	0.4944	0.4831	0.4793	0.3941
8 K	0.4852	0.5064	0.5569	0.5456	0.5418	0.4566
12 K	0.5477	0.5689	0.6194	0.6081	0.6043	0.5191
16 K	0.6102	0.6314	0.6819	0.6706	0.6668	0.5816

TABLE 12.4

Performance Comparison (in PSNR) of the Introduced Scheme and Shie et al.'s Scheme

Size of Secret Data (in Bits)	Shie et al.'s Scheme		Introduced Scheme	
	Lena	F16	Lena	F16
48 K	29.84	31.51	32.73	32.60
64 K	27.57	29.57	32.73	32.60
80 K	23.76	26.79	32.73	32.60

To compare the performance of the introduced scheme with Shie et al.'s scheme, we hide the same amounts of secret data in the compressed cover images. Before data hiding, the PSNR values of the compressed cover images, "Lena" and "F16," are 32.73 and 32.60 dB, respectively. For objective evaluation, Table 12.4 lists the performance comparison (in PSNR) of the introduced scheme and Shie et al.'s scheme, with the amounts of hidden data ranging from 48 to 80 kbit. As shown in Table 12.4, it demonstrates that the introduced scheme outperforms Shie et al.'s scheme when the same amounts of data are hidden in the cover images. For subjective evaluation by the human visual system, Figures 12.8 and 12.9 illustrate the original uncompressed cover images, the compressed cover images with different amounts of secret data hidden by the scheme of Shie et al. (2006) and the compressed cover images with secret data hidden by the scheme introduced in this section, respectively, for the test images "Lena" and "F16." Note that the introduced scheme induces no extra coding distortion in the compressed cover images. The experimental results also demonstrate that the visual quality of the cover images is quite acceptable after hiding large amounts of secret data with the introduced scheme. For comparing the compression performance of the introduced scheme and Shie et al.'s scheme, the bit rates with respect to the amount of hidden data are listed in Table 12.5. It reveals that the compression performance of the introduced scheme is not so good as Shie et al.'s scheme. However, the increase in bit rate is acceptable under the conditions that the visual quality is quite good even when the hiding capacity is large for the compressed cover images.

As illustrated in Figures 12.8 and 12.9, the introduced scheme outperforms Shie et al.'s scheme in terms of the visual quality of the compressed cover image. The introduced scheme outperforms Shie et al.'s scheme for the following reasons. In Shie et al.'s scheme, the secret data are hidden in the compressed cover image based on a modified encoding procedure of SMVQ. In traditional SMVQ, a small state codebook is adaptively generated from the master codebook for each image block. The codewords in the state codebook are similar, and each one of these codewords may be a good candidate for representing the currently encoded image block. Based on this concept, Shie et al. proposed a method to hide a set of secret bits in each SMVQ index by encoding the currently processed image block into the codeword index equivalent to the set of secret bits. Consequently, most of the cover image blocks are not encoded

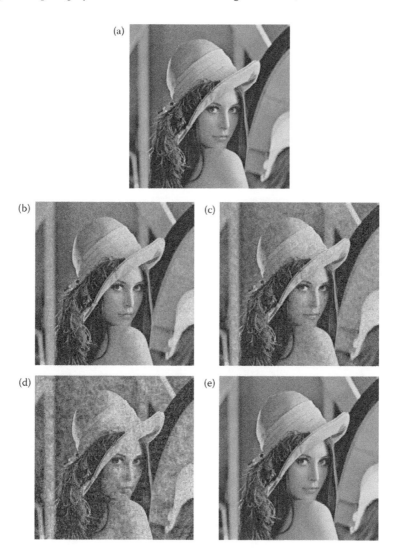

FIGURE 12.8   Visual quality of the cover image "Lena" under Shie et al.'s scheme and the introduced scheme. (a) The original uncompressed cover image, (b) the compressed cover image with 48K secret bits by Shie et al.'s scheme, (c) the compressed cover image with 64K secret bits by Shie et al.'s scheme, (d) the compressed cover image with 80K secret bits by Shie et al.'s scheme, and (e) the compressed cover image by the introduced scheme.

into the indexes that represent the best-match codewords. Under these circumstances, the visual quality of the compressed cover image could be distorted seriously. In contrast, the introduced scheme introduces no extra coding distortion to the compressed cover image during the data-hiding procedure. Therefore, the visual quality of the compressed cover image obtained by the introduced scheme is quite acceptable.

A novel image steganographic scheme, applied on the VQ compressed domain of host image, has been introduced in this section. The introduced scheme remains an

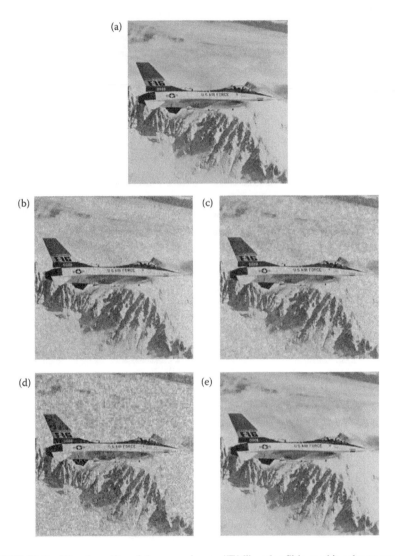

FIGURE 12.9 Visual quality of the cover image "F16" under Shie et al.'s scheme and the introduced scheme. (a) The original uncompressed cover image, (b) the compressed cover image with 48K secret bits by Shie et al.'s scheme, (c) the compressed cover image with 64K secret bits by Shie et al.'s scheme, (d) the compressed cover image with 80K secret bits by Shie et al.'s scheme, and (e) the compressed cover image by the proposed scheme.

acceptable bit rate for the compressed host image after a large amount of secret information have been embedded into the compressed codes of the host image. Furthermore, the hiding of secret information does not introduce any distortion on the visual quality for the compressed host image. The receiver can efficiently receive both the hidden information and the compressed host image simultaneously. The new steganographic scheme outperforms the earlier-proposed schemes for the following reasons: (1) The SOC algorithm, efficient lossless compression technique

TABLE 12.5

Performance Comparison (in Bits per Pixel) of the Introduced Scheme and Shie et al.'s Scheme

Capacity (in Bits)	Shie et al.'s Scheme		Introduced Scheme	
	Lena	F16	Lena	F16
48 K	0.4632	0.4413	0.502	0.496
64 K	0.4358	0.4257	0.565	0.559
80 K	0.4195	0.4052	0.627	0.621

for VQ index coding, is applied in the proposed scheme to reduce the bit rate of the cover image. (2) The secret data are hidden by technically embedding the secret bits into the compression codes of the cover image instead of modifying and translating these compression codes with other longer codes. (3) The hiding capacity of the cover image is not limited by the number of blocks partitioned from the cover image itself. (4) The proposed scheme does not introduce any visual distortion in the compressed cover image after secret data hiding.

12.5　VISUALLY IMPERCEPTIBLE IMAGE STEGANOGRAPHIC SCHEME BASED ON VQ

A novel visually imperceptible image steganographic scheme based on VQ is introduced in this section. Multiple secret images can be simultaneously and imperceptibly hidden into another cover image with the same image size by this scheme. In order to reduce the volume of secret images to be hidden, a codebook is first generated from the secret images and these images are encoded into binary indexes by the VQ technique. Then, the compressed data of secret images are embedded into the VQ codebook used in the encoding procedure by an adaptive LSB modification technique. For security purpose, the slightly modified codebook is further encrypted into a meaningless data stream by the DES cryptosystem. Finally, the encrypted codebook is embedded into the cover image using the greedy LSB modification technique. Simulation results show that this scheme provides a good improvement in the visual quality of the extracted secret images and the cover image at the receiver. In addition, it also provides a better hiding capacity for the cover image.

Visually imperceptible image steganography involves hiding one or more secret images into another noncritical image with minimal perceptible degradation. Recently, the technique of visually imperceptible image steganography has been widely studied and many researches have been proposed (Chen et al., 1998; Wang et al., 2001; Xuan et al., 2002; Chang et al., 2003; Hu, 2003; Thien and Lin, 2003; Chan and Cheng, 2004; Lin and Shie, 2004; Shie and Lin, 2009; Shie et al., 2010). Among these researches, a common technique for data embedding in image is on the basis of manipulating the LSB planes of image since LSB modification techniques usually achieve high-embedding capacity. Visually imperceptible image steganography can be used for covert communication. In the prior researches, raw images without any compression are directly embedded into the cover image. To solve the problem of inefficient hiding capacity, Chen et al. (1998) proposed an idea that secret images should be compressed by VQ and then encrypted before the hiding process, which is called virtual image cryptosystem. Hu (2003) proposed a revised algorithm of virtual image cryptosystem. In Hu's scheme, each pixel value of the cover image was split into two parts. The significant one is used for VQ codebook training, and the insignificant one is used for data embedding by the greedy substitution. Consequently, several secret images can be hidden into another cover image by Hu's scheme. Hu's scheme provides more hiding capacity, lower computational cost, and better image quality than those of Chen et al.'s scheme (1998). In order to improve the efficiency of image steganography, Lin and Shie (2004) proposed an idea to transmit a set of secret images via its corresponding VQ codebook. This scheme focuses on both the hiding capacity of the cover medium and the quality of extracted secret images at the receiver. To achieve a high-capacity and high-quality image steganographic scheme, the VQ technique is applied in Lin and Shie's scheme to compact the volume of secret images. Moreover, to guarantee the visual quality of extracted secret images at the receiver, the VQ codebook utilized in the encoding procedure is adopted as the cover medium. Lin and Shie's scheme provides a new and original approach to transmit a set of secret images via network, especially for limited-bandwidth communication channel. Although the appearance of the embedded and encrypted cover medium looks like a meaningless data stream to the possible interceptors, the cover medium of Lin and Shie's scheme (2004) is a VQ codebook, not a visually recognizable image. This may limit its practical utilization for the purpose of visual imperceptibility.

In this section, a novel visually imperceptible image steganographic scheme that improves Lin and Shie's scheme (2004) is presented. To design a high-capacity and high-quality image steganographic scheme, the VQ technique is utilized to compact the volume of secret images. Moreover, to adaptively guarantee the visual quality of extracted secret images at the receiver, the VQ codebook utilized in the encoding procedure is slightly modified and totally embedded into the cover image. This scheme provides a visually imperceptible image steganographic approach for delivering a set of secret images to the receiver.

12.5.1 Visually Imperceptible Image Steganographic Scheme Based on VQ

In the introduced visually imperceptible image steganographic scheme, the input of the transmitter is a set of secret images and a noncritical cover image, and the output of the transmitter is a stego-image with high visual quality. The goal of this scheme is to deliver a set of images secretly via a meaningful cover image of the same size.

Assume that there are t secret images to be delivered and these images are 8-bit gray-level images of $w \times h$ pixels. To compress the secret images with VQ, a codebook should be generated before the encoding procedure. Let the size of VQ codebook be N_c and the codeword be composed of $m \times n$ elements. These N_c codewords of codebook are generated based on the iterative LBG algorithm using the secret images as the training set. After the codebook C is generated, each of these secret images is partitioned into blocks of $m \times n$ pixels. Each image block is then encoded into a binary index of codeword. The compressed message of all secret images is obtained by merging these binary indexes. Therefore, the whole volume l (in the unit of bit) of the compressed information for secret images can be defined by the following equation:

$$l = t \times \lceil w/m \rceil \times \lceil h/n \rceil \times \log_2 N_c \qquad (12.10)$$

To embed multiple images into another cover image with minimal perceptible degradation and preserve good visual quality for secret images, the compressed information of secret images is first embedded into the VQ codebook associated with these images and the modified codebook is then embedded into the cover image. Let the number of modified LSBs for each codeword element and each cover image pixel be d and r, respectively. For 8-bit gray-level images, the following two formulas must be satisfied to accomplish the proposed image information-hiding scheme.

$$l \le d \times N_c \times m \times n \qquad (12.11)$$

$$8 \times N_c \times m \times n \le r \times w \times h \qquad (12.12)$$

The flowcharts of the introduced image steganographic scheme are given in Figure 12.10. Figure 12.10a describes the image-embedding procedures at the transmitter whereas the processes of image extraction at the receiver are presented in Figure 12.10b. The related parameters used in this scheme, t, w, h, m, n, N_c, d, and r,

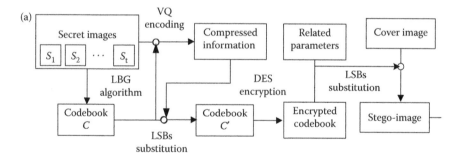

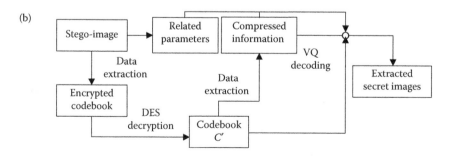

FIGURE 12.10 Flowcharts of the visually imperceptible image steganographic scheme. (a) Transmitter and (b) receiver.

have to be preserved well for future use at the receiver. These parameters can also be embedded into the cover image. Note that the compressed information, the codeword indices, of secret images can be further compressed on the basis of some famous lossless compression techniques such as Huffman coding (Huffman, 1952), arithmetic coding (Witten et al., 1987), and SOC (Hsieh and Tsai, 1996) algorithms. For more security, the modified codebook is encrypted by the DES cryptosystem (National Bureau of Standards, 1997) to make it an encrypted and meaningless message. Furthermore, the encrypted codebook can also be losslessly compressed before it is embedded into the cover image. Finally, a stego-image covering a set of secret images is generated.

The procedure of secret images' extraction is quite simple at the receiver. To reconstruct the secret images, the encrypted modified codebook and the related parameters are directly fetched from the stego-image. The encrypted codebook is first decrypted based on the DES decryption procedure. After the decryption process, the modified codebook C' is directly obtained. Therefore, the compressed information of secret images can be directly extracted from the codebook. With the parameters, including the number of secret images t, the image size w and h, the codeword size m and n, and the codebook size N_c, the compressed information of secret images can be decoded by performing table lookup operation on the modified codebook C'. Finally, all the secret images are reconstructed.

12.5.2 Experimental Results

In the computer experiments, the introduced image steganographic scheme is performed on a set of four images ("Airplane," "Lena," "Pepper," and "Toys"). "Airplane" is used as a cover image and the other three images are used as secret images. These test images are all with size 512×512 pixels and 256 levels (8 bits) per pixel. To reduce the overall volume of image information to be embedded into the cover image, these three secret images are divided into image blocks with 8×8 pixels and the VQ codebook size is 1024. Consequently, there are 65,536 ($1024 \times 8 \times 8$) elements and 524,288 ($1024 \times 8 \times 8 \times 8$) bits in this codebook. The length of each codeword index is 10 bits and the compressed information for each secret image is 40,960 ($10 \times (512 \times 512)/(8 \times 8)$) bits.

In this experiment, the compressed information of the three secret images can be simultaneously embedded into the codebook, without the need of any further compression, by adaptively modifying the first and the second least significant bits of each codeword element. Under this circumstances, 960 codewords of the codebook are enough to embed the compressed information of three secret images. For more security, the modified codebook is further encrypted by the DES cryptosystem. The encrypted codebook is given in Figure 12.11. To accomplish visual imperceptibility, the encrypted codebook is directly and totally embedded into the cover image "Airplane," without the need of compression, by greedily modifying the first and the second least significant bits of each pixel in "Airplane" ($512 \times 512 \times 2 = 5,24,288$).

To demonstrate the performance of the described image steganographic scheme, Figure 12.12 shows the original image "Airplane" and the stego-Airplane with three hidden secret images. In addition, the original images of "Lena," "Pepper," and "Toys,"

FIGURE 12.11   The DES-encrypted codebook with codebook size 1024 and codeword size 8×8.

and the corresponding secret images extracted at the receiver are given in Figures 12.13, 12.14, and 12.15, respectively.

To verify the performance of the introduced image steganographic scheme, the hiding capacity (the maximum number of secret images) with respect to the size of the codebook, the modified LSBs in a codeword element, and the modified LSBs in a cover image pixel are listed in Table 12.6. This table illustrates that more hiding capacity can be provided by reducing the codebook size, modifying more LSBs in each codeword element, or modifying more LSBs in each cover image pixel by the proposed scheme. However, to provide better visual quality for secret images at the receiver, it is suggested that the number of modified LSBs within each codeword element should not exceed three. Under these circumstances, in terms of mean-squared error, the degradation of image quality between the VQ-compressed secret image at the transmitter and the extracted secret image at the receiver is bounded by $(2^d - 1)^2$. Here, d is the number of modified LSBs within each codeword element, and r is the number of modified LSBs in each cover image pixel. Note that in Table 12.6, which the codebook size equals 1024, the data volume (in bit) of codebook exceeds the available number of LSBs that can be modified in the cover image when r equals 1.

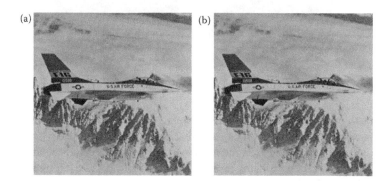

FIGURE 12.12   (a) Original "Airplane" and (b) stego-Airplane, PSNR = 45.883 dB.

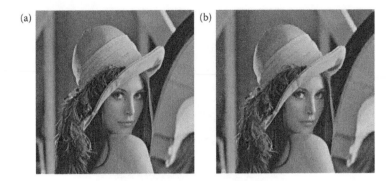

FIGURE 12.13   (a) Original "Lena" and (b) extracted secret "Lena," PSNR = 31.573 dB.

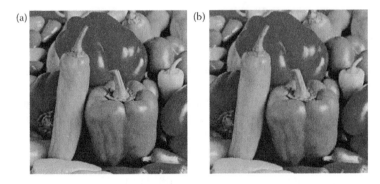

FIGURE 12.14   (a) Original "Pepper" and (b) extracted secret "Pepper," PSNR = 31.236 dB.

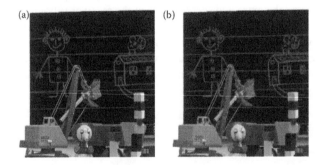

FIGURE 12.15   (a) Original "Toys" and (b) extracted secret "Toys," PSNR = 31.273 dB.

Consequently, the hiding capacity of our scheme is not available (NA). In addition, the hiding capacity is identical when r equals 2 and 3. This is because we can embed only one codebook with size 1024 into the cover image when r equals 3, although the third LSB bit-plane of cover image remains unused. Under these circumstances, some smaller codebooks can be embedded into the unused LSBs. Table 12.7 lists the hiding capacity of the introduced scheme when several codebooks with different sizes

TABLE 12.6

Hiding Capacity (Maximum Number of Secret Images) of the Introduced Scheme Using Codebooks with the Same Size

Codebook Size	Modified LSBs of Codeword	Modified LSBs of Cover Image		
		$r = 1$	$r = 2$	$r = 3$
64	$d = 1$	1	2	4
	$d = 2$	2	5	8
	$d = 3$	4	8	12
128	$d = 1$	1	2	3
	$d = 2$	2	4	6
	$d = 3$	3	6	10
256	$d = 1$	1	2	3
	$d = 2$	2	4	6
	$d = 3$	3	6	9
512	$d = 1$	0	1	2
	$d = 2$	1	3	5
	$d = 3$	2	5	8
1024	$d = 1$	NA	1	1
	$d = 2$	NA	3	3
	$d = 3$	NA	4	4

are embedded into a cover image. Table 12.8 lists the hiding capacity of the introduced image steganographic scheme using only one codebook and modifying the rest available LSBs of cover image with more compressed information (VQ indices) of secret images.

To compare the proposed scheme with Lin and Shie's scheme (2004), one example of the final appearance of cover medium, respectively, for these two schemes, is illustrated in Figure 12.16. The secret images are covered by a meaningless data

TABLE 12.7

Hiding Capacity (Maximum Number of Secret Images) of the Introduced Scheme Using Codebooks with Different Sizes

Codebook Sizes	Modified LSBs of Codeword	Modified LSBs of Cover Image		
		$r = 1$	$r = 2$	$r = 3$
One 512 and	$d = 1$	0	1	2
multiple 256	$d = 2$	1	3	5
	$d = 3$	2	5	8
One 1024 and	$d = 1$	NA	1	2
multiple 256	$d = 2$	NA	3	5
	$d = 3$	NA	4	7

TABLE 12.8

Hiding Capacity (Maximum Number of Secret Images) of the Introduced Scheme Using Only One Codebook and Modifying the Rest Available LSBs of Cover Image with More Compressed Information of Secret Images

Codebook Size	Modified LSBs of Codeword	Modified LSBs of Cover Image		
		$r = 1$	$r = 2$	$r = 3$
512	$d = 1$	0	8	15
	$d = 2$	1	8	16
	$d = 3$	2	9	16
1024	$d = 1$	NA	1	8
	$d = 2$	NA	3	9
	$d = 3$	NA	4	11

stream, as shown in Figure 12.16a, in Lin and Shie's scheme, whereas the secret images are under the protection of another high-quality image, as shown in Figure 12.16b, in the proposed scheme. The hiding capacity with respect to the codebook size and the modified LSBs in a cover medium (codeword) element provided in Lin and Shie (2004) is listed in Table 12.9.

A visually imperceptible image steganographic scheme capable of delivering multiple images secretly through another high-quality image has been introduced in this section. This scheme provides impressive improvement both in the visual quality of extracted secret images and in the hiding capacity of the cover image. Moreover, the visual quality of the stego-image is quite good when comparing it with the original cover image. This scheme outperforms Lin and Shie's scheme because of the following reasons: (1) The secret images are included in the training set in the codebook generation procedure. (2) The volume of secret images to be transmitted is greatly reduced by applying the VQ technique on these images. (3) The codebook

FIGURE 12.16 The final appearance of cover medium. (a) Lin and Shie's scheme and (b) the introduced scheme.

TABLE 12.9

Hiding Capacity (Maximum Number of Secret Images) in Lin and Shie's Scheme

Codebook Size	Data Stream Size	Modified LSBs of Cover Medium		
		$m = 1$	$m = 2$	$m = 3$
2048	128 kB	2	5	8
4096	256 kB	5	10	16

associated with these secret images is embedded into the cover image as well. (4) The cover medium is a meaningful image, not a meaningless data stream.

REFERENCES

Barni, M., Podilchuk, C. I., Bartolini, F., and Delp, E. J., Watermark embedding: Hiding a signal within a cover image, *IEEE Communications Magazine*, 39(8), 102–108, 2001.

Bender, W., Gruhl, D., Morimoto, N., and Lu, A., Techniques for data hiding, *IBM Systems Journal*, 35(3&4), 313–336, 1996.

Bourbakis, N. and Alexopoulos, C., Picture data encryption using scan patterns, *Pattern Recognition*, 25(6), 567–581, 1992.

Chan, C. K. and Cheng, L. M., Hiding data in images by simple LSB substitution, *Pattern Recognition*, 37, 469–474, 2004.

Chang, C. C., Chen, G. M., and Lin, M. H., Information hiding based on search-order coding for VQ indices, *Pattern Recognition Letters*, 25, 1253–1261, 2004.

Chang, C. C., Hsiao, J. Y., and Chan, C. S., Finding optimal least-significant-bit substitution in image hiding by dynamic programming strategy, *Pattern Recognition*, 36, 1583–1595, 2003.

Chen, T. S., Chang, C. C., and Hwang, M. S., A virtual image cryptosystem based upon vector quantization, *IEEE Transactions on Image Processing*, 7(10), 1485–1488, 1998.

Cox, I. J., Kilian, J., Leighton, T., and Shamoon, T., Secure spread spectrum watermarking for multimedia, *IEEE Transactions on Image Processing*, 6(12), 1673–1687, 1997.

Davis, R. M., *The Data Encryption Standard in Perspective*, Computer Security and the Data Encryption Standard, National Bureau of Standards Special Publication, February 1978.

Du, W. C. and Hsu, W. J., Adaptive data hiding based on VQ compressed images, *IEE Proceedings Vision, Image, and Signal Processing*, 150(4), 233–238, 2003.

Highland, H. J., Data encryption: A non-mathematical approach, *Computer & Security*, 16, 369–386, 1997.

Hsieh, C. H. and Tsai, J. C., Lossless compression of VQ index with search-order coding, *IEEE Transaction on Image Processing*, 5(11), 1579–1582, 1996.

Hu, Y. C., Grey-level image hiding scheme based on vector quantization, *IEE Electronics Letters*, 39(2), 202–203, 2003.

Huffman, D. A., A method for the construction of minimum redundancy codes, *Proceedings of the IRE*, 40, 1098–1101, 1952.

Jan, J. K. and Tseng, Y. M., On the security of image encryption method, *Information Processing Letters*, 60, 261–265, 1996.

Kohn, D., *The Codebreakers: The Story of Secret Writing*, Scribner, New York, 1996.

Lin, S. D. and Shie, S. C., Secret image communication scheme based on vector quantization, *IEE Electronics Letters*, 40(14), 859–860, 2004.

Linde, Y., Buzo, A., and Gray, R. M., An algorithm for vector quantizer design, *IEEE Transactions on Communications*, 28, 84–95, 1980.

Marvel, L. M. and Retter, C. T., Hiding information in images, *Proceedings of the IEEE International Conference on Image Processing*, vol. 2, pp. 396–398, 1998.

National Bureau of Standards (U.S.), *DES Encryption Standard*, Federal Information Processing Standards Publication, Vol. 46, National Technical Information Service, Springfield, VA, April 1997.

Petitcolas, F. A. P., Anderson, R. J., and Kuhn, M. G., Information hiding—A survey, *Proceedings of the IEEE*, 87(7), 1062–1078, 1999.

Rhee, M. Y., *Cryptography and Secure Communication*, McGraw-Hill Book Company, Singapore, 1994.

Rivest, R., Shamir, A., and Adleman, L., A method for obtaining digital signatures and public-key cryptosystems, *Communications ACM*, 21(2), 120–126, 1978.

Shie, S. C., Jiang, J. H., Chen, L. T., and Huang, Z. H., Secret image transmission scheme using secret codebook, *IEICE Transactions on Information and Systems*, E93-D(2), 399–402, 2010.

Shie, S. C., Lin, S. D., and Fang, C. M., Adaptive data hiding based on SMVQ prediction, *IEICE Transactions on Information and Systems*, E89-D(1), 358–362, 2006.

Shie, S. C. and Lin, S. D., Data hiding based on compressed VQ indices of images, *Computer Standards & Interfaces*, 31(6), 1143–1149, 2009.

Stallings, W., *Cryptography and Network Security*, Prentice Hall, New Jersey, 1999.

Thien, C. C. and Lin, J. C., A simple and high-hiding capacity method for hiding digit-by-digit data in images based on modulus function, *Pattern Recognition*, 36, 2875–2881, 2003.

Wang, R. Z., Lin, C. F., and Lin, J. C., Image hiding by optimal LSB substitution and genetic algorithm, *Pattern Recognition*, 34, 671–683, 2001.

Witten, I. H., Neal, R. M., and Cleary, J. G., Arithmetic coding for data compression, *Communications ACM*, 30(6), 520–540, 1987.

Wu, D. C. and Tsai, W. H., Spatial-domain image hiding using image differencing, *IEE Proceedings Vision, Image and Signal Processing*, 147(1), 29–37, 2000.

Xuan, G., Zhu, J., Chen, J., Shi, Y. Q., Ni, Z., and Su, W., Distortionless data hiding based on integer wavelet transform, *IEE Electronics Letters*, 38(25), 1646–1648, 2002.

13 Differential Evolution-Based Algorithm for Breaking the Visual Steganalytic System

Frank Y. Shih and Venkata Gopal Edupuganti

CONTENTS

13.1 INTRODUCTION

As digital information and data are transmitted over the Internet more often than ever before, new technology for protecting and securing sensitive messages needs to be discovered and developed. Digital steganography is the art and science of hiding information into covert channels, so as to conceal the information and prevent the detection of the hidden message. Information security research on covert channels is generally not a major player, but has been an extremely active topic in academia, industry, and government domains for the last 30 years.

The word steganography (Provos and Honeyman, 2003) is of Greek origin and means "covered/hidden writing." Steganography is the art and science of writing hidden messages in such a way that no one apart from the intended recipient knows the existence of the message. It is different from cryptography which does not hide the messages but encodes them in a secret manner, so that nobody can read it without the specific key. Another technique, watermarking (Cox et al., 2001; Shih, 2007) is used to guarantee authenticity and can be applied as proof that the content has not been altered since last insertion. Modern steganographic techniques include covert channel, invisible ink, microdot, null cipher, and spread-spectrum communication

(Shih, 2007). Although steganography is an ancient subject, computer technology provides a new aspect of applications by hiding messages in the multimedia content, such as audio, video, and images.

Steganography has been widely used in historic times. Examples include hidden messages in wax tablets. In ancient Greece, people wrote messages on wood and then covered it with wax, so it would look like an ordinary, unused tablet. Steganography was also used during World War II to send the messages secretly to the destination. In the field of radio and communication technology, the payload is the data to be transported secretly. The carrier is the signal, stream, or a data file into which the payload is hidden. The resulting signal, stream, or data file in which the payload is embedded is referred to as package, stego file, or covert message, respectively.

Steganography can be divided into three types: technical, linguistic, and digital. Technical steganography applies scientific methods to conceal the secret messages while linguistic steganography uses written natural language. Digital steganography, developed with the advent of computers, employs computer files or digital multimedia data. The fundamental requirement of the steganographic systems is that the stego-object must be perceptually similar to the cover (original) object. This means that the hidden message should introduce only a slight modification to the cover object. There are two ways of inspecting a stego-object: One is passive and the other is active. In the passive scheme, the stego-object is examined to determine whether it contains a hidden message or not. On the other hand, the active scheme alters the content of the stego-object even though it does not detect the hidden message.

Steganalytic systems can be categorized into two classes: spatial-domain steganalytic system (SDSS) (Westfeld and Pfitzmann, 1999a; Avcibas et al., 2003) and frequency-domain steganalytic system (FDSS) (Fridrich et al., 2003). The SDSS is used for checking the lossless compressed images by analyzing the spatial-domain statistic features. For lossy compression images such as Joint Photographic Experts Group (JPEG), the FDSS is used to analyze the frequency-domain statistic features. Variant steganalytic systems have been build based on least-significant bit (LSB) technique, image quality measure (IQM) (Avcibas et al., 2003), and discrete cosine transform (DCT). Wu and Shih (2006) proposed a genetic algorithm (GA)-based steganographic system to break the steganalytic system.

GA was used to generate the stego-image by artificially counterfeiting statistic features in order to break the corresponding steganalytic system. However, GA converges slowly to the optimal solution. There are other evolutionary approaches, such as differential evolution (DE) (Karboga and Okdem, 2004; Price et al., 2005) and evolution strategy (ES) (Beyer and Schwefel, 2004). One drawback of ES is that it is easier to get stuck on a local optimum. In this chapter, we use DE to increase the performance of the steganographic system because it is a simple and efficient adaptive scheme for global optimization over continuous space (Shih and Edupuganti, 2009). It can find the global optimum of a multidimensional, multimodal function with good probability.

The rest of the chapter is organized as follows. The DE-based methodology for breaking the visual steganalytic system (VSS) is presented in Section 13.2. The DE algorithm and its based image steganography are presented in Section 13.3. Experimental results are shown in Section 13.4. Finally, we draw conclusions in Section 13.5.

13.2   DE-BASED METHODOLOGY FOR BREAKING THE VSS

Kessler has presented an overview of steganography for the computer forensics examiner (Kessler, 2004). The steganalysis strategy usually follows the method that the steganography algorithm uses. A simple way is to visually inspect the carrier and steganography media. Many simple steganographic techniques embed secret messages in the spatial domain and select message bits in the carrier that are independent of its content (Wayner, 2002).

Most steganographic systems embed message bits in a sequential or pseudorandom manner. If the image contains certain connected areas of uniform color or saturated color (i.e., 0 or 255), we can use visual inspection to find suspicious artifacts. When the artifacts are not found, an inspection of the LSB plane is conducted.

Westfeld and Pfitzmann have presented a VSS that uses an assignment function of color replacement called the visual filter (Westfeld and Pfitzmann, 1999a,b). The idea is to remove all parts of the image that contain a potential message. The filtering process relies on the presumed steganographic tool and can produce the attacked carrier medium steganogram, the extraction of potential message bits, and the visual illustration.

As has been mentioned, in order to generate a stego-image to pass though the inspection of the SDSS, the messages should be embedded into the specific positions for maintaining the statistical features of a cover image. In the spatial-domain embedding approach, it is difficult to select such positions since the messages are distributed regularly. On the other hand, if the messages are embedded into specific positions of coefficients of a transformed image, the changes in the spatial domain are difficult to predict. We intend to find the desired positions on the frequency domain that produce minimum statistic features disturbance on the spatial domain.

Based on the DE, we generate the stego-image by adjusting the pixel values on the spatial domain using the following two criteria:

1. Evaluate the extracted messages obtained from the specific coefficients of a stego-image to ascertain that they are as close as possible to the embedded messages.
2. Evaluate the statistical features of the stego-image and compare them with those of the cover image such that the differences should be as few as possible.

Therefore, in order to break the VSS, the two results, VF^C and VF^S, of applying the visual filter on the cover- and the stego-images respectively, should be as identical as possible. The VSS was designed to detect the GIF format images by reassigning the color in the color palette of an image. Let the chromosome, ξ, be composed of 8×8 genes. Let C and S, respectively, denote a cover-image and a stego-image of size 8×8. The Analysis (ξ, C) to VSS indicating the sum of difference between VF^C and VF^S is defined as

$$\text{Analysis}(\xi, C) = \frac{1}{|C|} \sum_i (VF_i^C \oplus VF_i^S) \qquad (13.1)$$

where " $\oplus$ " denotes the exclusive-OR (XOR) operator. The algorithm is described below.

1. Divide a cover-image into a set of cover-images of size 8×8.
2. For each 8×8 cover-image, generate a stego-image based on the GA to perform the embedding procedure as well as to ensure that VF^C and VF^S are as identical as possible.
3. Combine all the 8×8 stego-images together to form a complete stego-image.

Let α_1 and α_2 denote weights, which can be adjusted according to the user's demand on the degree of distortion to the stego-image or the extracted message. The objective function is defined as:

$$\text{Evaluation}(\xi, C) = \alpha_1 \times \text{Analysis}(\xi, C) + \alpha_2 \times \text{BER}(\xi, C) \qquad (13.2)$$

where $\text{Analysis}(\xi, C)$ is the analysis function that maintains the statistic features between the stego-image and cover image by minimizing the difference defined in Equation 13.1, and $\text{BER}(\xi, C)$ is the bit error rate which sums up the bit differences between the embedded and extracted message as defined in Equation 13.3.

$$\text{BER}(\xi, C) = \frac{1}{|\text{Message}^H|} \sum_{i=0}^{\text{all pixels}} \left| \text{Message}_i^H - \text{Message}_i^E \right| \qquad (13.3)$$

where Message^H and Message^E, respectively, denote the embedded and the extracted binary secret messages, and $|\text{Message}^H|$ denotes the length of the message. For example, if $\text{Message}^H = 11111$ and $\text{Message}^E = 11101$, then $\text{BER}(\xi, C) = 0.2$.

13.3 DE ALGORITHM AND ITS BASED IMAGE STEGANOGRAPHY

DE is a kind of evolutionary algorithm (EA), which derives the name from natural biological evolutionary processes. EA is mimicked to obtain a solution to an optimization problem. One can use EA for problems that are difficult to solve by traditional optimization techniques, including the problems that are not well defined or are difficult to model mathematically. A popular optimization method, which belongs to the EA class of methods, is the GA (Holland, 1975). In recent years, GA has replaced traditional methods to be the preferred optimization tool, as several studies have conclusively proved. DE (Price et al., 2005) is a relative latecomer, but its popularity has been catching up. It is fast in numerical optimization and is more likely to find the true optimum.

13.3.1 CONCEPT

The method of DE mainly consists of four steps: initialization, mutation, recombination, and selection. In the first step, a random population of potential solutions is created within the multidimensional search space. To start with, we define the objective function $f(\mathbf{y})$ to be optimized, where $\mathbf{y} = (y^1, y^2, \ldots, y^n)$ is a vector of n decision variables. The aim is to find a vector $\mathbf{y}$ in the given search space, for which the value of the objective function is an optimum. The search space is first determined by providing the lower and upper bounds for each of the n decision variables of $\mathbf{y}$.

In the initialization step, a population of NP vectors, each of n dimensions, is randomly created. The parameters are encoded as floating point numbers. Mutation is basically a search mechanism, which directs the search toward potential areas of optimal solution together with recombination and selection. In this step, three distinct target vectors $\mathbf{y}_a$, $\mathbf{y}_b$, and $\mathbf{y}_c$ are randomly chosen from the NP parent population on the basis of three random numbers a, b, and c. One of the vectors $\mathbf{y}_c$ is the base of the mutated vector. This is added to the weighted difference of the remaining two vectors, that is, $\mathbf{y}_a - \mathbf{y}_b$, to generate a noisy random vector $\mathbf{n}_i$. The weighting is done using a scaling factor F, which is user-supplied and usually in the range from 0 to 1.2. This mutation process is repeated to create a mate for each member of the parent population.

In the recombination (crossover) operation, each target vector of the parent population is allowed to undergo the recombination by mating with a mutated vector. Thus, vector $\mathbf{y}_i$ is recombined with the noisy random vector, $\mathbf{n}_i$ to generate a trial vector, $\mathbf{t}_i$. Each element of the trial vector (t_i^m, where $i = 1, 2, \ldots, NP$ and $m = 1, 2, \ldots, n$) is determined by a binomial experiment whose success or failure is determined by the user-supplied crossover factor, CR. The parameter CR is used to control the rate at which the crossover takes place. The trial vector, $\mathbf{t}_i$, is thus the child of two parent vectors: noisy random vector $\mathbf{n}_i$ and the target vector $\mathbf{y}_i$. DE performs a non-uniform crossover, which determines the trial vector parameters to be inherited from which parent.

It is sometimes possible that some particular combinations of three target vectors from the parent population and the scaling factor F would result in noisy vector values, which are outside the bounds set for the decision variables. It is necessary, therefore, to bring such values within the bounds. For this reason, the value of each element of the trial vector is checked at the end of the recombination step. If it violates the bounds, it is heuristically brought back to lie within the bounded region. It is in the last stage of the selection, the trial vector is pitted against the target vector. The fitness is tested, and the fitter of the two vectors survives and proceeds to the next generation.

After NP competitions of this kind in each generation, one will have a new population, which is fitter than the population in the previous generation. This evolution procedure consisting of the above four steps is repeated over several generations until the termination condition is reached (i.e., when the objective function attains a prescribed optimum) or a specified number of generations are completed, whichever happens earlier. The flowchart is described in Figure 13.1.

13.3.2 Algorithm

1. The first step is the random initialization of the parent population. Each element of NP vectors in n dimensions is randomly generated as:

$$y_i^m = y_{min}^m + \text{rand}(0,1) \times (y_{max}^m - y_{min}^m) \tag{13.4}$$

where $i = 1, 2, \ldots, NP$ and $m = 1, 2, \ldots, n$.

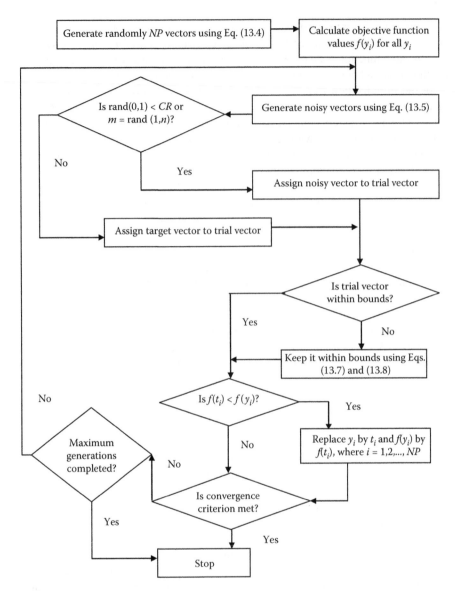

FIGURE 13.1 The DE flowchart.

2. Calculate the objective function values $f(\mathbf{y}_i)$ for all $\mathbf{y}_i$.
3. Select three random numbers (a, b, and c) within the range from 1 to NP. The weighted difference $(\mathbf{y}_a - \mathbf{y}_b)$ is used to perturb $\mathbf{y}_c$ to generate a noisy vector $\mathbf{n}_i$:

$$\mathbf{n}_i = \mathbf{y}_c + F \times (\mathbf{y}_a - \mathbf{y}_b) \tag{13.5}$$

where $i = 1, 2, \ldots, NP$.

4. Recombine each target vector $\mathbf{y}_i$ with the noisy random vector $\mathbf{n}_i$ to generate a trial vector $\mathbf{t}_i$:

$$\begin{cases} t_i^m = n_i^m & \text{if rand}(0,1) < CR \quad \text{or} \quad m = \text{rand}(1,n) \\ t_i^m = y_i^m & \text{otherwise} \end{cases} \tag{13.6}$$

where $i = 1, 2, \ldots, NP$ and $m = 1, 2, \ldots, n$.

5. Check whether each decision variable of the trail vector is within the bounds. If it is outside the bonds, force it to lie within the bounds using:

$$t_i^m = y_{\min}^m + 2 \times (p/q) \times (y_{\max}^m - y_{\min}^m) \tag{13.7}$$

where

$$\begin{cases} p = t_i^m - y_{\max}^m & \text{and} \quad q = t_i^m - y_{\min}^m & \text{if } t_i^m > y_{\max}^m \\ p = y_{\min}^m - t_i^m & \text{and} \quad q = y_{\max}^m - t_i^m & \text{otherwise} \end{cases} \tag{13.8}$$

6. Calculate the value of the objective function for the two vectors $\mathbf{t}_i$ and $\mathbf{y}_i$. The fitter of the two (i.e., the one with the lower objective function value) survives and proceeds to the next generation.
7. Check if the convergence criterion meets. If yes, stop; otherwise, go to step 8.
8. Check if the maximum number of generations is reached. If yes, stop; otherwise, go to step 3.

13.3.3 PROCEDURE OF APPLYING DE

Because the DE algorithm deals with floating point numbers, the procedure of generating the stego-image is slightly different from the GA algorithm. The evaluation function is the same as that of GA procedure. It uses the following rule to generate a stego-image: If $d_{ij} \geq 0.5$, $s_{ij} = c_{ij} + 1$; otherwise, $s_{ij} = c_{ij}$, where $1 \leq i, j \leq 8$, and s_{ij} and c_{ij}, respectively, denote the pixels in the stego-image and cover-image. This procedure is illustrated in Figure 13.2, where (a) is the generated 8×8 DE vector by the DE algorithm, (b) is the 8×8 cover-image, and (c) is the corresponding stego-image.

13.4 EXPERIMENTAL RESULTS

The GA- and the DE-based methodology for breaking the VSS was implemented in Java. JDEAL (Java Distributed Evolutionary Algorithms Library) was used to implement GA. Experiments were conducted on 50 cover images of size 256×256 and 10 secret messages of size 64×64. Some results are tabulated in Table 13.1 for iterations of 25, 50, and 100.

We apply the error measures, such as peak signal-to-noise Ratio (PSNR) and normalized correlation (NC), to compute the image distortion after embedding secret messages. The PSNR is often used in engineering to measure the signal ratio between the maximum power and the power of corrupting noise. Because signals possess a

(a)

0.43	0.23	0.36	0.78	0.82	0.46	0.78	0.65
0.98	0.14	0.09	0.36	0.49	0.61	0.48	0.24
0.96	0.84	0.98	0.92	0.87	0.88	0.94	0.64
0.14	0.19	0.03	0.78	0.86	0.94	0.42	0.56
0.64	0.54	0.13	0.26	0.37	0.48	0.87	0.64
0.94	0.64	0.44	0.24	0.92	0.51	0.66	0.16
0.24	0.45	0.56	0.64	0.46	0.24	0.76	0.43
0.24	0.94	0.48	0.65	0.78	0.92	0.93	0.46

(b)

12	67	98	245	129	67	230	11
119	142	231	12	65	98	114	167
24	46	78	134	145	64	25	142
98	211	198	46	98	34	12	9
97	74	96	43	123	213	56	78
87	46	215	87	13	58	34	156
98	46	16	178	126	214	101	98
19	36	47	58	36	24	126	48

(c)

12	67	98	246	130	67	231	12
120	142	231	12	65	99	114	167
25	47	79	135	146	65	26	143
98	211	198	47	99	35	12	10
98	75	96	43	123	213	57	79
88	47	215	87	14	59	35	156
98	46	17	179	126	214	102	98
19	37	47	59	37	25	127	48

FIGURE 13.2　(a) DE vector, (b) cover image, and (c) stego-image.

TABLE 13.1

Comparison of Image Steganography Using GA and DE Algorithms

			GA		DE	
Image	Watermark	Iteration	PSNR	NC	PSNR	NC
Lena	Chinese character	25	52.122	0.973	54.037	0.995
		50	52.588	0.979	55.502	1.0
		100	53.146	0.993	58.160	1.0
Barbara	Chinese character	25	51.133	0.967	53.963	0.999
		50	51.140	0.967	55.385	0.999
		100	51.146	0.993	57.972	1.0
Cameraman	Chinese character	25	51.143	0.918	54.048	0.976
		50	51.132	0.955	55.553	0.979
		100	51.151	0.974	58.130	0.987
Lena	Flower	25	51.125	0.8993	53.883	0.985
		50	51.133	0.965	55.228	0.993
		100	51.148	0.981	57.659	0.9926
Barbara	Flower	25	51.135	0.968	53.949	0.993
		50	51.134	0.985	55.317	0.997
		100	51.134	0.991	57.804	1.0
Cameraman	Flower	25	51.150	0.937	53.942	0.959
		50	51.138	0.933	55.282	0.985
		100	51.151	0.961	57.549	0.990

largely wide dynamic range, we apply the logarithmic decibel scale to limit its variation. This measures the quality of reconstruction in image compression; however, it just provides a rough quality measure. In comparing two video files, the mean PSNR is often calculated. The PSNR is thus defined:

$$\text{PSNR} = 10 \times \log_{10}\left(\frac{\sum_{i=1}^{N}\sum_{j=1}^{N}[h^{Wa}(i,j)]^2}{\sum_{i=1}^{N}\sum_{j=1}^{N}[h(i,j) - h^{Wa}(i,j)]^2}\right) \quad (13.9)$$

where h^{Wa} is the stego-image, h is the cover image, and $N \times N$ is the size of the image. The correlation between two images is often used in feature detection. Normalized correlation can be used to locate a pattern on a target image that best matches the specified reference pattern from the registered image base. The NC is defined as:

$$\text{NC} = \frac{\sum_{i=1}^{N}\sum_{j=1}^{N} W(i,j) \times W^F(i,j)}{\sum_{i=1}^{N}\sum_{j=1}^{N}[W(i,j)]^2} \quad (13.10)$$

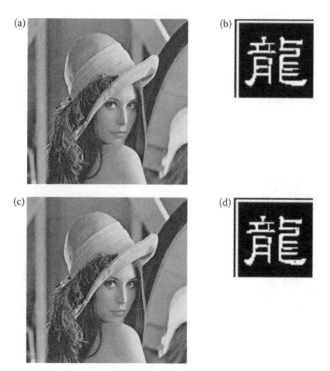

FIGURE 13.3 Images before and after the application of DE methodology to break VSS. (a) The Lena cover-image, (b) secret message to be embedded, (c) stego-image after 25 iterations of each block using DE, and (d) retrieved secret message from (c).

where W^F is the extracted secret message, W is the embedded secret message, and $N \times N$ is the size of the message.

From Table 13.1, it is clearly observed that for the same number of iterations, DE algorithm not only improves the PSNR values of the stego-image, but also promotes the NC values of the secret messages. The percentage increase in PSNR values ranges from 5 to 13%, and that of NC values ranges from 0.8 to 3%.

Figure 13.3 shows the results obtained by embedding a 64×64 Chinese character into a 256×256 Lena image for 25 iterations of each 8×8 block using DE. Figure 13.4 shows the results obtained by embedding a 64×64 flower image into a 256×256 Barbara image for 100 iterations of each 8×8 block using GA and DE. Figure 13.5 shows the results obtained by embedding a 64×64 Chinese character into a 256×256 Cameraman image for 50 iterations of each 8×8 block using GA and DE. Comparing Figures 13.4d and f and comparing Figures 13.5d and f, we can observe that the extracted secret messages from the DE-generated stego-images have higher quality than those from the GA-generated stego-images.

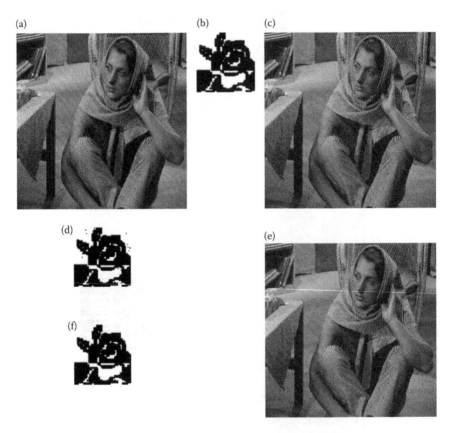

FIGURE 13.4 The images before and after the applications of GA and DE algorithms. (a) Barbara 256×256 cover-image, (b) flower 64×64 watermark, (c) Barbara stego-image after 100 iterations using GA, (d) extracted watermark from (c), (e) Barbara stego-image after 100 iterations using DE, and (f) extracted watermark from (e).

FIGURE 13.5 The images before and after the applications of GA and DE algorithms. (a) Cameraman 256 × 256 cover-image, (b) Chinese character 64 × 64 watermark, (c) cameraman stego-image after 50 iterations using GA, (d) extracted watermark from (c), (e) cameraman stego-image after 50 iterations using DE, and (f) extracted watermark from (e).

13.5 CONCLUSIONS

In this chapter, a different evolutionary approach, named DE, is used for image steganography to break the VSS. It is a simple and efficient adaptive scheme for global optimization over continuous space. Experimental results show that the DE-based steganography is superior to the GA-based steganography. The application of DE algorithm not only improves the PSNR values of the stego-images, but also upgrades the NC values of the extracted secret messages for the same number of iterations as in that of GA algorithm.

REFERENCES

Avcibas, I., Memon, N., and Sankur, B., Steganalysis using image quality metrics, *IEEE Trans Image Process*, 12(2), 221–229, 2003.

Beyer, H. and Schwefel, H., Evolution strategies—a comprehensive introduction, *Nat Comput*, 1, 3–52, 2004.

Cox, I. J., Miller, M., and Bloom, J., *Digital Watermarking: Principles & Practice*, San Francisco: Morgan Kaufmann, 2001.

Fridrich, J., Goljan, M., and Hogea, D., New methodology for breaking steganographic techniques for JPEGs, *Proc of SPIE Electronic Imaging*, Santa Clara, CA, pp. 143–155, 2003.

Holland, J. H., *Adaptation in Natural and Artificial Systems*, Ann Arbor, Michigan: The University of Michigan Press, 1975.

Karboga, D. and Okdem, S., A simple and global optimization algorithm for engineering problems: differential evolution algorithm, *Turkish J Electr Eng Comput Sci*, 12(1), 53–60, 2004.

Kessler, G. C., An overview of steganography for the computer forensics examiner, *Forensic Sci Commun*, 6(3), 1–29, 2004.

Price, K., Storn, R., and Lampinen, J., *Differential Evolution: A Practical Approach to Global Optimization*, Berlin: Springer, 2005.

Provos, N. and Honeyman, P., Hide and seek: An introduction to steganography, *IEEE Sec Priv Mag*, 1(3), 32–44, 2003.

Shih, F. Y., *Digital Watermarking and Steganography: Fundamentals and Techniques*, Boca Raton, FL: CRC Press, 2007.

Shih, F. Y. and Edupuganti, V. G., A differential evolution based algorithm for breaking the visual steganalytic system, *Soft Comput*, 13(4), 345–353, 2009.

Wayner, P., *Disappearing Cryptography: Information Hiding: Steganography and Watermarking*, Second Ed., San Francisco: Morgan Kaufmann, 2002.

Westfeld, A. and Pfitzmann, A., Attacks on steganographic systems breaking the steganographic utilities Ezstego, Jsteg, Steganos and S-Tools and some lessons learned, *Proc Third Intl Workshop on Information Hiding*, Dresden, Germany, pp. 61–76, Sep. 1999a.

Westfeld, A. and Pfitzmann, A., Attacks on steganographic systems, *Proc Third Intl Workshop on Information Hiding*, Lecture Notes in Computer Science, Berlin: Springer-Verlag, vol. 1768, pp. 61–76, 1999b.

Wu, Y. and Shih, F. Y., Genetic algorithm based methodology for breaking the steganalytic systems, *IEEE Trans Syst, Man Cybern*, 36(1), 24–29, 2006.

14 Genetic Algorithm-Based Methodology for Breaking the Steganalytic Systems

Yi-Ta Wu and Frank Y. Shih

CONTENTS

14.1 INTRODUCTION

Steganography is an ancient art of hiding secret messages within another apparently innocuous carrier in such a way that only the receiver knows the existence of the messages. It is different from cryptography which encodes messages, so that nobody can read it without the specific key. Another technique, digital watermarking (Cox et al., 2001; Wu and Shih, 2004), is concerned with issues related to copyright protection and intellectual property and therefore a watermark usually contains the information regarding the carrier and the owner. The well-known steganographic methods include covert channel, invisible ink, microdot, and spread-spectrum communication (Norman, 1973; Kahn, 1996). Although steganography is an ancient subject, computer technology provides a new aspect of applications by hiding messages in the multimedia materials, such as audio and images.

A famous classic steganographic model presented by Simmons (1984) is the prisoners' problem that Alice and Bob in a jail plan to escape together. All communications between them are monitored by Wendy, a warden. Therefore, they must hide the messages in other innocuous-looking media (cover-object) in order to obtain the stego-object. Then, the stego-object is sent through the public channel. Wendy is free to inspect all the messages between Alice and Bob with two options, *passive* or *active*. The passive way is to inspect the message in order to determine whether it contains a hidden message and then to conduct a proper action. On the other hand, the active way is always to alter messages though Wendy may not perceive any trace of a hidden message. Note that, in this chapter we focus on the passive warden.

For steganographic systems, the fundamental requirement is that the stego-object should be perceptually indistinguishable to the degree that it does not raise suspicion. In other words, the hidden information introduces only slight modification to the cover-object. Most passive warden distinguishes the stego-images by analyzing their statistical features.

In general, the steganalytic systems can be categorized into two classes: spatial-domain steganalytic system (SDSS) and frequency-domain steganalytic system (FDSS). The SDSS (Westfeld and Pfitzmann, 1999; Avcibas et al., 2003) is adopted for checking the lossless compressed images by analyzing the spatial-domain statistical features. For the lossy compression images such as JPEG, FDSS (Farid, 2001; Fridrich et al., 2003) is used to analyze the frequency-domain statistical features. Westfeld and Pfitzmann (1999) presented two SDSSs based on visual and chi-square attacks. The visual attack uses human eyes to inspect stego-images by checking their lower bit-planes. The chi-square attack can automatically detect the specific characteristic generated by the least-significant bit (LSB) steganographic technique. Avcibas and Sankur (2002) proposed image quality measure (IQM) which is based on a hypothesis that the steganographic systems leave statistical evidences that can be exploited for detection using IQM and multivariate regression analysis. Fridrich et al. (2003) presented an FDSS for detecting the JPEG stego-images by analyzing their discrete cosine transformation (DCT) with cropped images. The above steganalytic systems (Westfeld and Pfitzmann, 1999; Avcibas et al., 2003; Fridrich et al., 2003) are used in our experiments to test the correctness of our GA-based methodology.

Since the steganalytic system analyzes certain statistical features of an image, the idea of developing a robust steganographic system is to generate the stego-image by avoiding changing the statistical features of the cover-image. In literature, several papers have presented the algorithms for steganographic and steganalytic systems. Very few papers have discussed the algorithms for breaking the steganalytic systems. Recently, Chu et al. (2004) presented a DCT-based steganographic system by using the similarities of DCT coefficients between the adjacent image blocks where the embedding distortion is spread. Their algorithm can allow random selection of DCT coefficients in order to maintain key statistical features. However, the drawback of their approach is that the capacity of the embedded message is limited, that is, only 2 bits for an 8×8 DCT block.

In this chapter, we present a new method for breaking steganalytic systems (Wu and Shih, 2006). The emphasis is shifted from traditionally avoiding the change of statistical features to artificially counterfeiting them. Our paper is based on the methodology: in order to manipulate the statistical features for breaking the inspection of steganalytic systems, the genetic algorithm (GA)-based approach is adopted to counterfeit several stego-images (candidates) until one of them can break the inspection of steganalytic systems. We are the first to utilize the evolutionary algorithms in the field of steganographic systems.

This chapter is organized as follows. The overview of our GA-based breaking methodology is described in Section 14.2. We present our GA-based breaking algorithm on SDSS and FDSS in Sections 14.3 and 14.4, respectively. Experimental results are shown in Section 14.5. The complexity analysis is provided in Section 14.6. Finally, we draw conclusions in Section 14.7.

14.2 OVERVIEW OF OUR GA-BASED BREAKING METHODOLOGY

The GA, introduced by Holland (1975) in his seminal work, is commonly used as an adaptive approach that provides a randomized, parallel, and global search based on the mechanics of natural selection and genetics in order to find solutions to a problem.

In general, the GA starts with some randomly selected genes as the first generation, called *population*. Each individual in the population corresponding to a solution in the problem domain is called *chromosome*. An objective, called *fitness function*, is used to evaluate the quality of each chromosome. The chromosomes of high quality will survive and form a new population of the next generation. Using the three operators: reproduction, crossover, and mutation, we can recombine a new generation to find the best solution. The process is repeated until a predefined condition is satisfied or a constant number of iterations are reached. The predefined condition in this chapter is the situation when we can correctly extract the desired hidden message.

In order to apply the GA for embedding messages into the frequency domain of a cover-image to obtain the stego-image, we use the chromosome, ξ, consisting of n genes as $\xi = g_0, g_1, g_2, \ldots, g_n$. Figure 14.1 gives an example of a chromosome

(a)

g_0	g_1	g_2	g_3	g_4	g_5	g_6	g_7
g_8	g_9	g_{10}	g_{11}	g_{12}	g_{13}	g_{14}	g_{15}
g_{16}	g_{17}	g_{18}	g_{19}	g_{20}	g_{21}	g_{22}	g_{23}
g_{24}	g_{25}	g_{26}	g_{27}	g_{28}	g_{29}	g_{30}	g_{31}
g_{32}	g_{33}	g_{34}	g_{35}	g_{36}	g_{37}	g_{38}	g_{39}
g_{40}	g_{41}	g_{42}	g_{43}	g_{44}	g_{45}	g_{46}	g_{47}
g_{48}	g_{49}	g_{50}	g_{51}	g_{52}	g_{53}	g_{54}	g_{55}
g_{56}	g_{57}	g_{58}	g_{59}	g_{60}	g_{61}	g_{62}	g_{63}

(b)

0	0	1	0	0	2	0	1
1	2	0	0	1	1	0	1
2	1	1	0	0	-2	0	1
3	1	2	5	0	0	0	0
0	2	0	0	0	1	0	0
0	-1	0	2	1	1	4	2
1	1	1	0	0	0	0	0
0	0	0	0	1	-2	1	3

FIGURE 14.1 The numbering positions corresponding to 64 genes. (a) The distribution order of a chromosome in an 8×8 block and (b) an example of the corresponding chromosome.

($\xi \in Z^{64}$) containing 64 genes ($g_i \in Z$ (integers)). Figure 14.1a shows the distribution order of a chromosome in an 8×8 block, and Figure 14.1b shows an example of the corresponding chromosome.

The chromosome is used to adjust the pixel values of a cover-image to generate a stego-image, so that the embedded message can be correctly extracted and at the same time the statistical features remain in order to break steganalytic systems. A fitness function is used to evaluate the embedded message and statistical features.

Let C and S, respectively, denote the cover- and the stego-images of size 8×8. We generate the stego-images by adding the cover-image and the chromosome as

$$S = \{s_i \mid s_i = c_i + g_i, \text{ where } 0 \le i \le 63\} \quad (14.1)$$

14.2.1 FITNESS FUNCTION

In order to embed messages into DCT-based coefficients and avoid the detection of steganalytic systems, we develop a fitness function to evaluate the following two terms:

1. Analysis(ξ, C)—The analysis function evaluates the difference between the cover-image and the stego-image in order to maintain the statistical features. It is related to the type of the steganalytic systems used and will be explained in Sections 14.3 and 14.4.
2. BER(ξ, C)—The bit error rate (BER) sums up the bit differences between the embedded and the extracted messages. It is defined as

$$\text{BER}(\xi, C) = \frac{1}{|\text{Message}^H|} \sum_{i=0}^{\text{all pixels}} |\text{Message}_i^H - \text{Message}_i^E| \quad (14.2)$$

where Message^H and Message^E denote, respectively, the embedded and the extracted binary messages, and $|\text{Message}^H|$ denotes the length of the message. For example, if $\text{Message}^H = 11111$ and $\text{Message}^E = 10101$, then BER (ξ, C) = 0.4.

We use a linear combination of the analysis and the BER to be the fitness function as

$$\text{Evaluation}(\xi, C) = \alpha_1 \times \text{Analysis}(\xi, C) + \alpha_2 \times \text{BER}(\xi, C) \quad (14.3)$$

where α_1 and α_2 denote weights. The weights can be adjusted according to the user's demand on the degree of distortion to the stego-image or the extracted message.

14.2.1.1 Reproduction

$$\text{Reproduction}(\Psi, k) = \{\xi_i \mid \text{Evaluation}(\xi_i, C) \le \Omega \text{ for } \xi_i \in \Psi\} \quad (14.4)$$

where Ω is a threshold for sieving chromosomes, and $\Psi = \{\xi_1, \xi_2, \ldots, \xi_n\}$. It is used to reproduce k better chromosomes from the original population for higher qualities.

14.2.1.2 Crossover

$$\text{Crossover}(\Psi, l) = \left\{ \xi_i \Theta \xi_j \,\middle|\, \xi_i, \xi_j \in \Psi \right\} \qquad (14.5)$$

where Θ denotes the operation to generate chromosomes by exchanging genes from their parents: ξ_i and ξ_j. It is used to gestate l better offspring by inheriting good genes (i.e., higher qualities in the fitness evaluation) from their parents. The often-used crossovers are one-point, two-point, and multipoint crossovers. The criteria of selecting a suitable crossover depend on the length and structure of chromosomes. We adopt the one- and two-point crossovers as shown in Shih and Wu (2005) through this chapter.

14.2.1.3 Mutation

$$\text{Mutation}(\Psi, m) = \{ \xi_i \circ j \mid 0 \le j \le |\xi_i| \text{ and } \xi_i \in \Psi \} \qquad (14.6)$$

where $\circ$ denotes the operation to randomly select a chromosome ξ_i from Ψ and change the j-th bit from ξ_i. It is used to generate m new chromosomes. The mutation is usually performed with a probability p ($0 < p \le 1$), meaning only p portion of the genes in a chromosome will be selected to be mutated. Since the length of a chromosome is 64 in this chapter and there are only one or two genes to be mutated when the GA mutation operator is performed, we select p to be 1/64 or 2/64.

Note that in each generation, the new population is generated by the above three operations. The new population is actually the same size of $k + l + m$ as the original population. Note that, to break the inspection of the steganalytic systems, we use a straightforward GA-selection method that the new generation is generated based on the chromosomes having superior evaluation values in the current generation. For example, only top 10% of the current chromosomes will be considered for the GA operations, such as reproduction, crossover, and mutation.

The mutation tends to be more efficient than the crossover if a candidate solution is close to the real optimum solution. In order to enhance the performance of our GA-based methodology, we generate a new chromosome with desired minor adjustment when the previous generation is close to the goal. Therefore, the strategy of dynamically determining the ratio of three GA operations is used. Let R_P, R_M, and R_C, respectively, denote the ratios of production, mutation, and crossover. In the beginning, we select a small R_P and R_M and a large R_C to enable the global search. After certain iterations, we will decrease R_C and increase R_M and R_P to shift focuses on local search if the current generation is better than the earlier one; otherwise, we will increase R_C and decrease R_M and R_P to enlarge the range of global search. Note that, this property must be satisfied: $R_P + R_M + R_C = 100\%$.

The recombination strategy of our GA-based algorithm is presented below. We apply the same strategy in recombining the chromosome through this chapter except that the fitness function is differently defined with respect to the properties of individual problem.

The Algorithm for Recombining the Chromosomes:

1. Initialize the base population of chromosomes, R_P, R_M, and R_C.
2. Generate candidates by adjusting pixel values of the original image.
3. Determine the fitness value of each chromosome.
4. If a predefined condition is satisfied or a constant number of iterations are reached, the algorithm will stop and output the best chromosome to be the solution; otherwise, go to the following steps to recombine the chromosomes.
5. If a certain number of iterations are reached, go to step 6 to adjust R_P, R_M, and R_C; otherwise, go to step 7 to recombine the new chromosomes.
6. If 20% of the new generation are better than the best chromosome in the preceding generation, then $R_C = R_C - 10\%$, $R_M = R_M + 5\%$, and $R_P = R_P + 5\%$; otherwise, $R_C = R_C + 10\%$, $R_M = R_M - 5\%$, and $R_P = R_P - 5\%$.
7. Obtain the new generation by recombining the preceding chromosomes using production, mutation, and crossover.
8. Go to step 2.

14.3 GA-BASED BREAKING ALGORITHM ON SDSS

As aforementioned, in order to generate a stego-image to pass though the inspection of SDSS, the messages should be embedded into the specific positions for maintaining the statistical features of a cover-image. In the spatial-domain embedding approach, it is difficult to select such positions since the messages are distributed regularly. On the other hand, if the messages are embedded into specific positions of coefficients of a transformed image, the changes in the spatial domain are difficult to predict. In this section, we intend to find the desired positions on the frequency domain that produce minimum statistical features disturbance on the spatial domain. We apply the GA to generate the stego-image by adjusting the pixel values on the spatial domain using the following two criteria:

1. Evaluate the extracted messages obtained from the specific coefficients of a stego-image to be as close as to the embedded messages.
2. Evaluate the statistical features of the stego-image and compare them with those of the cover-image such that the differences should be as small as possible.

14.3.1 GENERATING THE STEGO-IMAGE ON THE VISUAL STEGANALYTIC SYSTEM

Westfeld and Pfitzmann (1999) presented a visual steganalytic system (VSS) that uses an assignment function of color replacement, called the *visual filter*, to efficiently detect a stego-image by translating a gray-scale image into binary. Therefore, in order to break the VSS, the two results, VF^C and VF^S, of applying the visual filter on the cover- and the stego-images, respectively, should be as identical as possible. The VSS was originally designed to detect the GIF format images by reassigning the color in the color palette of an image. We extend their method to detect the BMP format images as well by setting the odd- and even-numbered gray scales to black and white, respectively. The Analysis(ξ, C) to VSS indicating the sum of difference between LSB^C and LSB^S is defined as

$$\text{Analysis}(\xi, C) = \frac{1}{|C|} \sum_{i=0}^{\text{all pixels}} (VF_i^C \oplus VF_i^S) \tag{14.7}$$

where $\oplus$ denotes the exclusive-OR (XOR) operator. Our algorithm is described below.

The Algorithm for Generating a Stego-image on VSS:

1. Divide a cover-image into a set of cover-images of size 8×8.
2. For each 8×8 cover-image, we generate a stego-image based on the GA to perform the embedding procedure as well as to ensure that LSB^C and LSB^S are as identical as possible.
3. Combine all the 8×8 stego-images together to form a complete stego-image.

14.3.2 GENERATING THE STEGO-IMAGE ON THE IQM-BASED STEGANALYTIC SYSTEM (IQM–SDSS)

Avcibas et al. (2003) proposed a steganalytic system by analyzing the IQMs (Avcibas and Sankur, 2002) of the cover- and the stego-images. The IQM–SDSS consists of two phases: training and testing. In the training phase, the IQM is calculated between an image and its filtered image using a low-pass filter based on the Gaussian kernel. Suppose there are N images and q IQMs in the training set. Let x_{ij} denote the score in the i-th image and the j-th IQM, where $1 \leq i \leq N$, $1 \leq j \leq q$. Let y_i be the value of -1 or 1 indicating the cover- or stego-image, respectively. We can represent all the images in the training set as

$$
\begin{aligned}
y_1 &= \beta_1 x_{11} + \beta_2 x_{12} + \cdots + \beta_q x_{1q} + \varepsilon_1 \\
y_2 &= \beta_1 x_{21} + \beta_2 x_{22} + \cdots + \beta_q x_{2q} + \varepsilon_2 \\
&\vdots \\
y_N &= \beta_1 x_{N1} + \beta_2 x_{N2} + \cdots + \beta_q x_{Nq} + \varepsilon_N
\end{aligned}
\tag{14.8}
$$

where ε_1, ε_2, ..., ε_N denote random errors in the linear regression model (Rencher, 1995). Then the linear predictor $\beta = [\beta_1, \beta_2, ..., \beta_q]$ can be obtained from all the training images.

In the testing phase, we use the q IQMs to compute y_i to determine whether it is a stego-image. If y_i is positive, the test image is a stego-image; otherwise, it is a cover-image.

We first train the IQM–SDSS in order to obtain the linear predictor, that is, $\beta = [\beta_1, \beta_2, ..., \beta_q]$, from our database. Then, we use β to generate the stego-image by our GA-based algorithm, so that it can pass through the inspection of IQM–SDSS.

Note that the GA procedure is not used in the training phase. The Analysis(ξ, C) to the IQM–SDSS is defined as

$$\text{Analysis}(\xi, C) = \beta_1 x_1 + \beta_2 x_2 + \cdots + \beta_q x_q \tag{14.9}$$

The Algorithm for Generating a Stego-image on IQM–SDSS:

1. Divide a cover-image into a set of cover-images of size 8×8.
2. Adjust the pixel values in each 8×8 cover-image based on GA to embed messages into the frequency domain and ensure that the stego-image can pass through the inspection of the IQM–SDSS. The procedures for generating the 8×8 stego-image are presented next after this algorithm.
3. Combine all the 8×8 embedded-images together to form a completely embedded image.
4. Test the embedded-image on IQM–SDSS. If it passes, it is the desired stego-image; otherwise, repeat steps 2–4.

The Procedure for Generating an 8×8 Embedded-image on IQM–SDSS:

1. Define the fitness function, the number of genes, the size of population, the crossover rate, the critical value, and the mutation rate.
2. Generate the first generation by a random selection.
3. Generate an 8×8 embedded-image based on each chromosome.
4. Evaluate the fitness value for each chromosome by analyzing the 8×8 embedded-image.
5. Obtain the better chromosome based on the fitness value.
6. Recombine new chromosomes by crossover.
7. Recombine new chromosomes by mutation.
8. Repeat steps 3–8 until a predefined condition is satisfied or a constant number of iterations are reached.

14.4   GA-BASED BREAKING ALGORITHM ON FDSS

Fridrich et al. (2003) presented a steganalytic system for detecting JPEG stego-images based on the assumption that the histogram distributions of some specific AC DCT coefficients of a cover-image and its cropped image should be similar. Note that, in the DCT coefficients, only the zero frequency (0,0) is the DC component, and the remaining frequencies are the AC components. Let $h_{kl}(d)$ and $\bar{h}_{kl}(d)$, respectively, denote the total number of AC DCT coefficients in the 8×8 cover and its corresponding 8×8 cropped images with the absolute value equal to d at location (k, l), where $0 \le k, l \le 7$. Note that, the 8×8 cropped images, defined in Fridrich et al. (2003), are obtained using the same way as the 8×8 cover images with a horizontal shift by four pixels.

The probability, ρ_{kl}, of the modification of a nonzero AC coefficient at (k, l) can be obtained by

$$\rho_{kl} = \frac{\overline{h}_{kl}(1)[h_{kl}(0) - \overline{h}_{kl}(0)] + [h_{kl}(1) - \overline{h}_{kl}(1)][\overline{h}_{kl}(2) - \overline{h}_{kl}(1)]}{[\overline{h}_{kl}(1)]^2 + [\overline{h}_{kl}(2) - \overline{h}_{kl}(1)]^2} \qquad (14.10)$$

Note that, the final value of the parameter ρ is calculated as an average over the selected low-frequency DCT coefficients $(k, l) \in \{(1, 2),(2, 1),(2, 2)\}$, and only 0, 1, and 2 are considered when checking the coefficient values of the specific frequencies between a cover and its corresponding cropped image.

Our GA-based breaking algorithm on the JFDSS is intended to minimize the differences between the two histograms of a stego-image and its cropped-image. It is presented below.

Algorithm for Generating a Stego-image on JFDSS:

1. Compress a cover-image by JPEG and divide it into a set of small cover-images of size 8×8. Each is performed by DCT.
2. Embed the messages into the specific DCT coefficients and decompress the embedded image by IDCT.
3. We select a 12×8 working window and generate an 8×8 cropped-image for each 8×8 embedded-image.
4. Determine the overlapping area between each 8×8 embedded-image and its cropped-image.
5. Adjust the overlapping pixel values by making that the coefficients of some specific frequencies (k, l) of the stego-image and its cropped-image are as identical as possible and the embedded messages are not altered.
6. Repeat steps 3–6 until all the 8×8 embedded-images are generated.

Let $\text{Coef}^{\text{Stego}}$ and $\text{Coef}^{\text{Crop}}$ denote the coefficients of each 8×8 stego-image and its cropped-image. The Analysis(ξ, C) to the JFDSS is defined as

$$\text{Analysis}(\xi, C) = \frac{1}{|C|} \sum_{i=0}^{\text{all pixels}} (\text{Coef}_i^{\text{Stego}} \otimes \text{Coef}_i^{\text{Crop}}) \qquad (14.11)$$

where $\otimes$ denotes the operator defined by

$$
\begin{cases}
\text{Coef}_i^{\text{Stego}} \otimes \text{Coef}_i^{\text{Crop}} = 1 & \text{if } (\text{Coef}_i^{\text{Stego}} = 0 \text{ and } \text{Coef}_i^{\text{Crop}} \neq 0) \text{ or} \\
& \quad (\text{Coef}_i^{\text{Stego}} = 1 \text{ and } \text{Coef}_i^{\text{Crop}} \neq 1) \text{ or} \\
& \quad (\text{Coef}_i^{\text{Stego}} = 2 \text{ and } \text{Coef}_i^{\text{Crop}} \neq 2) \text{ or} \\
& \quad (\text{Coef}_i^{\text{Stego}} \neq 0,1,2 \text{ and } \text{Coef}_i^{\text{Crop}} = 0,1,2) \\
\text{Coef}_i^{\text{Stego}} \otimes \text{Coef}_i^{\text{Crop}} = 0 & \text{otherwise}
\end{cases}
\qquad (14.12)
$$

Note that, 0, 1, and 2 denote the values in the specific frequencies obtained by dividing the quantization table. We only consider the values of the desired frequencies

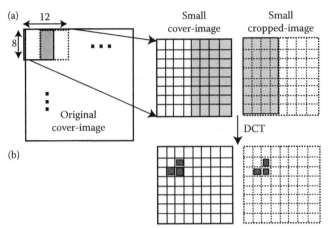

The frequency domain of cover-image and cropped-images

FIGURE 14.2 An example of our GA-based algorithm on the JFDSS. (a) A 128 working window; (b) the desired locations.

to be 0, 1, 2, or some values in Equation 14.12 because of the strategy of JFDSS in Equation 14.10.

Figure 14.2 shows an example of our GA-based algorithm on the JFDSS. In Figure 14.2a, we select a 12×8 working window for each 8×8 stego-image, and generate its 8×8 cropped-image. Note that, the shaded pixels indicate their overlapping area, and the three black boxes in Figure 14.2b are the desired locations.

14.5 EXPERIMENTAL RESULTS

In this section, we provide experimental results to show that our GA-based steganographic system can successfully break the inspection of steganalytic systems. For testing our algorithm, we use a database of 200 gray-scale images of size 256×256. All the images were originally stored in the BMP format.

14.5.1 GA-Based Breaking Algorithm on VSS

We test our algorithm on VSS. Figure 14.3a and f shows a stego-image and a message-image of sizes 256×256 and 64×64, respectively. We embed four bits into the 8×8 DCT coefficients on frequencies (0,2), (1,1), (2,0), and (3,0) for avoiding distortion. Note that, the stego-image in Figure 14.3a is generated by embedding Figure 14.3f into the DCT coefficients of the cover-image using our GA-based algorithm. In Figure 14.3b–j, respectively, display the bit-planes from 7 to 0. Figure 14.4 shows the stego-image and its visual filtered result. It is difficult to determine that Figure 14.4a is a stego-image.

Figure 14.5 shows the relationship of the average iteration for adjusting an 8×8 cover-image versus the correct rate of the visual filter. The correct rate is the percentage of similarity between the transformed results of the cover- and the stego-images using the visual filter. Note that, the BERs in Figure 14.5 are all 0%.

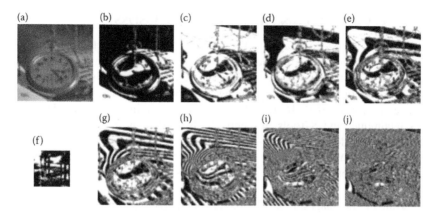

FIGURE 14.3 A stego-image generated by our GA-based algorithm and its 8 bit-planes. (a) The embedding of (f) into the DCT coefficients of the cover-image using the proposed GA-based algorithm, (b)–(e) and (g)–(j), respectively, display the bit-planes from 7 to 0.

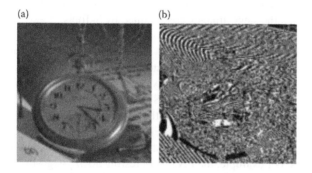

FIGURE 14.4 (a) A stego-image and (b) its visual filtered result.

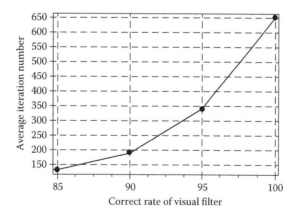

FIGURE 14.5 The relationship of the average iteration versus the correct rate of the visual filter.

14.5.2   GA-Based Breaking Algorithm on IQM–SDSS

We generate three stego-images as the training samples for each cover-image by Photoshop plug-in Digimarc (PictureMarc, 2006), Cox's technique (Cox et al., 1997), and S-Tools (Brown, 1994). Therefore, there are totally 800 images of 256×256 size, including 200 cover-images and 600 stego-images. The embedded message sizes were 1/10, 1/24, and 1/40 of the cover image size for Digimarc, Cox's technique, and S-Tools, respectively. Note that the IQM–SDSS (Avcibas et al., 2003) can detect the stego-images containing the message size of 1/100 of the cover-image. We develop the following four training strategies to obtain the linear predictors as:

1. Train all the images in our database to obtain the linear predictor β^A.
2. Train 100 cover-images and 100 stego-images to obtain the linear predictor β^B, in which the stego-images are obtained by Cox's technique.
3. Train 100 cover-images and 100 stego-images to obtain the linear predictor β^C, in which the stego-images are obtained by Photoshop plug-in Digimarc.
4. Train 100 cover-images and 100 stego-images to obtain the linear predictor β^D, in which the stego-images are obtained by S-Tools.

In the testing phase, we generate 50 stego-images for each linear predictor. Therefore, we will obtain four types of stego-images SI_A, SI_B, SI_C, and SI_D corresponding to β^A, β^B, β^C, and β^D, respectively. It is obvious that all the SI_A images will pass through the inspection of IQM–SDSS with the linear predictor β^A, but may fail with other linear predictors, similarly for the SI_B, SI_C, and SI_D.

Table 14.1 shows the false-negative (FN; Type I Error) and false-positive (FP; Type II Error) rates of testing each type of stego- and cover-images under different linear predictors. For example, the FN rate of testing 50 SI_A stego-images under β^B is 94% indicating that the β^B-IQMSS falsely decides that 94% of our stego-images belong to the cover-image. On the other hand, the FP rate of testing 50 cover-images under β^B is 10% indicating that the β^B-IQMSS decides that 10% of cover-images belong to the stego-images. Note that the cover-images used in the testing phase are different to those used in the training phase.

TABLE 14.1
The Experimental Results of GA-Based Algorithm
on IQM–SDSS

		β^A (%)	β^B (%)	β^C (%)	β^D (%)	Average (%)
50 SI_A images	FN		94	94	100	96
50 SI_B images	FN	84		84	80	83
50 SI_C images	FN	82	86		88	85
50 SI_D images	FN	82	86	82		83
50 cover images	FP	6	10	16	16	12

14.5.3 GA-BASED BREAKING ALGORITHM ON JFDSS

Figure 14.6 shows an example of adjusting an 8×8 embedded-image to obtain an 8×8 stego-image for breaking the JFDSS. Figure 14.6a shows a 12×8 working window, where the enclosed is the overlapping area. Figures 14.6b and c show the original 8×8 embedded- and cropped-images, respectively. We embed "1" on (1,2), (2,1), and (2,2) by compressing Figure 14.6b using JPEG under 70% compression quality to obtain Figure 14.6d. Note that, the top-left pixel is (0,0) and the messages can be embedded into any frequency of the transformed domain, so that the embedding capacity could be sufficiently high. Due to this, the JFDSS only checks frequencies (1,2), (2,1), and (2,2), we show an example of embedding three bits into these three frequencies. Similarly, Figure 14.6e is obtained by compressing Figure 14.6c using JPEG under the same compression quality. By evaluating the frequencies (1,2), (2,1), and (2,2), the JFDSS can determine whether the embedded-image is a stego-image. Therefore, in order to break the JFDSS, we obtain the stego-image as in Figure 14.6f. Figures 14.6g and h respectively show the new embedded- and cropped-images. Similarly, we obtain Figures 14.6i and j by compressing Figures 14.6g and h, respectively, using JPEG under 70% compression quality. Therefore, the JFDSS cannot distinguish from the frequencies (1,2), (2,1), and (2,2).

Let $Q\text{Table}(i, j)$ denote the standard quantization table, where $0 \le i, j \le 7$. The new quantization table, $\text{NewTable}(i, j)$, with $x\%$ compression quality can be obtained by

$$\text{NewTable}(i, j) = \frac{Q\text{Table}(i, j) \times \text{factor} + 50}{100} \qquad (14.13)$$

(a)

169	166	162	158	157	158	161	163	157	158	161	163
166	164	162	159	159	159	161	162	159	159	161	162
162	162	161	161	161	161	161	161	161	161	161	161
158	159	161	162	162	162	160	160	162	162	160	160
157	159	161	163	163	161	160	158	163	161	160	158
158	159	161	162	162	160	158	157	162	160	158	157
160	161	161	160	160	158	157	156	160	158	157	156
162	162	161	159	158	157	156	156	158	157	156	156

(f)

169	166	162	158	164	163	164	159	157	158	161	163
166	164	162	159	162	162	162	161	159	159	161	162
162	162	161	161	161	162	161	161	161	161	161	161
158	159	161	162	160	162	161	161	162	162	160	160
157	159	161	163	160	158	159	160	163	161	160	158
158	159	161	162	160	160	157	159	162	160	158	157
160	161	161	160	161	159	161	158	160	158	157	156
162	162	161	159	161	158	159	159	158	157	156	156

(b) / (d)

169	166	162	158	157	158	161	163		127	1	0	0	0	0	0	0
166	164	162	159	159	159	161	162		1	0	1	0	0	0	0	0
162	162	161	161	161	161	161	161		0	1	1	0	0	0	0	0
158	159	161	162	162	162	160	160		0	0	0	0	0	0	0	0
157	159	161	163	163	161	160	158		0	0	0	0	0	0	0	0
158	159	161	162	162	160	158	157		0	0	0	0	0	0	0	0
160	161	161	160	160	158	157	156		0	0	0	0	0	0	0	0
162	162	161	159	158	157	156	156		0	0	0	0	0	0	0	0

(g) / (i)

169	166	162	158	164	163	164	159		127	1	0	0	0	0	0	0
166	164	162	159	162	162	162	161		1	0	1	0	0	0	0	0
162	162	161	161	161	162	161	161		1	1	1	0	0	0	0	0
158	159	161	162	160	162	161	161		0	0	0	0	0	0	0	0
157	159	161	163	160	158	159	160		0	0	0	0	0	0	0	0
158	159	161	162	160	160	157	159		0	0	0	0	0	0	0	0
160	161	161	160	161	159	161	158		0	0	0	0	0	0	0	0
162	162	161	159	161	158	159	159		0	0	0	0	0	0	0	0

(c) / (e)

157	158	161	163	157	158	161	163		126	0	0	0	0	0	0	0
159	159	161	162	159	159	161	162		1	-1	0	-1	0	0	0	0
161	161	161	161	161	161	161	161		-1	0	0	0	0	0	0	0
162	162	160	160	162	162	160	160		0	0	0	0	0	0	0	0
163	161	160	158	163	161	160	158		0	0	0	0	0	0	0	0
162	160	158	157	162	160	158	157		0	0	0	0	0	0	0	0
160	158	157	156	160	158	157	156		0	0	0	0	0	0	0	0
158	157	156	156	158	157	156	156		0	0	0	0	0	0	0	0

(h) / (j)

164	163	164	159	157	158	161	163		127	1	0	0	0	0	0	0
162	162	162	161	159	159	161	162		1	0	1	0	0	0	0	0
161	162	161	161	161	161	161	161		1	1	1	0	0	0	0	0
160	162	161	161	162	162	160	160		0	0	0	0	0	0	0	0
160	158	159	160	163	161	160	158		0	0	0	0	0	0	0	0
160	160	157	159	162	160	158	157		0	0	0	0	0	0	0	0
161	159	161	158	160	158	157	156		0	0	0	0	0	0	0	0
161	161	158	159	158	157	156	156		0	0	0	0	0	0	0	0

FIGURE 14.6 An example of adjusting an 8×8 embedded-image.

(a)

16	11	10	16	24	40	51	61
12	12	14	19	26	58	60	55
14	13	16	24	40	57	69	56
14	17	22	29	51	87	80	62
18	22	37	56	68	109	103	77
24	35	55	64	81	104	113	92
49	64	78	87	103	121	120	101
72	92	95	98	112	100	103	99

(b)

10.1	7.1	6.5	10.1	14.9	24.5	31.1	37.1
7.7	7.7	8.9	11.9	16.1	35.3	36.5	33.5
8.9	8.3	10.1	14.9	24.5	34.7	41.9	34.1
8.9	10.7	13.7	17.9	31.1	52.7	48.5	37.7
11.3	13.7	22.7	34.1	41.3	65.9	62.3	46.7
14.9	21.5	33.5	38.9	49.1	62.9	68.3	55.7
29.9	38.9	47.3	52.7	62.3	73.1	72.5	61.1
43.7	55.7	57.5	59.3	67.7	60.5	62.3	59.9

FIGURE 14.7 (a) The standard quantization table and (b) the quantization table using JPEG under 70% compression quality.

where the factor is determined by

$$
\begin{cases}
\text{factor} = \dfrac{5000}{x} & \text{if } x \le 50 \\
\text{factor} = 200 - 2x & \text{otherwise}
\end{cases}
\tag{14.14}
$$

Figures 14.7a and b show the quantization tables of the standard and 70% compression quality, respectively.

14.6 COMPLEXITY ANALYSIS

In general, the complexity of our GA-based algorithm is related to the size of the embedded message and the position of embedding (Shih and Wu, 2003). Figure 14.8 shows the relationship between the embedded message and the required iterations in which the cover-image is of size 8×8 and the message is embedded into the LSB of

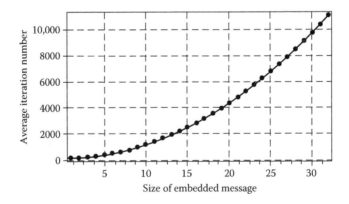

FIGURE 14.8 The relationship between the size of the embedded message and the required iterations.

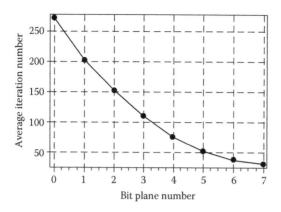

FIGURE 14.9   Embedding a 4-bit message into the different bit-plane.

the cover-image by the zigzag order starting from the DC (zero-frequency) component. We observe that the more messages embedded in a stego-image, the more iterations are required in our GA-based algorithm. Figure 14.9 shows an example when we embed a 4-bit message into the different bit-plane of DCT coefficients. We observe that the lower bit-plane used for embedding, the more iterations are required in our GA-based algorithm.

14.7   CONCLUSIONS

In this chapter, a new GA-based algorithm of generating a stego-image is used to break the detection of the SDSS and the FDSS by artificially counterfeiting statistical features. We design a fitness function to evaluate the quality of each chromosome in order to generate the stego-image that can pass through the inspection of steganalytic systems. Experimental results show that our GA-based algorithm can not only successfully break the detection of the current steganalytic systems, but also correctly embed messages into the frequency domain of a cover-image. Moreover, our GA-based algorithm can be used to enhance the quality of stego-images.

REFERENCES

Avcibas, I., Memon, N., and Sankur, B., Steganalysis using image quality metrics, *IEEE Trans Image Process*, 12(2), 221–229, 2003.

Avcibas, I. and Sankur, B., Statistical analysis of image quality measures, *J Electron Imag*, 11, 206–223, 2002.

Brown, A., *S-Tools for Windows, Shareware 1994*, ftp://idea.sec.dsi.unimi.it/pub/security/crypt/code/s-tools4.zip.

Chu, R., You, X., Kong, X., and Ba, X., A DCT-based image steganographic method resisting statistical attacks, *Int Conf on Acoustics, Speech, and Signal Processing*, Montreal, Quebec, Canada, May 2004.

Cox, I. J., Bloom, J., Miller, M., and Cox, I., *Digital Watermarking: Principles & Practice*, San Francisco: Morgan Kaufmann, 2001.

Cox, I. J., Kilian, J., Leighton, T., and Shamoon, T., Secure spread spectrum watermarking for multimedia, *IEEE Trans Image Process*, 6, 1673–1686, 1997.

Farid, H., Detecting steganographic message in digital images, Technical Report, TR2001–412, Computer Science, Dartmouth College, 2001.

Fridrich, J., Goljan, M., and Hogea, D., New methodology for breaking steganographic techniques for JPEGs, *Proc EI SPIE*, Santa Clara, CA, pp. 143–155, Jan. 2003.

Holland, J. H., *Adaptation in Natural and Artificial Systems*, Ann Arbor, Michigan: The University of Michigan Press, 1975.

Kahn, D., *The Codebreakers*, 2nd edition, New York: Macmillan, 1996.

Norman, B., *Secret Warfare*, Washington, DC: Acropolis Books, 1973.

PictureMarc, Embed Watermark, v 1.00.45, Digimarc Corporation, 2006.

Rencher, A. C., *Methods of Multivariate Analysis*, New York: John Wiley, 1995, Ch. 6, 10.

Shih, F. Y. and Wu, Y.-T., Combinational image watermarking in the spatial and frequency domains, *Pattern Recognit*, 36(4), 969–975, 2003.

Shih, F. Y. and Wu, Y.-T., Enhancement of image watermark retrieval based on genetic algorithm, *J Visual Commun Image Repre*, 16, 115–133, 2005.

Simmons, G. J., Prisoners' problem and the subliminal channel, *Proc Int Conf on Advances in Cryptology*, 51–67, 1984.

Westfeld, A. and Pfitzmann, A., Attacks on steganographic systems breaking the steganographic utilities EzStego, Jsteg, Steganos, and S-Tools and some lessons learned, *Proc Third Int Workshop on Information Hiding*, Dresden, Germany, pp. 61–76, Sep. 1999.

Wu, Y. and Shih, F. Y., An adjusted-purpose digital watermarking technique, *Pattern Recognit*, 37(12), 2349–2359, 2004.

Wu, Y. and Shih, F. Y., Genetic algorithm based methodology for breaking the steganalytic systems, *IEEE Trans Syst, Man Cybern*, 36(1), 24–29, 2006.

Part IV

Forensics

15 Image Inpainting Using an Enhanced Exemplar-Based Algorithm

I-Cheng Chang and Chia-We Hsu

CONTENTS

15.1 INTRODUCTION

The growth of digital images has led to an increasing need for image editing techniques. Image inpainting is one such authoring technique for repairing or adjusting on-hand images. Image inpainting focuses on automatically repairing the missed regions of an image in a visually plausible way. Original image inpainting algorithms have been presented to repair small missed or damaged areas (Bertalmio et al., 2000; Ballester et al., 2001; Telea, 2004), such as speckles, scratches, and text in image. The texture synpaper method (Efros and Leung, 1999; Drori et al., 2003.) can effectively repair the damaged image, but is more suitable for repairing the image single or simple texture. Since texture synpaper is based on a known texture sample, it is inappropriate for repairing images involving complex and unknown textures.

The most important issue for image inpainting of large missing area is to determine the structure information of these areas, and repair them seamlessly from existing surrounding information. Previous studies address this issue by solving two problems: analyzing the structural information from the remainder image (e.g., the contour and shape) and keeping the texture information in order to reduce the visual artifact of repaired image. Ballester et al. (2001) and Bertalmio et al. (2001) adopted a method that utilizes partial differential equations to propagate the structure of the missing area, but which produces a vague repaired result if the missed area is large. Jia and Tang (2003) addressed color and texture segmentation, applying the tensor voting algorithm to segment the image for propagating the structure of the missed area. The resulting structure is used to identify the searching range and the filling order. Since two parameters control the smoothness of the propagation curve, the algorithm can repair not only linear structures but also nonlinear structures in a missed area. However, it takes a long time for color segmentation and structure propagation. To reduce the time taken, the missed area can be filled manually according to the structure drawn. Wang et al. (2005) and Rares et al. (2005) presented similar algorithms to Jia and Tang (2003). Wang et al. (2005) adopted the constrained texture synthesis (Wang et al., 2004) to analyze the texture distribution, and then defined the searching range for the filling image. Due to the visual artifacts in the repaired image, an algorithm was presented to detect the artifacts and refill these regions again. Rares et al. (2005) proposed an algorithm that first extracts the edges or corners of an image, then defines an order for edge permutation. However, the algorithm is only suitable for some structure distribution, and produces a blur effect in the repaired image.

To repair an image effectively without apparent artifacts, Criminisi et al. (2003, 2004) adopted the magnitude of the gradient and undamaged area of an image to define the filling order and searching direction for finding the best patches. The advantage of this algorithm is that it is very suitable for repairing linear structures in the missed area, and it smoothly connects the repaired region with the nearby undamaged region. However, the filling order is random and unreliable. Hence, further works (Cheng et al., 2005; Li et al., 2005; Nie et al., 2006; Shao et al., 2006) have been conducted to improve the result by redefining the filling order. In addition to the structure information, texture information is also an important factor for repairing images. Some studies focus on analyzing texture information (Bertalmio et al., 2003; Shao et al., 2006). Most related studies use an exemplar-based method to keep the texture information of an image. Shao et al. (2006) presented an algorithm to repair images based on the Poisson equation. The algorithm decomposes an image into structure and texture images, and assigned the weights for both images to compose a patch to fill in the missing region. Their method successfully keeps the texture information, but may produce blurred artifacts.

In this chapter, we will introduce a robust image inpainting algorithm to determine a new filling order based on the structure priority value. The algorithm can fill the missing area with more accurate structure information. Besides, the dynamic searching range is utilized to improve the efficiency of finding the best patches in the source region. The rest of this chapter is organized as follows. Section 15.2 describes the concept of image inpainting and the related problems. Section 15.3 presents the details of the proposed algorithm. Section 15.4 presents the experimental results and discussions. Finally, conclusions are drawn in Section 15.5.

15.2 IMAGE INPAINTING

The difficulty of repairing a large missing area is how to define the filling order and keep the structure information and texture information from the surrounding of the area. Criminisi et al. (2004) developed a very effective and popular algorithm that provides a good solution for this problem. However, the algorithm has some problems. This section reviews the algorithm and discusses the associated problems.

15.2.1 REVIEW OF CRIMINISI'S ALGORITHM

Criminisi's algorithm is an isophote-driven image-sampling process. The filling order in the repairing process is determined by the gradient information of an exemplar. Figure 15.1 illustrates a single iteration in filling the target region centered at point p. An image I is divided into disjoint regions, the target region Ω and the source region Φ. The target region is the selection area that needs to be filled in, and the

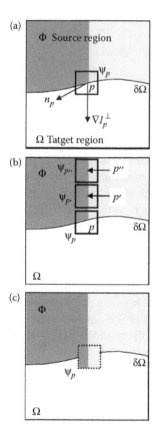

FIGURE 15.1 Illustration of a single iteration of Criminisi's algorithm in filling the target region centered at point p. (a) Initial status. (b) Searching the best matching patch. (c) The result of iteration 1.

corresponding contour is denoted as $\delta \Omega$. An image excluding the target region and the contour of Ω is defined as the source region, that is, $\Phi = I - \Omega$. Ψ_p denotes a square patch centered at a point p, where p is located at $\delta \Omega$.

A key part of the inpainting process is to determine the filling order for the target region. The priority value of each square patch $\Psi_p \in \Omega$ is adopted to find the filling order for all unfilled pixels. The priority value $P(p)$ is defined as the product of two terms:

$$P(p) = C(p) \cdot D(p) \tag{15.1}$$

where $C(p)$ and $D(p)$ are called confidence term and data term, respectively. These terms are defined as follows:

$$C(p) = \frac{\sum_{q \in \Psi_p \cap (I-\Omega)} C(q)}{\left|\Psi_p\right|} \tag{15.2}$$

$$D(p) = \frac{\left|\nabla I_p^{\perp} \cdot \vec{n}_p\right|}{\alpha} \tag{15.3}$$

where $\left|\Psi_p\right|$ denotes the area of Ψ_p and α is a normalized parameter. For gray-level image, α is assigned to 255. In the initial step, the value of $C(p)$ is set to be 0 if $p \in \Omega$ and 1 if $p \in \Phi$. In Equation 15.3, n_p is the normal vector of the $\delta \Omega$ at point p, ∇ is denoted as a gradient operator, and $\perp$ is an isophote (direction and intensity) orthogonal to the gradient. When all priority values of the patches along $\delta \Omega$ have been computed, the patch with the maximum priority value would be selected as the first one to be filled. If $\Psi_{\hat{p}}$ is with the maximum priority value after the filling order determination, the next step is to search for a patch around the source region that is most similar to $\Psi_{\hat{p}}$ according to the following equation:

$$\Psi_{\hat{q}} = \arg\min_{\Psi_q \in \Phi} d(\Psi_{\hat{p}}, \Psi_q) \tag{15.4}$$

where the distance $d(\Psi_{\hat{p}}, \Psi_q)$ between two patches $\Psi_{\hat{p}}$ and Ψ_q is defined as the sum of squared differences (SSD) of the already repaired pixels in these two patches. After a patch has been filled, the confidence $C(p)$ is updated as follows:

$$C(p) = \gamma C(\hat{p}) \quad \forall p \in \Psi_{\hat{p}} \cap \Omega \tag{15.5}$$

In general, $\gamma < 1$, and the new $C(p)$ inherits a certain portion of the value corresponding to the last $C(p)$. Additionally, to reduce the time of finding best match patch, the search path will follow the isophote of point p. The filling process proceeds until all pixels in the target region have been filled.

15.2.2 PROBLEMS WITH IMAGE INPAINTING ALGORITHM

The priority $P(p)$ contains two parameters, the data term $D(p)$ and the confidence term $C(p)$. The data term contains the gradient information and stores the structure part when the missed area is being filled. If the data term is a large value, then the linear structures are synthesized first. However, if the patch contains too few pixels of the source image, then the corresponding data term is not reliable. To avoid this mistake, the confidence term is added as a measurement of the reliable information around the pixel p. Using the two factors, the algorithm tries to retain not only the structure information of an image but also the texture information. The method indeed achieves good performance, but it still suffers from three problems.

1. The structure information may be broken by inappropriate values of $C(p)$ and $D(p)$. Assume there are two patches, Ψ_p and Ψ_q, and the algorithm needs to determine the first one to be filled. If the vectors of $\bar{n}_p$ and $\bar{n}_q$ are the same, for example, $(0,1)$, and the patch size is $n \times n$, then the gradient and the confidence term of Ψ_p and Ψ_q are $\nabla \bar{I}_p = (x_p, y_p)$, $\nabla \bar{I}_q = (x_q, y_q)$, $C(p) = s_p / n \times n$ and $C(q) = s_q / n \times n$. Hence, the isophotes of Ψ_p and Ψ_q are $\nabla \bar{I}_p^{\perp} = (y_p, -x_p)$ and $\nabla \bar{I}_q^{\perp} = (y_q, -x_q)$, respectively. Assume that $(x_p, y_p) = (4.5, 7)$, $(x_q, y_q) = (8, 7)$, $s_p = 70$, $s_q = 40$, and $n = 9$. Then the priority value of Ψ_p is larger than that of Ψ_q, and so the patch Ψ_p will be filled first. However, in the case, the data value of Ψ_q is larger than that of Ψ_p. Experimental results indicate that once the confidence values are above a particular value, the patch with the larger data value should be chosen, even if it has the smaller priority value; otherwise, the filling process loses the structure information of the image. Thus, the confidence and data values should have different weighting factors to generate the priority value, but adequate weighting factors are hard to obtain.

2. The updating rule of the confidence term, that is, Equation 15.5, makes the priority value to decay to zero, and causes the patch selection to be random. If the inherited weight α is set between 0 and 1, then the confidence value decays as the iteration number of filling process increases. The value of $C(p)$ approaches zero after a series of filling operations.

$$C^n(p) = \alpha^n \times C(\hat{p}) \quad \text{and} \quad \lim_{n \to \infty} C^n(p) = 0 \tag{15.6}$$

Here, n denotes the iteration number of filling operation, and $\alpha \in (0,1)$. Once the confidence value drops below a small value, the priority value is dominated by the confidence value, and also drops to a small value, regardless of the data value of a patch. Cheng et al. (2005) also observed the same problem, which causes the filling process to fill the target region randomly. Therefore, the inpainting result does not properly represent the structure information. The dropping effect causes the random filling process to destroy the structure information. Figure 15.2a shows the distribution of the priority value during the filling process of an image. The priority is very small after a number of iterations due to the decay of the confidence value (Figure 15.2b). Figure 15.2c shows the corresponding data value distribution.

The above-mentioned problem still exists if the value of the inherited weight is increased to delay the confidence value dropping to zero, as long as the target region is large enough. An excessive inherited weight also forces the filling direction along some specific path, on the other hand, while a smaller weight enhances the impact of the dropping effect. Selecting an appropriate inherited weight is difficult, while an unsuitable weight would not retain the structure information effectively.

3. The searching range defined by the isophote is not so reliable. Criminisi defined the isophote as the searching hint to determine the best matched patch. The method works well if the texture around the patch is simple; however, the related gradient cannot be calculated accurately when a patch contains two or more materials or textures. The corresponding isophote

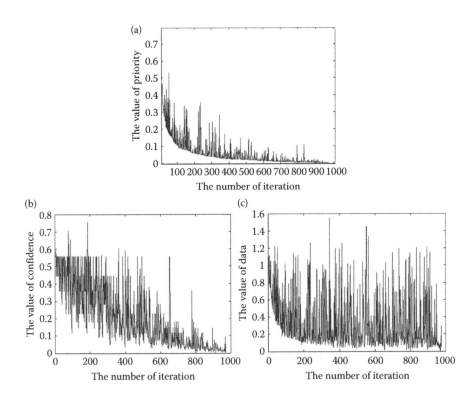

FIGURE 15.2 Illustration of the filling processing of an image. (a) The distribution of the priority value. (b) The distribution of the confidence value. (c) The distribution of the data value.

then induces the wrong searching direction, and probably finds an unsuitable patch.

15.3 THE PROPOSED IMAGE INPAINTING ALGORITHM

The inpainting algorithm concerns two major issues: (1) appropriate filling order and (2) adequate searching range. The proposed inpainting algorithm defines the new parameter, structure priority value, which can efficiently keep the structure information of an image. Besides, it adopts the example-based method and dynamic searching range to maintain the texture information. Different from the algorithms of Criminisi et al. (2004) and Cheng et al. (2005), we do not need to adjust the parameters for the damaged images with a different scene, for example, the weights of the confidence and data terms, and can still keep the structure information effectively.

15.3.1 Structure Priority Value

The notation adopted here is similar to that defined in Section 15.2.1. Figure 15.3 illustrates the meanings of the notation. If some region is removed, the removed region is denoted as target region Ω; the contour of the target region is represented as $\delta\Omega$ and the area excluding the target region is marked as source region ($\Phi = I - \Omega$).

The study adopts the edge ratio instead of the data value used in Criminisi's algorithm, to preserve the structure information. The new term is denoted as the *structure priority (SP) value*, and is defined as follows:

$$SP(p) = C(p) \times ER(p) \tag{15.7}$$

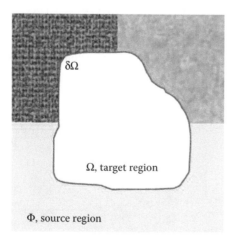

FIGURE 15.3 Notation diagram of our algorithm. The symbol Ω is the target region and its corresponding contour is denoted as $\delta\Omega$.

where $C(p)$ is the confidence value that determines the reliability of the corresponding patch, and ER(p) denotes the edge value. The term ER(p) is defined as the ratio of edge points relative to the source region inside the patch, and the equation is as follows:

$$\text{ER}(p) = \frac{\sum_{q \in \Psi_p \cap \Phi} e(q)}{\sum_{q \in \Psi_p \cap \Phi} C(q)} \tag{15.8}$$

$$e(q) = \begin{cases} 1 & \text{if } q \in \text{edge} \\ 0 & \text{otherwise} \end{cases} \tag{15.9}$$

where $e(q)$ denotes whether pixel p is an edge point. A higher SP indicates a higher priority for filling the corresponding patch. This definition of filling order can avoid the disadvantage of the previous algorithms (Criminisi et al., 2004; Li et al., 2005; Nie et al., 2006). Although the priority function in (Cheng et al., 2005; Li et al., 2005; Nie et al., 2006; Shao et al., 2006) is modified by combining the confidence term and the data term, the confidence term still has a negative impact on the determination of the priority values. Moreover, the confidence value is computed at each filling iteration rather than the inherited rule (Equation 15.5), thus avoiding the second disadvantage in the previous chapter. For instance, in Figure 15.4a–c, the patch Ψ_p is filled in the first iteration, and the confidence value is then recalculated by Equation 15.2. If $C(p') < C(q)$, then Ψ_q is filled in the second iteration. However, it still cannot reduce the confidence term impact on the priority value. The influence of the confidence term on the priority value is still not lowered even if more structure information is kept.

The definition of the filling order in our algorithm does not majorly depend on the confidence term. In our experiments, we observed that the $C(p)$ of most selected patches are higher than 0.5. The dropping effect does not occur in the filling process. Figure 15.5 demonstrates one example of the filling process. Figure 15.5a is the original image, and 96% of the values of $C(p)$ are higher than 0.5 during repairing process (Figure 15.5b). Figure 15.5c is the corresponding ER distribution.

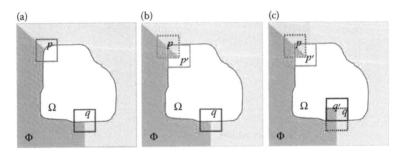

FIGURE 15.4 The filling process. (a) Initial status. (b) The result of iteration 1. (c) The result of iteration 2.

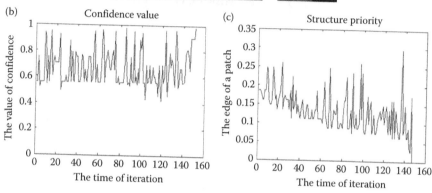

FIGURE 15.5 Illustration of the filling processing of an image. (a) Input image. (b) The distribution of the confidence value. (c) The distribution of the edge ratio.

Calculating the ER(p) of $\delta\Omega$ involves deriving the edge map of the corresponding image using the Canny edge detector. The benefit of the Canny edge detector is that it does not blur the edge in image like the Gaussian filter, and therefore produces accurate gradient information for the input image. Moreover, Canny edge can produce continuous edges, unlike some other detectors, for example, Sobel and Laplacian.

15.3.2 Dynamic Searching Range

Besides the filling order, the searching range is another important factor which influences the result of the repaired image. The determination of search range also affects the time taken to perform the inpainting process.

A straightforward way to search the best match is to regard the entire image as the searching range; however, this approach takes a long time, and an excessively large search range creates many errors. One fast and easy way to obtain the best matching patch is to define the minimum boundary rectangle (MBR) of the target region, and expand the area of MBR to obtain the searching range. Since the shapes of the target region may be different, for example, a mixture of convex or concave shapes, expanding MBR with a fixed width, does not work. Instead, the searching range is dynamically adjusted based on the shape of the target region, checking whether the ratio between the target region and the source region is beyond a predefined bound. The expansion process of the searching range is described as follows:

$$\text{Initial condition: } SR^0(I) = 0, \quad \text{and} \quad E\_MBR^0(I) = MBR(I) \qquad (15.10)$$

while

$$\left(\frac{\sum_{x,y \in \Phi} E\_MBR^t(x,y)}{\sum_{x,y \in \Omega} E\_MBR^t(x,y)} < \beta \right)$$

$$E\_MBR^{t+1} = \text{Expand}(E\_MBR^t(I))$$

$$SR^{t+1}(I) = E\_MBR^{t+1}(I) - MBR(I)$$

where $MBR(I)$ denotes the MBR of the target region in an image I, E_MBR^t is the expansion of MBR, and the initial state is set to $MBR(I)$. The function of Expand($\bullet$) is to expand E_MBR by one pixel width. E_MBR is expanded until the ratio of the pixels belonging to the target and source regions in the E_MBR is higher than β. A typical suitable value of β is 1.5. Figure 15.6a shows the searching range of a convex target region and Figure 15.6b illustrates that of a concave target region. The expansion of MBR for a concave shape is normally smaller than that for a convex shape, since the initial MBR already includes enough of the source region.

In addition to the reduction of the searching time compared to the full search, the proposed algorithm can also avoid the drawback of Criminisi's algorithm. Criminisi's algorithm only searches along the particular direction (isophote) of the image, but introduces an error when the patch contains complex textures. Additionally, the searching range needs to exclude the area filled in previous filling iterations, because

FIGURE 15.6 Dynamic searching range. (a) Target region of convex shape. (b) Target region of concave shape. The searching range is the area between the red and blue rectangles.

FIGURE 15.7 Reduce the artifact in the repaired image. (a) Original image: the car is set as the target region. (b) Inpainting result with artifact in the red rectangle: the searching range includes the previous filled area. (c) Inpainting result: the searching range excludes the previous filled area.

the patch found in the current iteration may have been filled patch in a previous iteration. Filling such a patch may result in appearance artifacts in the repaired image, as shown in Figure 15.7. If the car on the left side of Figure 15.7a is the target region, then Figure 15.7b shows the inpainting result when the searching range includes the filled area. The area marked by the red rectangle is the artifact, and the filled area refers to the same patch. However, the artifact disappears if the searching range is limited to the original source region (Figure 15.7c).

15.4 EXPERIMENTAL RESULTS

This section evaluates and discusses the proposed algorithm in three subsections. Section 15.4.1 demonstrates the robustness under the different shapes of target regions. Section 15.4.2 evaluates the performance of the process on natural images. The experimental results are examined to determine whether the filling results can keep the structure information. Section 15.4.3 compares the results between the proposed method and the texture synthesis method. Section 15.4.4 describes the processed results of the man-made images.

15.4.1 ANALYSIS OF THE INFLUENCE OF THE SHAPE OF TARGET REGION

To compare the performance of different image inpainting algorithms, Figure 15.8 illustrates several inpainting results where the jump man is removed. Figure 15.8a is the original image and Figure 15.8b shows the result of applying Criminisi's algo-

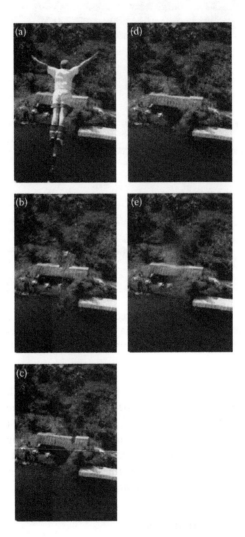

FIGURE 15.8 Example of repairing jump man. (a) Original image. (b) The result of Criminisi et al. (2004). (Adapted from Criminisi, A., Pérez, P., and Toyama, K., *IEEE Transactions on Image Processing*, 13(9), 1200–1212, 2004.) (c) The result of Wang et al. (2005) without artifact detection. (d) The result of Wang et al. (2005) with artifact detection. (Adapted from Wang, J.-F., Hsu, H.-J., and Liao, S.-C. *IEEE International Conference on Image Processing*, vol. 2, pp. 73–76, September 2005.) (e) The result of Shao et al. (2006). (Adapted from Shao, X., Liu, Z., and Li, H., *IEEE International Conference on Document Image Analysis for Libraries*, pp. 368–372, April 2006.)

rithm. Figure 15.8c shows the result of applying Wang et al. (2005) without artifact detection, where the eave structure contains a crack. Figure 15.8d illustrates the result of artifact detection and refilling. However, the eaves in Figure 15.8d are still sunken. Figure 15.8e shows the result of applying Shao et al. (2006).

Figure 15.9 shows the inpainting results for the jump man image, demonstrating that the proposed algorithm is robust to a change in the target region. The size and shape of the target region significantly affect the inpainting result since the

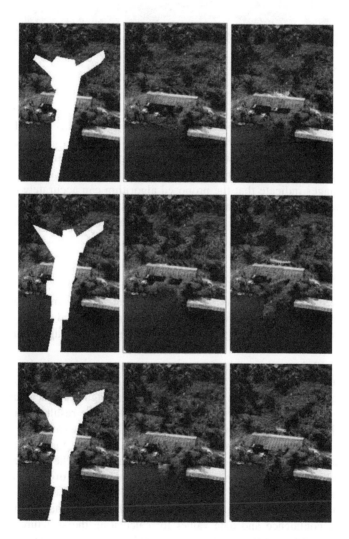

FIGURE 15.9 The repaired jump man images for different target regions. The left column shows four removed regions; the results of the proposed algorithm are shown in the middle column; and the right column presents the results using Criminisi et al. (2004) algorithm. (Adapted from Criminisi, A., Pérez, P., and Toyama, K., *IEEE Transactions on Image Processing*, 13(9), 1200–1212, 2004.)

FIGURE 15.9 Continued.

reference source image is different after the target region is changed. Figure 15.9 indicates that the proposed approach can maintain good quality even if the target region varies. The left column of Figure 15.9 shows the image for a different target region; the middle column shows the results of the proposed method; and the right column displays those of the Criminisi's algorithm for the same target region. The resulting images indicate that the proposed algorithm can accurately reconstruct the eaves with the natural scene. Although the result in the fourth row is not as good as that of the upper three, the eaves of the results of Criminisi et al. (2004) are distorted.

15.4.2 Inpainting Results of Natural Images

The section presents the results of several different natural images, and checks whether the distribution of the confidence value is almost beyond a level. The differences between the proposed approach and other inpainting algorithms are also described.

Figure 15.10 shows the result of repairing a tiny area, such as the leg of the dog. Figure 15.10a shows an image of the riverside and Figure 15.10b shows the target region, where the sign is removed from the picture. Figure 15.10c illustrates the result using Criminisi's algorithm and Figure 15.10d is the result of the proposed algorithm. The repaired leg of the dog in Figure 15.10d is more natural and smoother than that in Figure 15.10c. Figure 15.10e and f shows the corresponding distributions of the confidence value and the edge ratio. Notably, almost all the confidence values are greater than 0.5, and the confidence values do not have the dropping effect as in the previous method.

Cheng et al. (2005) adopts a good approach, which uses the linear combination to generate the confidence and data values for calculating the structure value. The method assigns different weights to these two values based on the structure or texture for each image, but requires manual trial and error to determine the appropriate weights. Figure 15.11b and c shows the results with different weighting policies. Figure 15.11d illustrates the result of our approach, where the inner and outer structure of the sand tower is more complete, and we do not need to be assigned the

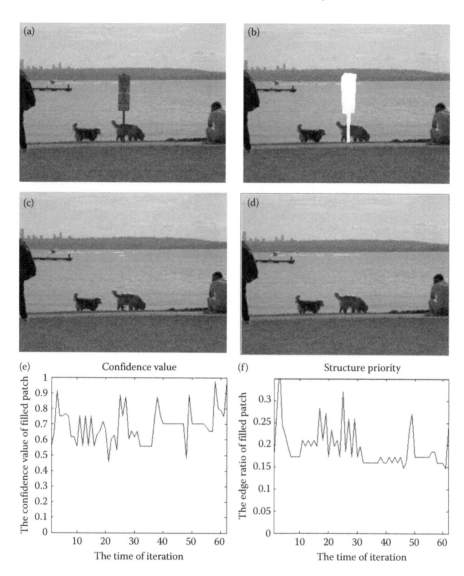

FIGURE 15.10 Repair the tiny area of an image. (a) Original image. (b) The target region is marked with white. (c) The result of Criminisi et al. (2004). (d) The result of the proposed algorithm. (e) The distribution of the confidence value. (f) The distribution of the edge ratio where the iteration number is 62. (Adapted from Criminisi, A., Pérez, P., and Toyama, K., *IEEE Transactions on Image Processing*, 13(9), 1200–1212, 2004.)

weights for each input image. Figure 15.11e shows the distribution of confidence values during the process, which remains above 0.5 during almost the entire iteration process. Figure 15.11f is the corresponding distribution of edge ratios.

The algorithm of Shao et al. (2006) applies the technique of texture analysis to fill the image, as indicated in Figures 15.12b and 15.13b. Although the proposed

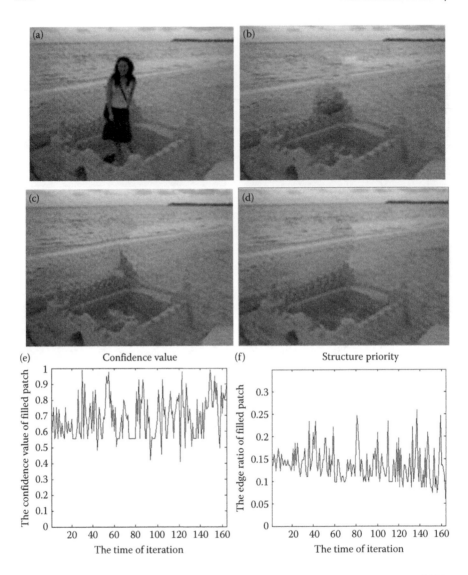

FIGURE 15.11 Comparison between the proposed algorithm and Cheng et al. (2005). (a) Original image. (b) and (c) Reconstruction with different weighting factors of Cheng et al. (2005). (d) Result of the proposed algorithm, where the texture and structure information of an image is successfully kept. (e) The distribution of the confidence value. (f) The distribution of the edge ratio where the iteration number is 165. (Adapted from Cheng, W.-H., et al., *IEEE International Conference on Computer Graphics, Imaging and Vision*, pp. 26–29, July, 2005.)

algorithm does not analyze the texture information before the inpainting process, the repaired texture of Figures 15.12c and 15.13c is as natural as the results of Shao et al. (2006).

Figure 15.14a is an image of a girl standing before a wall and a small garden. The girl was selected to be removed, and the target region covered four materials from

FIGURE 15.12 The removed region is the bicycle men. (a) Original image. (b) The result of Shao et al. (2006). (c) The result of the proposed algorithm. (Adapted from Shao, X., Liu, Z., and Li, H., IEEE *International Conference on Document Image Analysis for Libraries*, pp. 368–372, April 2006.)

FIGURE 15.13 Remove texts of the image. (a) Original image. (b) The result of Bornard et al. (2002). (c) The result of the proposed algorithm. (Adapted from Bornard, R. et al., Missing date correction in still images and image sequences, *In Proceedings of the Tenth ACM International Conference on Multimedia*, December 2002.)

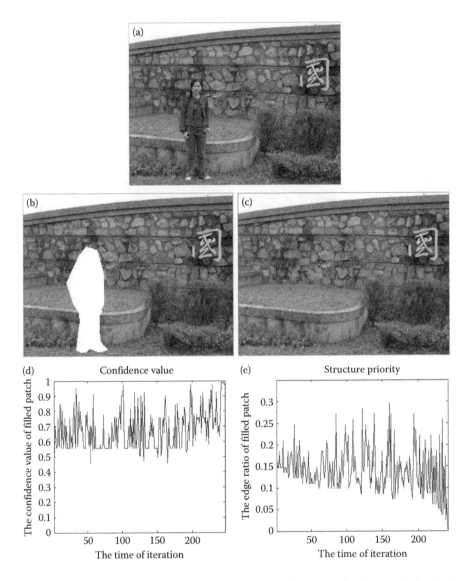

FIGURE 15.14 An example of keeping the structure information of an image. (a) Original image. (b) The target region. (c) The result of the proposed algorithm. (d) The distribution of the confidence values. (e) The distribution of the edge ratio.

top to down: rock wall, grass, cement wall, and ground (Figure 15.14b). Figure 15.14c shows the inpainting result. The proposed approach successfully rebuilt the rock wall–grass, grass–cement wall, and cement wall–ground boundary areas. Figure 15.14d and e illustrates the distributions of the confidence values and edge ratios.

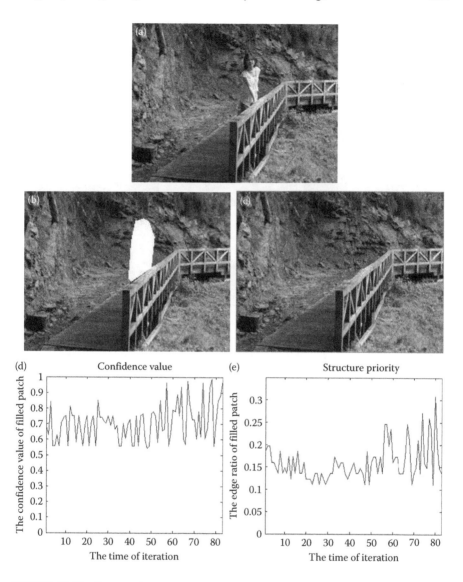

FIGURE 15.15 Remove the person of an image. (a) Original image. (b) The target region. (c) The result of the proposed algorithm. (d) The distribution of the confidence value. (e) The distribution of the edge ratio.

Figure 15.15a shows a girl standing on a bridge, and a rock in the background. The girl was removed and the target region is shown in Figure 15.15b. Figure 15.15c shows the inpainting result, which indicates that the background of grass and rock was successfully reconstructed. The confidence values remained above 0.5 during the iteration process (see Figure 15.15d). Figure 15.15e illustrates the distribution of the edge ratio.

15.4.3 EVALUATION OF MAN-MADE IMAGES

Finally, the performance on the man-made images was evaluated, as shown in Figures 15.16a and 15.17a. The key point is to observe whether the boundary between adjacent textures can be restored smoothly. Figures 15.16b and 15.17b illustrate the removed regions, and Figures 15.16c and 15.17c show the reconstructed images. The texture elements and the boundaries were successfully rebuilt in both cases. Figure 15.16d and e illustrate the corresponding distributions of the confidence value and the edge ratio.

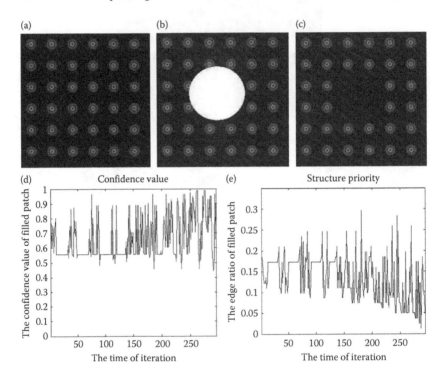

FIGURE 15.16 Circle dot image. (a) Original image. (b) The target region. (c) The result of the proposed algorithm. (d) The distribution of confidence value. (e) The distribution of the edge ratio where the iteration number is 296.

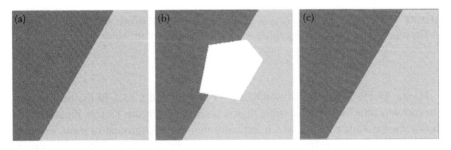

FIGURE 15.17 Single-texture images. (a) Original image. (b) Target region. (c) Inpainting result.

15.5 CONCLUSIONS

This work presents a new method for determining the filling order of inpainting. The proposed approach is robust in filling missing or removed regions of an image. It also performs well for different kinds of man-made images and natural images, regardless of whether their textures are simple or complex. Experimental results demonstrate that the proposed algorithm can fill the damaged image with visually plausible backgrounds.

REFERENCES

Ballester, C., Caselles, V., Verdera, J., Bertalmio, M., and Sapir, G., A variational model for filling-in gray level and color images, In *Eighth IEEE International Conference on Computer Vision*, vol. 1, pp. 10–16, July 2001.

Bertalmio, M., Bertozzi, A. L., and Sapiro, G., Navier-Stokes, Fluid dynamics, and image and video inpainting, In *Proceedings of IEEE Conference on Computer Vision and Pattern Recognition*, vol. 1, pp. 355–362, December 2001.

Bertalmio, M., Sapiro, G., Caselles, V., and Ballester, C., Image inpainting, In *Proceeding of ACM Conference on Computer Graphics (SIGGRAPH)*, pp. 417–424, July 2000.

Bertalmio, M., Vese, L., Sapiro, G., and Osher, S., Simultaneous structure and texture image inpainting, *IEEE Transactions on Image Processing*, 12(8), 882–889, 2003.

Bornard, R., Lecan, E., Laborelli, L., and Chenot, J.-H., Missing date correction in still images and image sequences, In *Proceedings of the Tenth ACM International Conference on Multimedia*, France, December 2002.

Cheng, W.-H., Hsieh, C.-W., Lin, S.-K., Wang, C.-W., and Wu, J.-L., Robust algorithm for examplar based image inpainting, In *IEEE International Conference on Computer Graphics, Imaging and Vision*, pp. 26–29, July, 2005.

Criminisi, A., Pérez, P., and Toyama, K., Object removal by exemplar-based inpainting, In *IEEE Conference on Computer Vision and Pattern Recognition*, vol. 2, pp. 721–728, June 2003.

Criminisi, A., Pérez, P., and Toyama, K., Region filling and object removal by exemplar-based image inpainting, *IEEE Transactions on Image Processing*, 13(9), 1200–1212, 2004.

Drori, I., Cohen-Or, D., and Yeshurun, H., Fragment-based image completion, *ACM Transactions on Graphics*, 22(3), 2003.

Efros, A. A. and Leung, T. K., Texture synpaper by non-parametric sampling, In *The Proceedings of the Seventh IEEE International Conference on Computer Vision*, vol. 2, pp. 1033–1038, September 1999.

Jia, J. and Tang, C.-K., Image repairing: Robust image synpaper by adaptive ND tensor voting, In *IEEE Computer Society Conference on Computer Vision and Pattern Recognition*, vol. 1, pp. 643–650, June 2003.

Li, B.-R., Qi, Y., and Shen, X.-K., An image inpainting method, In *IEEE Ninth International Conference on Computer Aided Design and Computer Graphics*, pp. 531–536, December 2005.

Nie, D., Ma, L., and Xiao, S., Similarity based image inpainting method, In *Proceedings of IEEE International Conference on Multi-Media Modelling*, January 2006.

Rares, A., Reinders J., Marcel T., and Biemond, J., Edge-based image restoration, *IEEE Transactions on Image Processing*, 14(10), 1454–1468, 2005.

Shao, X., Liu, Z., and Li, H., An image inpainting approach based on Poisson equation, In *IEEE International Conference on Document Image Analysis for Libraries*, pp. 368–372, April 2006.

Telea, A., An image inpainting technique based on the fast marching method, *Journal of Graphics Tools*, 9(1), 25–36, 2004.

Wang, J.-F., Hsu, H.-J., and Liao, S.-C. A novel framework for object removal from digital photograph, In *IEEE International Conference on Image Processing*, vol. 2, pp. 73–76, September 2005.

Wang, J.-F., Hsu, H.-J., and Wang, H.-M., Constrained texture synpaper by scalable sub-patch algorithm, In *IEEE International Conference on Multimedia and Expo*, vol. 1, pp. 635–638, June 2004.

16 A Comparison Study on Copy–Cover Image Forgery Detection

Frank Y. Shih and Yuan Yuan

CONTENTS

16.1 INTRODUCTION

It was a very difficult task in old times without digital cameras and computers to create a good splicing photograph, which requires sophisticated skill of darkroom masking. Due to rapid advances in powerful image-processing software, digital images are easy to manipulate and modify. This makes it more difficult for humans to check the authenticity of a digital image. Nowadays, modifying the content of digital images becomes much easier with the help of sophisticated software such as Adobe Photoshop™. It was reported that there were five million registered users of Adobe Photoshop up to the year 2004 (Hafner, 2004). Image editing software is generally available, and some of them are even free, such as GIMP™ (the GNU Image Manipulation Program) and Paint.Net™ (the free image editing and photo manipulation software were designed to be used in computers that run Microsoft Windows™). The ease of creating fake digital images with a realistic quality makes us think twice before accepting an image as authentic. For the news photographs and the electronic check clearing systems, image authenticity becomes extremely critical.

As an example of image forgery after the U.S. Civil War, a photograph of Lincoln's head was superimposed onto a portrait of the southern leader John Calhoun, as shown in Figure 16.1. Another example of image forgery appeared in a video of Osama bin Laden issued on Friday, September 7, 2007 before the sixth anniversary of 9/11. According to Neal Krawetz of Hactor Factor, an expert on digital image forensics, this video contained many visual and audio splices, and all of the modifications were of very low quality.

Checking the internal consistencies within an image, such as whether the shadow is consistent with the lighting or the objects in an image are in a correct perspective view, is one method to examine the authenticity of images. Minor details of fake images are likely to be overlooked by forgers, and thus it can be used to locate possible inconsistency. However, minor or ambiguous inconsistencies can be easily argued unless there are major and obvious inconsistencies. Moreover, it is not difficult for a professional to create a digital photomontage without major inconsistencies.

In this chapter, we describe and compare the techniques of copy–cover image forgery detection (Shih and Yuan, 2010). It is organized as follows. Section 16.2 reviews watermarking technique for image authentication. Section 16.3 introduces image splicing detection. Section 16.4 presents four copy–cover detection methods, including principal component analysis (PCA), discrete cosine transform (DCT), spatial domain, and statistical domain. Section 16.5 compares the four copy–cover detection methods, and provides the effectiveness and sensitivity under variant additive noises and lossy Joint Photographic Experts Group (JPEG) compressions. Finally, we draw conclusions in Section 16.6.

FIGURE 16.1 The 1860 portrait of (a) President Abraham Lincoln and (b) Southern politician John Calhoun. (Courtesy of Hoax Photo Gallery.)

16.2 WATERMARKING FOR IMAGE AUTHENTICATION

Watermarking is not a brand new phenomenon. For nearly 1000 years, watermarks on papers have been often used to visibly indicate a particular publisher and to discourage counterfeiting in currency. A watermark is a design impressed on a piece of paper during production and used for copyright identification. The design may be a pattern, a logo, or an image. In the modern era, as most of the data and information are stored and communicated in a digital form, proving authenticity plays an increasingly important role. As a result, digital watermarking is a process whereby arbitrary information is encoded into an image in such a way that the additional payload is imperceptible to image observers. Figure 16.2 shows the general procedure of watermarking.

Digital watermarking has been proposed as a tool to identify the source, creator, owner, distributor, or authorized consumer of a document or an image. It can also be used to detect a document or an image that has been illegally distributed or modified. In a digital world, a watermark is a pattern of bits inserted into a digital media that can identify the creator or authorized users. The digital watermark, unlike the printed visible stamp watermark, is designed to be invisible to viewers. The bits embedded into an image are scattered all around to avoid identification or modification. Therefore, a

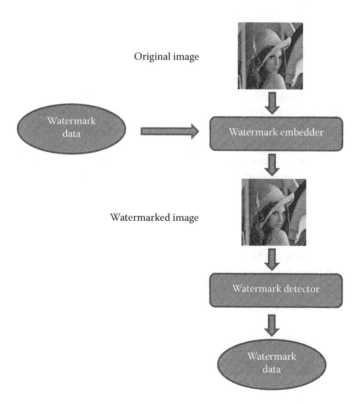

FIGURE 16.2 The general procedure of watermarking.

digital watermark must be robust enough to survive the detection, compression, and operations that are applied.

In general, watermarking techniques, such as fragile watermarking (Shih, 2007), semifragile watermarking (Wu and Shih, 2007), or content-based watermarking (Bas et al., 2002), are often used for the image authentication application. However, watermarking techniques have some drawbacks. Fragile watermark is not suitable for such applications involving compression of images, which is a common practice before sharing images on the Internet. Even though the compression operation is content preserving, fragile watermarking techniques would probably declare a compressed image as unauthentic. Although semifragile watermark can be designed to tolerate a specific set of content-preserving operations such as JPEG compression (Lin and Chang, 1998), designing such a watermark that can meet the complex requirements of real-life applications is very challenging. It is indeed not easy to develop a watermarking algorithm that can resist such errors produced from scanning and transmission, as well as can tolerate the intensity and size adjustments.

Several watermarking methods have been proposed. Yuan et al. (2006) and Huang et al. (2006) put forward integer wavelet-based multiple logo-watermarking schemes, in which a visual meaningful logo is embedded in the wavelet domain. The watermark is inserted in different bands of wavelet coefficients to make it robust to different attacks. Wu and Cheung (2010) presented a reversible watermarking algorithm, which exploits the redundancy in the watermarked object to save the recovery information. Kalantari et al. (2010) proposed a robust watermarking algorithm in the ridgelet domain by modifying the amplitude of the ridgelet coefficients to resist additive white noise and JPEG compression. Luo et al. (2010) developed a watermarking algorithm using interpolation techniques to restore the original image without any distortion after the hidden data are extracted. Kang et al. (2010) proposed a watermarking algorithm which is resilient to both print–scan process and geometric distortion by adopting a log-polar mapping.

Since the watermark generation and embedding techniques are closely coordinated in the process of watermarking, the overall success of detection relies upon the security of the watermark generation and embedding. There are several issues to be considered: (1) how easy it is to disable the embedding of watermark, (2) how easy it is to hack the embedding procedure, and (3) how easy it is to generate a valid watermark or embed a valid watermark into a manipulated image. Unfortunately, the embedded watermark can be removed by exploiting the weak points of a watermarking scheme. When a sufficient number of images with the same secret watermark key are obtained, a watermarking scheme can be hacked. There is still not a fully secure watermarking scheme available till date.

16.3   IMAGE SPLICING DETECTION

Image splicing is a fundamental step in digital photomontage, which makes a paste-up image by editing images using digital tools such as Adobe Photoshop. Ng et al. (2004) investigated the prospect of using bicoherence features for blind image splicing detection. Image splicing is an essential operation for digital photomontaging, which in turn is a technique for creating image forgery. They fed bicoherence features of images to support vector machine for classification, and the detection accuracy reaches 70%.

Based on the observation that digitally processed image forgery makes the digital image data highly correlated, Gopi et al. (2006) exploited this property by using autoregressive (AR) coefficients as the feature vector for identifying the location of digital forgery in an image. There are 300 feature vectors extracted from different images to train an artificial neural network (ANN). The accuracy rate of identifying digital forgery is 77.67%.

Although image splicing is a very powerful technique, one still can detect the trace of splicing. Image splicing generally introduces a step-jump transition that does not usually happen at crossing boundaries of two objects in a natural scene image. Figure 16.3 shows the difference between the boundaries introduced by image splicing and natural boundary.

In general, the camera-captured images do not usually contain step-like edge transitions based upon the following reasons that make edges blurred:

1. Most output images from digital cameras are in the JPEG compressed format, which could cause loss of fine details.
2. Noise is usually unavoidable.
3. When people take a picture, the hand shaking is unavoidable.
4. Digital cameras often use a color filter array interpolation to produce RGB images.
5. The output image is often resized or resampled for application purposes.

Based on the above observations, we present an algorithm to detect the trace of image splicing:

1. Let Z_2 and Z_3 denote the horizontal and vertical neighbors of Z_1, respectively. We calculate the first-order derivative of the input image by

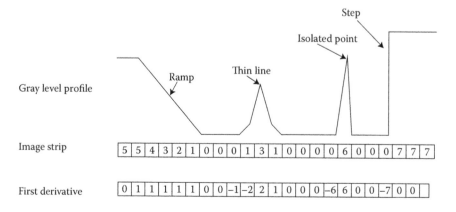

FIGURE 16.3 In a camera-captured image, there are ramp, isolated point, flat segment, and thin line. On the other hand, "step" is produced by image splicing.

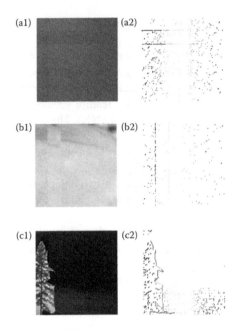

FIGURE 16.4 Some examples, where the left-hand images are spliced images and the right-hand images are our resulting images, in which the lines indicate the traces of splicing. (a1), (b1), and (c1) are spliced images and (a2), (b2), and (c2) are the respectively resulting images, in which the lines indicate the traces of splicing.

$$\nabla f = \sqrt{(Z_2 - Z_1)^2 + (Z_3 - Z_1)^2} \tag{16.1}$$

2. Perform thresholding on the first-order derivatives to obtain the binary image BW1.
3. Apply a binary morphological erosion with a 2×2 structuring element on BW1 to obtain a new binary image BW2.
4. Calculate the difference image, that is, BW3 = BW1 − BW2.
5. Obtain the connected components in BW3. The noisy components whose sizes are less than a given threshold are removed. The components in BW3 are the traces of image splicing.

We apply this algorithm on the image splicing dataset images from Columbia University (Columbia, 2010). Figure 16.4 shows some examples, where the left-hand images are spliced images and the right-hand images are the resulting images, in which the lines indicate the traces of splicing.

16.4   COPY–COVER IMAGE FORGERY DETECTION

The copy–cover technique is the most popular technique for making image forgery. Copy–cover means that one portion of a given image is copied and then used to cover some object in the given image. If the processing is properly done, most people would not notice that there are identical (or nearly identical) regions in an image. Figure 16.5 shows an example of copy–cover image forgery, where a region of wall background in the left image is copied and then used to cover two boxes on the wall.

Several researchers have explored the copy–cover image forgery detection. Mahdian and Saic (2007) proposed a method to automatically detect and localize duplicated regions in digital images. Their method is based on blur moment invariants, allowing the successful detection of copy–cover forgery even when blur degradation, additional noise, or arbitrary contrast changes are present in the duplicated regions. These modifications are commonly used techniques to conceal traces of copy–cover forgery.

Fridrich et al. (2003) presented an effective copy–cover forgery detection algorithm using DCT and quantization techniques. Popescu and Farid (2004) used the PCA domain representation to detect the forged part, even when the copied area is corrupted by noise. Ju et al. (2007) adopted PCA for small image blocks with fixed sizes, and generates the degraded dimensions representation for copy–cover detection.

Although both DCT and PCA representation methods are claimed to be successful in copy–cover forgery detection, there is a lack of performance comparisons. We evaluate the two methods in terms of time complexity, efficiency, and robustness, as well as evaluate two other methods—one based on spatial-domain representation and the other on statistical domain representation.

Given an image of N pixels, our goal is to identify the location and shape of duplicated regions resulting from copy–cover forgery. The general copy–cover detection method is described below. First, a given image is split into small overlapping blocks; each block is transformed into another domain, such as DCT or PCA domain. A two-dimensional matrix is constructed in the way that the pixels in a block are placed in a row by a raster scan and the total number of rows corresponds to the total number

FIGURE 16.5 The image on the left is original, and the image on the right is forged, in which a region of wall background is copied and then used to cover two boxes on the wall.

of blocks in the given image. By lexicographically sorting all the rows, identical or nearly identical blocks can be detected since they are adjacent to each other. The computational cost of this method is the lexicographic sorting (with time complexity $O(N \log N)$) and domain transformation.

We use four different domain representations for copy–cover detection. The first method is based on PCA domain. The dimension of each block is reduced, and only a number of principal coefficients are preserved. The second method is based on the DCT domain. Only a number of most significant DCT coefficients are preserved. Both PCA and DCT methods are in general robust to noise introduced in the process of forgery and can reduce the time consumption in the lexicographical sorting.

The third method is based on the spatial domain. All the small blocks are sorted directly according to their pixel values. It saves time since no transformation is involved. However, the lexicographical sorting consumes much more time. The fourth method is based on statistical domain, which uses the mean value and standard deviation of each block for sorting. These four methods are described in more detail below.

Before describing the four methods, we need to define the parameter notations. Let N be the total number of pixels in a square grayscale or color image (i.e., the image has $\sqrt{N} \times \sqrt{N}$ pixels in dimensions). Let b denote the number of pixels in a square block (i.e., the block has $\sqrt{b} \times \sqrt{b}$ pixels in dimensions); there are totally $N_b = (\sqrt{N} - \sqrt{b} + 1)^2$ blocks. Let Nc be the number of principal components preserved in the PCA domain, and let Nt be the number of significant DCT coefficients preserved in the DCT domain. Let Nn denote the number of neighboring rows to search for in the lexicographically sorted matrix. Let Nd be the minimum offset threshold.

16.4.1 PCA Domain Method

PCA is known as the best data representation in the least-square sense for classical recognition (Jolliffe, 2002). It is commonly used to reduce the dimensionality of images and retain most information. The idea is to find the orthogonal basis vectors or the eigenvectors of the covariance matrix of a set of images, with each image being treated as a single point in a high-dimensional space. Since each image contributes to each of the eigenvectors, the eigenvectors resemble ghost-like images when displayed. The PCA domain method for copy–cover detection is described below.

1. Using PCA, we can compute the new $N_b \times Nc$ matrix representation, in which each row is composed of the first Nc principal components in each block. If a color image is used, we convert the color image into a grayscale image or analyze each color channel separately.
2. Sort the rows of the above matrix in a lexicographic order to yield a matrix S. Let $\vec{s}_i$ denote the rows of S, and let (x_i, y_i) denote the position of the block's image coordinates (top-left corner) that corresponds to $\vec{s}_i$.
3. For every pair of rows $\vec{s}_i$ and $\vec{s}_j$ from S such that $|i - j| < Nn$, place the pair of coordinates (x_i, y_i) and (x_j, y_j) in a list.

4. For all elements in this list, compute their offset, as defined by $(x_i - x_j, y_i - y_j)$.
5. Discard all the pairs whose offset magnitude $\sqrt{(x_i - x_j)^2 + (y_i - y_j)^2}$ is less than Nd.
6. Find out the pairs of coordinates with highest offset frequency.
7. From the remaining pairs of blocks, build a duplication map by constructing a zero image, whose size is the same as the original, and coloring all pixels in a duplicated region by a unique grayscale intensity value.

16.4.2 DCT DOMAIN METHOD

Many video and image compression algorithms apply the DCT to transform an image to the frequency domain and perform quantization for data compression (Yip and Rao, 1990). One of its advantages is the energy compaction property, that is, the signal energy is concentrated on a few components while most other components are zero or negligibly small. This helps separate an image into parts (or spectral sub-bands) of hierarchical importance (with respect to the image's visual quality). The popular JPEG compression technology uses the DCT to compress an image.

We replace step 1 of the aforementioned PCA domain method by DCT to compute the new $N_b \times Nt$ matrix representation, where each row of the matrix is composed of Nt significant coefficients by zigzag ordering of DCT coefficients in each block.

16.4.3 SPATIAL DOMAIN METHOD

We replace step 1 of the PCA domain method by the $N_b \times b$ matrix representation, where each row is the column-wise concatenation of the b pixels in each block.

16.4.4 STATISTICAL DOMAIN METHOD

We replace step 1 of the PCA domain method by an $N_b \times 2$ matrix, where each row contains the mean value and the standard deviation of each block.

16.5 EXPERIMENTAL RESULTS

To compare the performance of the aforementioned four methods, we create an image database composed of 500 images for use in our experiment. The image resolution is of size 256×256. The content of those images includes landscape, buildings, flowers, human, animals, and so on. For each image, we randomly copy a region of size 81×81 and paste it to another location to form a tempered image. Since the detection results depend somewhat on the content of image and the region selected, we conduct the experiment for all the images and use the average value for comparisons. The following parameters are preset: $b = 256$, $Nn = 1$, $Nd = 10$, $Nc = 26$, and $Nt = 26$.

To compare the robustness of the four methods with respect to JPEG compression and Gaussian blurring, we use XNview (a software for viewing and converting graphic files, which is a freeware available at http://www.xnview.com), to accomplish the JPEG compression and Gaussian blurring. When conducting the JPEG

FIGURE 16.6   (a) The original image and (b) the copy–cover forgery image.

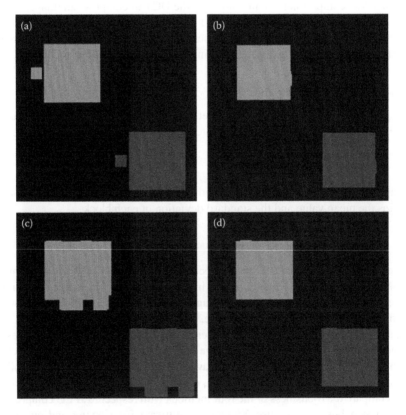

FIGURE 16.7   Output of copy–cover forgery detection by (a) PCA, (b) DCT, (c) spatial domain, and (d) statistical domain detection methods. Note that the matched blocks are displayed in light and dark gray.

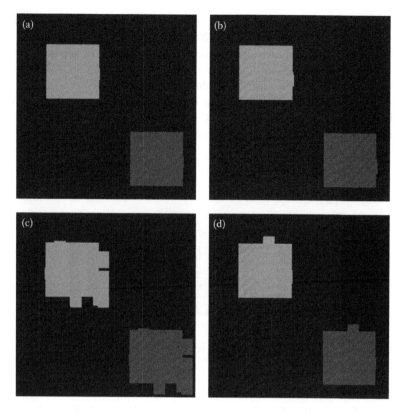

FIGURE 16.8   Output of copy–cover forgery detection by (a) PCA, (b) DCT, (c) spatial domain, and (d) statistical domain detection methods when the given image is corrupted by Gaussian blurring with block size 7.

compression, XNview allows us to choose the compression ratio between 0 and 100. The smaller the number, the smaller the output file will be. When conducting the Gaussian blurring, XNview allows us to choose the Gaussian blurring filter size. The larger the size, the more the given image will become blurred.

Figure 16.6 shows an example of a copy–cover forgery, in which the tampering consists of copying and pasting a region to cover a significant content. Figure 16.7 shows the output images of copy–cover forgery detection when the given image is compressed with JPEG quality 50, where Figures a–d are respectively obtained by PCA, DCT, spatial domain, and statistical domain detection methods. Figure 16.8 shows the output copy–cover forgery detection when the given image is corrupted by Gaussian blurring of block size 7.

16.5.1   ROBUSTNESS TO JPEG COMPRESSION

Since most images available are JPEG compressed, we apply JPEG compression ratios from 50 to 100 to compress the test images for comparing the robustness of the four methods under JPEG compression. The obtained true positive rates related to JPEG quality are shown in Figure 16.9. We observe that the performances of DCT domain and PCA domain methods are very similar, and are better than those of spatial domain and statistical domain methods. Moreover, the statistical domain method is slightly better than the spatial-domain method.

The true negative rates with respect to JPEG quality are shown in Figure 16.10. We observe that the performance of DCT domain, PCA domain, and statistical domain methods are very similar. However, the performance of spatial domain is worse than the other three.

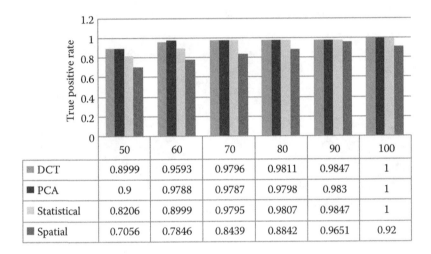

	50	60	70	80	90	100
▪ DCT	0.8999	0.9593	0.9796	0.9811	0.9847	1
▪ PCA	0.9	0.9788	0.9787	0.9798	0.983	1
Statistical	0.8206	0.8999	0.9795	0.9807	0.9847	1
▪ Spatial	0.7056	0.7846	0.8439	0.8842	0.9651	0.92

FIGURE 16.9   True positive rates with respect to JPEG compression ratio.

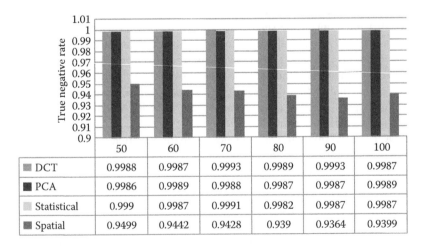

	50	60	70	80	90	100
▪ DCT	0.9988	0.9987	0.9993	0.9989	0.9993	0.9987
▪ PCA	0.9986	0.9989	0.9988	0.9987	0.9987	0.9989
Statistical	0.999	0.9987	0.9991	0.9982	0.9987	0.9987
▪ Spatial	0.9499	0.9442	0.9428	0.939	0.9364	0.9399

FIGURE 16.10   True negative rates with respect to JPEG compression ratio.

16.5.2 Robustness to Gaussian Blurring

Since copy–cover image forgery will produce two identical regions in an image, one method to conceal the forgery is to apply Gaussian blurring on the composite image to conceal the forgery. We apply Gaussian blurring with different block sizes from 1×1 to 11×11 on the test images and then perform the detection. Note that the image using 1×1 Gaussian blurring is the same as the original image.

The true positive rates with respect to Gaussian blurring are shown in Figure 16.11. We observe that the performances of DCT, PCA, and statistical domain methods are similar, which are slightly better than that of the spatial-domain method.

The true negative rates with respect to PSNR are shown in Figure 16.12. We observe that the true negative rate of the spatial domain method is the lowest. The true negative rates of DCT, PCA, and statistical domain methods are similar.

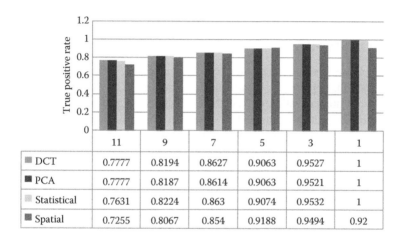

FIGURE 16.11   True positive rates with respect to Gaussian blurring.

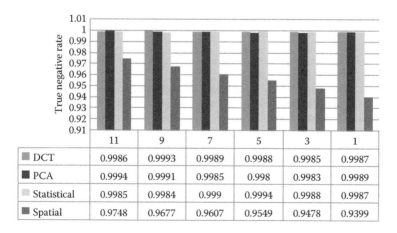

FIGURE 16.12   True negative rates with respect to Gaussian blurring.

TABLE 16.1
Running Time of the Four Detection Methods

Method	Running Time (s)
DCT domain	29.8594
PCA domain	20.5313
Spatial domain	17.0156
Statistical domain	20.8281

The experiment is performed on a Dell notebook computer with a 1.70 GHz Intel Pentium Mobile Processor and 512 MB of RAM running Windows XP™. The program is coded in MATLAB®. For a given image of size 256×256, the average running time of the four methods is shown in Table 16.1.

16.6   CONCLUSIONS

In this chapter, we discuss the techniques of watermarking for authentication and the four methods for copy–cover identification, including PCA, DCT, spatial domain, and statistical domain. We evaluate their effectiveness and sensitivity under the influences of Gaussian blurring and lossy JPEG compressions. We conclude that the PCA domain method outperforms the other methods in terms of time complexity and detection accuracy. Our future work is to extend the capability of copy-and-paste modification from different images. Furthermore, we will explore more complicated spatially distributed copy-and-paste modification; that is, instead of copying a consecutive area, one may copy one pixel here and another pixel there in a random-like distribution.

REFERENCES

Bas, P., Chassery, J. M., and Macq, B., Image watermarking: An evolution to content based approaches, *Pattern Recognition*, 35, 545–561, 2002.

Columbia Image Splicing Detection Evaluation Dataset, The DVMM Laboratory of Columbia University, 2010.

Fridrich, J., Soukal, D., and Lukáš. J., Detection of copy-move forgery in digital images, in *Digital Forensic Research Workshop*, 2003.

Gopi, E. S., Lakshmanan, N., Gokul, T., Ganesh, S., and Shah, P. R., Digital image forgery detection using artificial neural network and auto regressive coefficients, *Proceedings of the Canadian Conference on Electrical and Computer Engineering*, pp. 194–197, May 2006.

Hafner, K., The camera never lies, but the software can, *New York Times*, 2004.

Huang, D., Yuan, Y., and Lu, Y., Novel multiple logo-watermarking algorithm based on integer wavelet, *Chinese Journal of Electronics*, 15, 857–860, 2006.

Jolliffe, I. T., *Principal Component Analysis*, Springer, New York, 2002.

Ju, S., Zhou, J., and He, K., An authentication method for copy areas of images, in *International Conference on Image and Graphics*, pp. 303–306, 2007.

Kalantari, N. K., Ahadi, S. M., and Vafadust, M., A robust image watermarking in the ridgelet domain using universally optimum decoder, *IEEE Transactions on Circuits and Systems for Video Technology*, 20, 396–406, 2010.

Kang, X., Jiwu, H., and Wenjun, Z., Efficient general print-scanning resilient data hiding based on uniform log-polar mapping, *IEEE Transactions on Information Forensics and Security*, 5, 1–12, 2010.

Lin, C. Y. and Chang, S. F., A robust image authentication method surviving JPEG lossy compression, in *SPIE Conference on Storage and Retrieval of Image/Video Database*, pp. 296–307, 1998.

Luo, L., Chen, Z., Chen, M., Zeng, X.n, and Xiong, Z., Reversible image watermarking using interpolation technique, *IEEE Transactions on Information Forensics and Security*, 5, 187–193, 2010.

Mahdian, B. and Saic, S., Detection of copy-move forgery using a method based on blur moment invariants, *International Journal of Forensic Science*, 171, 180–189, 2007.

Ng, T.-T., Chang, S.-F., and Sun, Q., Blind detection of photomontage using higher order statistics, *Proceedings of the IEEE International Symposium on Circuits and Systems*, Vancouver, Canada, pp. 23–26, May 2004.

Popescu, A. C. and Farid, H., Exposing digital forgeries by detecting duplicated image regions, Technical Report, TR2004-515, Dartmouth College, 2004.

Shih, F. Y., *Digital Watermarking and Steganography: Fundamentals and Techniques*, CRC Press, Boca Raton, FL, 2007.

Shih, F. Y. and Yuan, Y., A comparison study on copy-cover image forgery detection, *The Open Artificial Intelligence Journal*, 4, 49–54, 2010.

Wu, H. T. and Cheung, Y. M., Reversible watermarking by modulation and security enhancement, *IEEE Transactions on Instrumentation and Measurement*, 59, 221–228, 2010.

Wu, Y. and Shih, F. Y., Digital watermarking based on chaotic map and reference register, *Pattern Recognition*, 40, 3753–3763, 2007.

Yip, P. and Rao, K., *Discrete Cosine Transform: Algorithms, Advantages, and Applications*, Academic Press, San Diego, CA, 1990.

Yuan, Y., Decai, H., and Duanyang, L., An integer wavelet based multiple logo-watermarking scheme, in *First International Multi-Symposiums on Computer and Computational Sciences*, 2, 175–179, 2006.

17 Chaos-Based Hash Function with Both Modification Detection and Localization Capabilities

Di Xiao, Frank Y. Shih, and Xiaofeng Liao

CONTENTS

17.1 INTRODUCTION

As one of the cores of Cryptography, hashing is a basic technique for information security (Schneier, 1996; Stinson, 1995). A cryptographic hash function is used to compress the message data to a fixed-size hash value in such a way that any alternation to the data will generate a different hash value. Chaos is a kind of deterministic random-like process provided by nonlinear dynamic systems. Chaotic systems are systems that are random-like, but in fact are not random. They are governed by physical laws to make the accurate prediction almost impossible. It is a promising direction to utilize the characteristics of chaos to design hash function.

Wong (2003) developed a combined encryption and hashing scheme based on the iteration of Logistic Map and the dynamical update of a look-up table, which is further improved by Xiao et al. (2006). Xiao et al. (2005) proposed an algorithm for one-way hash function construction based on a Piecewise Linear Chaotic Map (PWLCM) with changeable-parameter. Yi (2005) proposed a hash function algorithm based on Tent Maps. Zhang et al. (2007) proposed a novel chaotic keyed hash algorithm by using a Feedforward–Feedback Nonlinear Filter. Arumugam et al. (2007) studied the suitability of Logistic Map and Lorenz Map in generating Message Authentication Codes. However, the above algorithms have a common point that they are only able to verify whether there is modification, but unable to locate where the modification takes place. Besides, their iterative hash structures are all in a sequential mode. The processing of the current message unit, that is, a character as in Wong (2003) and Xiao et al. (2005, 2006) or a block as in Yi (2005), Zhang et al. (2007), and Arumugam et al. (2007), cannot start until the previous one has been processed. These limitations restrict their applications.

In this chapter, we propose an algorithm for both modification detection and localization, whose structure can support parallel processing mode (Xiao et al., 2010). The mechanism of both changeable-parameter and self-synchronization is utilized to achieve all the performance requirements of hash function. The rest of this chapter is arranged as follows. Section 17.2 introduces the brief preliminaries about the chaotic maps used in the proposed algorithm. In Section 17.3, the hash function algorithm is described in detail. Performance analyses are given in Section 17.4. Finally, this chapter is concluded in Section 17.5.

17.2 PRELIMINARIES

The inherent merits of chaos, such as the sensitivity to tiny changes in initial conditions and parameters, mixing property, ergodicity, unstable periodic orbits with long periods, and one-way iteration, form the potential foundation for excellent hash function construction. In the proposed algorithm, PWLCM will be used.

PWLCM is defined as

$$X_{t+1} = F_P(X_t) = \begin{cases} X_t/P, & 0 \le X_t < P \\ (X_t - P)/(0.5 - P), & P \le X_t < 0.5 \\ (1 - X_t - P)/(0.5 - P), & 0.5 \le X_t < 1 - P \\ (1 - X_t)/P, & 1 - P \le X_t \le 1 \end{cases} \tag{17.1}$$

where $X_t \in [0, 1]$ and $P \in (0, 0.5)$ denote the iteration trajectory value and the parameter of PWLCM, respectively. According to Baranousky and Daems (1995), $\{X_t\}$ is ergodic and uniformly distributed in $[0, 1]$, and the auto-correlation function of $\{X_t\}$ is δ-like.

17.3 PROPOSED ALGORITHM

17.3.1 Algorithm Description

17.3.1.1 Padding, Division, and Preprocessing

In our algorithm, a lookup table with 16×16 blocks is built in advance, as shown in Figure 17.1. The original message M is padded such that its length is a multiple of 256 characters (2048 bits): let m be the length of the original message M; the padding bits $(100 \cdots 0)_2$ with length n (such that $(m + n)\mathrm{mod}2048 = 2048 - 64 = 1984$, $1 \le n \le 2048$) are appended. The left 64-bit is used to denote the length of the original message M. If m is greater than 2^{64}, then $m\mathrm{mod}2^{64}$. After padding, M is constituted by s divisions with 256 characters (2048 bits), $M = (M_1, M_2, \ldots, M_s)$. For each division, its 256 characters are filled into the 256 blocks of the lookup table in turn. When all the divisions have been processed in this way, there will be s characters within each block of the lookup table. Without loss of generality, let ith block of the lookup table hold character array-$c_1, c_2, \ldots, c_s$. The detailed preprocessing of the ith block ($i = 1, 2, \ldots, 256$) in the lookup table is described in Figure 17.1.

1. Translates the pending character array-$c_1, c_2, \ldots, c_s$ to the corresponding ASCII numbers, then maps these ASCII numbers into a number array-C_1, $C_2, \ldots, C_s$ by means of linear transform, in which the element is a number $\in [0, 1]$ and the length is the character number s in the block of the lookup table.
2. The iteration process of PWLCM is as follows:

1st: $P_1 = (C_1 + P_0)/4 \in (0, 0.5)$, $\quad X_1 = F_{P_1}(X_0) \in (0, 1)$,
2nd–sth: $P_k = (C_k + X_{k-1})/4 \in (0, 0.5)$, $\quad X_k = F_{P_k}(X_{k-1}) \in (0, 1)$,
$(s + 1)$th: $P_{s+1} = (C_s + X_s)/4 \in (0, 0.5)$, $\quad X_{s+1} = F_{P_{s+1}}(X_s) \in (0, 1)$,
$(s + 2)$th–$2s$th: $P_k = (C_{(2s-k+1)} + X_{k-1})/4 \in (0, 0.5)$, $\quad X_k = F_{P_k}(X_{k-1}) \in (0, 1)$.

Block 1	Block 2	$\cdots$	Block 16	H_Group 1
Block 17	Block 18	$\cdots$	Block 32	H_Group 2
. . .	. . .	.	.	
Block 241	Block 242	$\cdots$	Block 256	H_Group 16
V_Group 1	V_Group 2		V_Group 16	

FIGURE 17.1 Look-up table.

Here, the initial condition $X_0 \in [0, 1]$ and initial parameter $P_0 \in (0, 1)$ of PWLCM are used as the secret key of the algorithm. If a certain iteration value X_k is equal to 0 or 1, then an extra iteration is carried out. The property of chaos can ensure that this kind of extra iteration time is very less.

3. X_{2s} obtained in Step (ii) is the set as the representative value of the ith block $(i = 1, 2, \ldots, 256)$ in the block of the lookup table.

17.3.1.2 Grouping and Processing in a Parallel Mode

The 256 blocks of the lookup table are assigned in turn into 16 horizontal groups and 16 vertical groups for further processing in a parallel mode, respectively. As shown in Figure 17.1, the groups are:

Horizontal Direction: H_Group 1-Block1, Block2, ..., Block16; H_Group 2-Block17, Block18, ..., Block32; ..., H_Group 16-Block241, Block242, ..., Block256;

Vertical Direction: V_Group 1-Block1, Block17, ..., Block241; V_Group 2-Block2, Block18, ..., Block242; ..., V_Group 16-Block16, Block32, ..., Block256.

Without loss of generality, let $CB_{i1}, CB_{i2}, \ldots, CB_{i16} \in (0, 1)$ be the corresponding representative values of the blocks in the ith horizontal group, the detailed processing of the ith horizontal group $(i = 1, 2, \ldots, 16)$ is described as follows:

1. The iteration process of PWLCM is as follows:
   1st: $P_1 = (CB_{i1} + P_0 + i/16)/6 \in (0,0.5)$, $X_1 = F_{P1}(X_0) \in [0,1]$, where the initial condition $X_0 \in [0, 1]$ and initial parameter $P_0 \in (0, 1)$ of PWLCM are the secret key of the algorithm, and "i" is the order of each Horizontal Group.
   2nd–16th: $P_k = (CB_{ik} + X_{k-1})/4 \in (0,0.5)$, $X_k = F_{P_k}(X_{k-1}) \in [0,1]$,
   17th: $P_{17} = (CB_{i16} + X_{16})/4 \in (0,0.5)$, $X_{17} = F_{P_{17}}(X_{16}) \in [0,1]$,
   18th–32nd: $P_k = (C_{(32-k+1)} + X_{k-1})/4 \in (0,0.5)$, $X_k = F_{P_k}(X_{k-1}) \in [0,1]$,
   33rd~34th: $X_k = F_{P_{32}}(X_{k-1}) \in [0,1]$.

2. We transform the iteration values-X_{32}, X_{33}, X_{34} to the corresponding binary format, extract 40, 40, 48 bits after the decimal point, respectively, and juxtapose them from left to right to get a 128-bit DHK_i, the keystream of the ith horizontal group $(i = 1, 2, \ldots, 16)$. At the same time, we extract 4 bits after the decimal point from the binary format of X_{32} to get a 4-bit LHK_i $(i = 1, 2, \ldots, 16)$.

Similar operations have been performed on the jth vertical group $(j = 1, 2, \ldots, 16)$. Then we will obtain a 128-bit DVK_j, the keystream of the jth Vertical Group $(j = 1, 2, \ldots, 16)$, and a 4-bit LVK_j $(j = 1, 2, \ldots, 16)$.

17.3.1.3 Obtaining the Detection Hash and Localization Hash Values

The final 256-bit hash value of the message M is jointly composed of DHASH and LHASH, which are used to accomplish the detection and the localization, respectively. The 128-bit detection hash value can be obtained by $DHASH = DHK_1 \oplus DHK_2 \cdots \oplus DHK_{16} \oplus DVK_1 \oplus DVK_2 \cdots \oplus DVK_{16}$, where $\oplus$ denotes XOR (Exclusive-OR) operation. At the same time, all 4-bit LHK_i ($i = 1, 2, ..., 16$) and 4-bit LVK_j ($j = 1, 2, ..., 16$) are juxtaposed from left to right to get a 128-bit localization hash value-LHASH.

Note that the 16 horizontal groups and the 16 vertical groups of the lookup table can be processed in a parallel mode, respectively. Therefore, the efficiency of the proposed algorithm is promising. The most important point is that the generation of the keystream in each group must be under the control of the corresponding algorithm key as well as the order and the content of the current group, which can guarantee the security of the algorithm.

17.3.2 CHARACTERISTICS OF THE ALGORITHM CONSTRUCTION

17.3.2.1 Both Modification Detection and Localization Capabilities

The proposed algorithm has the capabilities of both modification detection and localization. During the application of the hash function, we first verify whether there are modifications on the pending message by computing the new detection hash value and comparing it with the former one. If the comparison indicates some modifications, then we may further realize modification localization by using the localization hash value. This kind of function is very useful for information authentication or communications with resource constraints, which cannot be provided by most of other hash algorithms.

17.3.2.2 Parallel Mode

The parallel mode of the proposed algorithm is embodied in two aspects. One is that the preprocessing among 256 blocks of the lookup table can be performed in a parallel mode, and the other from the whole structural point of view, that the processing among 16 horizontal groups and 16 vertical groups can be performed in a parallel mode (see Figure 17.1). Actually, the processing of the 16 blocks within each horizontal or vertical group can also be adapted to a parallel mode if necessary.

17.3.2.3 Changeable-Parameter and Self-Synchronizing Keystream

In steps 1 and 2, the message unit at different positions will cause the parameter of chaotic maps to change dynamically during the iteration process of PWLCM. In step 2, the iteration process of PWLCM is also related to the order of each message group. On one hand, perturbation is introduced in a simple way to avoid the dynamical degradation of chaos, and on the other hand, self-synchronizing stream is realized, which ensures that the generated keystream is closely related to the algorithm key as well as the content and the order of each message group. The mechanism of both changeable-parameter and self-synchronization provides the foundation for the security of the proposed algorithm.

17.4 PERFORMANCE ANALYSIS

To implement the proposed algorithm for performance analyses, PWLCM given by Equation 17.1 is chosen. Its initial condition $X_0 = 0.232323$ and initial parameter $P_0 = 0.858485$ are set as the algorithm key. The randomly chosen original paragraph of a message is

"As a ubiquitous phenomenon in nature, chaos is a kind of deterministic random-like process generated by nonlinear dynamic systems. The properties of chaotic cryptography includes: sensitivity to tiny changes in initial conditions and parameters, random-like behavior, unstable periodic orbits with long periods and desired diffusion and confusion properties, etc. Furthermore, benefiting from the deterministic property, the chaotic system is easy to be simulated on the computer. Unique merits of chaos bring much promise of application in the information security field."

17.4.1 DETECTION HASH PERFORMANCE

In this section, we focus on the performance of the 128-bit detection hash value.

17.4.1.1 Distribution of Hash Value

The uniform distribution of hash value is one of the most important requirements of hash functions related to security. Simulation experiment has been conducted on the above paragraph of message. Two-dimensional graphs are used to demonstrate the differences between the original message and the final detection hash value. In Figure 17.2, the ASCII codes of the original message are localized within a small area; while in Figure 17.3, the hexadecimal detection hash value spreads around very uniformly. This simulation result indicates that no information (including the statistic information) of the original message can be left over after the diffusion and confusion.

17.4.1.2 Sensitivity of Hash Value to the Message and the Secret Key

In order to evaluate the sensitivity of hash value to the message and the secret key, hash simulation experiments have been performed using the following seven conditions:

C1: The original message;
C2: Change the first character *A* in the original message into *B*;
C3: Change the word *unstable* in the original message into *anstable*;
C4: Change the full stop at the end of the original message into comma;

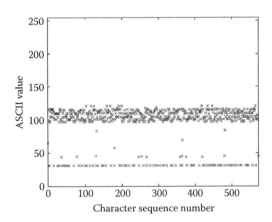

FIGURE 17.2 Distribution of the original message in ASCII.

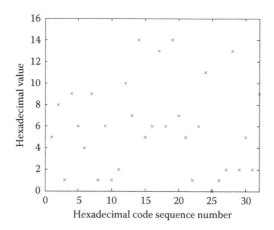

FIGURE 17.3 Distribution of the hash values in hexadecimal format.

C5: Exchange the message blocks in the 1st message group M_1 with the corresponding message blocks in the 2nd message group M_2, respectively—"*As a ubiquitous*" with "*phenomenon in na*," "e behavior, unst" with "able periodic or," and "ch promise of ap" with "plication in the";

C6: Change the secret key X_0 from 0.232323 to 0.2323230000000001;

C7: Change the secret key P_0 from 0.858485 to 0.8584850000000001.

The corresponding hash values in hexadecimal format are obtained as follows:

C1: 58196491612A7E56D6E7516B012D2529
C2: 580BDDEAE5730A88421E5A5375C16043
C3: 28BBDC5395CB941E90DC8448F6147A5F
C4: D1B7BC19F4D3AD0F6B01298EB687D797
C5: E62DF1C64672D8198154D35D1923B7CF
C6: E81515BD73C0097EB71A4FAF37279EE9
C7: 58196491612A7E56D6E7516B012D2529

The graphical display of binary sequences is shown in Figure 17.4.

The simulation result indicates that the sensitivity property of the proposed algorithm is so perfect that any least difference of the message or key will cause huge changes in the final detection hash value.

17.4.1.3 Statistic Analysis of Diffusion and Confusion

In order to hide message redundancy, Shannon (1949) introduced diffusion and confusion, which are two general principles to the practical cipher design, including hash functions. Since the hash value is in a binary format, that is, each bit is only 1 or 0, the ideal diffusion effect should be that any tiny changes in original conditions lead to the 50% changing probability for each bit of the hash value.

We have performed the following diffusion and confusion test. The detection hash value of the above paragraph of the original message is generated. Then a bit in

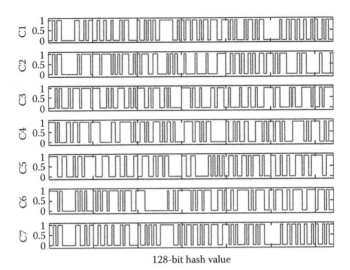

128-bit hash value

FIGURE 17.4   Hash values under different conditions.

the message is randomly selected and toggled, and a new detection hash value is generated. Two hash values are compared and the number of changed bits is counted as B_i. This kind of test is performed J-time, and the corresponding distribution of changed bit number is shown as Figure 17.5, where $J = 2048$. Obviously, the changed bit number corresponding to 1-bit changed message concentrates around the ideal changed bit number of 64 bits. It indicates that the algorithm has very strong capability for diffusion and confusion.

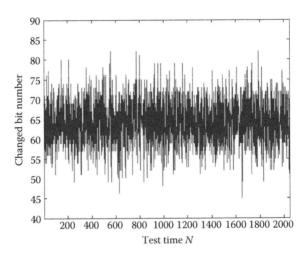

FIGURE 17.5   Distribution of changed bit number.

Usually, four statistics are defined as follows:

Mean changed bit number $\bar{B} = 1/N \sum_{i=1}^{N} B_i$

Mean changed probability $P = (\bar{B}/128) \times 100\%$

$$\Delta B = \sqrt{\frac{1}{N-1} \sum_{i=1}^{N} (B_i - \bar{B})^2}$$

$$\Delta P = \sqrt{\frac{1}{N-1} \sum_{i=1}^{N} (B_i/128 - P)^2} \times 100\%$$

Through the tests with $J = 256, 512, 1024, 2048$, respectively, the corresponding data are listed in Table 17.1.

Based on the analysis of the data in Table 17.1, we can draw the conclusion: the mean changed bit number $\bar{B}$ and the mean changed probability P are both very close to the ideal value 64 bit and 50%. While ΔB and ΔP are very little, which indicates the capability for diffusion and confusion is very stable.

17.4.1.4   Analysis of Collision Resistance

The mechanism of both changeable-parameter and self-synchronization expedites the avalanche effect. We have performed the following test to conduct quantitative analysis on collision resistance (Wong, 2003; Xiao et al., 2005, 2006; Zhang et al., 2007): First, the detection hash value of the above paragraph of original message is generated and stored in ASCII format. Then a bit in the message is selected randomly and toggled. A new detection hash value is then generated and stored in ASCII format. Two hash values are compared, and the number of ASCII character with the same value at the same location in the hash value, namely the number of hits, is counted. Moreover, the absolute difference of two hash values is calculated using the formula: $d = \sum_{i=1}^{N} | t(e_i) - t(e_i') |$, where e_i and e_i' be the ith ASCII character of the original and the new hash value, respectively, and the function $t( )$ converts the entries to their equivalent decimal values. This kind of collision test has been

TABLE 17.1
Statistics of Number of Changed Bit B_i

	$N = 256$	$N = 512$	$N = 1024$	$N = 2048$	Mean
$\bar{B}$	64.5430	64.3184	63.7266	64.1777	64.1914
$P/\%$	50.42	50.25	49.79	50.14	50.15
ΔB	5.6359	5.7108	5.7475	5.5859	5.6700
$\Delta P/\%$	4.40	4.46	4.49	4.36	4.405

TABLE 17.2

Absolute Differences of Two Hash Values

Maximum	Minimum	Mean	Mean/Character
2224	573	1401.1	87.5625

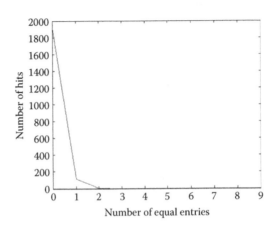

FIGURE 17.6 Distribution of the number of ASCII characters with the same value at the same location in the hash value.

performed 2048 times, with the secret key $X_0 = 0.232323$ and $P_0 = 0.858485$. Within them, the number of 0-hit is 1932, the number of 1-hit is 114, and the number of 2-hit is 2. The maximum, mean, minimum values of d, and the mean per character are listed in Table 17.2. A plot of the distribution of the number of hits is given in Figure 17.6. It should be noted that the maximum number of equal character is only 2 and the collision is very low.

17.4.2 LOCALIZATION HASH PERFORMANCE

In this section, we will focus on the performance of the 128-bit localization hash part. In the proposed algorithm, the 256 blocks of the lookup table are assigned into 16 horizontal groups and 16 vertical groups, respectively, as shown in Figure 17.1. The processing of each horizontal group or vertical group will generate the corresponding 4-bit LHK_i ($i = 1, 2, ..., 16$) or LVK_j ($j = 1, 2, ..., 16$), which constitute the final 128-bit localization hash value-LHASH. Note that each block belongs to a particular horizontal group and a particular vertical group at the same time. Therefore, any tiny modification in one block, actually in one character of the original message in a certain block from the lower level's point of view, will very likely lead to the changes in the corresponding LHK_i and LVK_j. Theoretically, if a particular modification happens, the miss probability in one direction will be $1/2^4$, and that in two directions will be $1/2^8$, which are quite small.

We have performed the following modification localization test. The chosen original message is the same as above, with the length of 572 characters. The tiny modifications include: the 1st character-"A" to "B"; the 6th character-"u" to "v"; the 136th character-"p" to "q"; the 256th character-"k" to "l." In the look-up table, the 1st character has been filled in Block 1, which belongs to H_Group 1 and V_Group 1; the 6th character has been filled in Block 6, which belongs to H_Group 1, and V_Group 6; the 136th character has been filled in Block 136, which belongs to H_Group 9 and V_Group 8; the 256th character has been filled in Block 256, which belongs to H_Group 16 and V_Group 16.

The comparison between the new localization hash value and the former one indicates that there are modifications within H_Group 1, H_Group 9, H_Group 16, V_Group 1, V_Group 6, V_Group 8, and V_Group 16. The possible modification blocks can be located within the intersection between the above groups, namely within 12 blocks (all the four modified blocks have been correctly located, but other eight innocent blocks have also been identified with false alarm), while the modification-free blocks are the rest 244 ones. From another perspective, it is more convenient to identify the modification-free area. On the condition that only a few modifications scatter in the message, this kind of localization is meaningful.

Theoretical analysis and simulation indicate that the proposed algorithm can realize modification localization, although some limitations still remain, such as miss probability, false alarm, and so on. It is our future work to find out more suitable solutions to overcome them, for example, by extending the length of 4-bit LHK_i and LVK_j, increasing the grouping time, or introducing a random number to grouping, and so on.

17.4.3 SECURITY OF KEY

The security of key includes two aspects. One is the key nonrecovery property. It must be computationally infeasible to recover key, given one or more message-hash value pairs. The other is the size of the key space, which characterizes the capability of resisting brute-force attack.

In the hashing process of our algorithm, the sensitivity to tiny changes in initial conditions and parameters as well as the mechanism of both changeable-parameter and self-synchronization are fully utilized. It makes the algorithm to possess strong one-way property, and there exist complicated nonlinear and sensitive dependence among message, hash value, and secret key. Therefore, it is immune from key recovery attack.

To investigate the key space size, the following evaluations are performed. Let the tiny change of the initial value X_0 of PWLCM be larger than 10^{-16}. For example, when X_0 is changed from 0.232323 to 0.2323230000000001, the corresponding changed bit number of detection hash value obtained is around 64. Similarly, let the tiny change of the initial parameters P_0 of PWLCM be larger than 10^{-16}. For example, when P_0 is changed from 0.858485 to 0.8584850000000001, the corresponding changed bit number of hash value obtained is also around 64. Readers can refer to Section 3.1.1. However, if the tiny changes of X_0 and P_0 are set as 10^{-17}, no corresponding bit of the hash value changes. Therefore, the sensitivities to X_0 and P_0 are

TABLE 17.3

Number of Required Multiplicative Operations for Each Character of Algorithms

	Schneier (1996)	Stinson (1995)	Wong (2003)	This Chapter
Multiplication	6	11.7	32 (by software) /8 (by hardware)	$4 + 33/(8\ s)$

Source: Adapted from Schneier, B., *Applied Cryptography: Protocols, Algorithms, and Source Code in C*, 2nd ed., Wiley, New York, 1996; Stinson, D. R., *Cryptography: Theory and Practice*, CRC Press, Boca Raton, FL, 1995; Wong, K. W., *Phys. Lett. A*, 307, 292–298, 2003.

s represents the number of 256-character (2048-bit) divisions of message after padding, namely the number of characters within each block.

both considered as 10^{-16}. Considering the value ranges of components, $X_0 \in [0, 1]$ and $P_0 \in (0, 1)$, it can be derived that the size of the key space is approximately larger than 2^{106}, which is large enough to resist the brute-force attack.

17.4.4 ANALYSIS OF SPEED

For speed comparison among different algorithms, the numbers of required multiplicative operations for each ASCII character (8-bit) message during the hash process are listed in Table 17.3. Since each multiplicative operation consumes much more time than each additive operation, this kind of comparison is objective, in spite of different implementing platforms. Obviously, our proposed algorithm is the fastest one.

The complexity calculation of our proposed algorithm is as follows. Each group has 16 blocks, and there are s characters within each block. First, step 1, in each block, the preprocessing of each character needs 4-time multiplicative operations. Therefore, all the 16 blocks in each group need 64 s-time multiplicative operations. Second, step 2, in each group, the processing of all the 16 blocks needs $4 \times 16 + 2 = 66$-time multiplicative operations. In all, the number of required multiplicative operations for each character in the proposed algorithm is $(64\ s + 66)/(16\ s) = 4 + 33/(8\ s)$. As the character number of message increases, the required multiplicative operation for each character becomes only slightly larger than 4.

Furthermore, since the proposed algorithm can support the parallel mode, its efficiency is predominant, especially compared to other hash algorithm in the sequential mode.

17.4.5 IMPLEMENTATION AND FLEXIBILITY

In the proposed hash algorithm, double precision floating-point arithmetic will be involved. Since IEEE 754 floating point standard is available on virtually almost all the computing platforms produced since 1980 (Goldberg and Priest, 1991), the proposed algorithm is suggested to implement with IEEE 754 double-precision floating-point arithmetic. Clearly, two hash values of a message with the same secret key produced on two computing platforms will be the same as long as IEEE 754 floating-point standard and the same operation order are implemented on both platforms, in spite of different O.S. or different program languages.

Through simply modifying the way to process X_{32}, X_{33}, and X_{34} in step 2 of the proposed algorithm, the length of the final hash value can be extended. Compared to the conventional hash algorithm such as MD5 with fixed 128-bit length, the proposed algorithm can adapt to the actual demand better.

17.5   CONCLUSIONS

In this chapter, a new hash function is proposed for both modification detection and localization. Its structure can also support the parallel processing mode. By means of the mechanism of both changeable-parameter and self-synchronization, the key-stream establishes a close relation with the algorithm key, the content, and the order of each message unit. The proposed algorithm fulfills the performance requirements of hash functions. It is simple, efficient, practicable, and reliable.

REFERENCES

Arumugam, G., Praba, V. L., and Radhakrishnan, S., Study of chaos functions for their suitability in generating message authentication codes, *Appl. Soft Comput.*, 7, 1064–1071, 2007.

Baranousky, A. and Daems, D., Design of one-dimensional chaotic maps with prescribed statistical properties, *Int. J. Bifurcation Chaos*, 5(6), 1585–1598, 1995.

Goldberg, D. and Priest, D., What every computer scientist should know about floating-point arithmetic, *ACM Comput. Surv.*, 23(1), 5–48, 1991.

Schneier, B., *Applied Cryptography: Protocols, Algorithms, and Source Code in C*, 2nd ed., Wiley, New York, 1996.

Shannon, C. E., Communication theory of secrecy systems, *Bell Syst. Tech. J.*, 28(4), 656–715, 1949.

Stinson, D. R., *Cryptography: Theory and Practice*, CRC Press, Boca Raton, FL, 1995.

Wong, K. W., A combined chaotic cryptographic and hashing scheme, *Phys. Lett. A*, 307, 292–298, 2003.

Xiao, D., Liao, X. F., and Deng, S. J., One-way hash function construction based on the chaotic map with changeable-parameter, *Chaos Solitons Fractals*, 24(1), 65–71, 2005.

Xiao, D., Liao, X. F., and Wong, K. W., Improving the security of a dynamic look-up table based chaotic cryptosystem, *IEEE Trans. Circuits Syst. II*, 53(6), 502–506, 2006.

Xiao, D., Shih, F. Y., and Liao, X., A chaos-based hash function with both modification detection and localization, *Commun. Nonlinear Sci. Numer. Simul.*, 15(9), 2254–2261, 2010.

Yi, X., Hash function based on chaotic tent maps, *IEEE Trans. Circuits Syst. II*, 52(6), 354–357, 2005.

Zhang, J. S., Wang, X. M., and Zhang, W. F., Chaotic keyed hash function based on feedforward–feedback nonlinear digital filter, *Phys. Lett. A*, 362, 439–448, 2007.

18 Video Forensics

Hsiao-Rong Tyan and Hong-Yuan Mark Liao

CONTENTS

18.1 INTRODUCTION

Computational forensics (Bijhold et al., 2007; Franke and Srihari, 2007, 2008; Lambert et al., 2007) is an emerging research field. Franke and Srihari (2007) discussed the growing trend of using mathematical, statistical, and computer science methods to develop new procedures for the investigation and prosecution of crimes as well as for law enforcement. Computational methods are becoming increasingly popular and important in forensic science because they provide powerful tools for representing human knowledge as well as for realizing the recognition and reasoning capabilities of machines. Another reason is that forensic testimony presented in court is often criticized by defense lawyers because they argue that it lacks a scientific basis (Saks and Koehler, 2005). However, Franke and Srihari point out that computational methods can systematically help forensic practitioners to (1) analyze and identify traces in an objective and reproducible manner; (2) assess the quality of an examination method; (3) report and standardize investigative procedures; (4) search large volumes of data efficiently; (5) visualize and document the results of analysis; (6) assist in the interpretation and presentation of results; and (7) reveal previously unknown patterns/links, to derive new rules and contribute to the generation of new knowledge.

A survey of the literature shows that computational forensics has become a very popular research area in recent years because it provides crucial crime scene investigation tools. Indeed, recent advances in computational forensics could benefit a number of research fields that are closely related to criminal investigations, for example, DNA analysis, blood spatter analysis, bullet trajectory analysis, voiceprint matching, fingerprint recognition, and video surveillance. In this chapter, we focus on video forensics, which we define as follows: Techniques that utilize computers to process video content and then apply the results to crime investigations can be called video forensics. Sun et al. (2005, 2006) developed an on-road vehicle detection system that can be used to alert drivers about driving environments and possible collision with other vehicles. Robust and reliable computer vision-based detection is a critical component of the system. To utilize the system in crime investigations, it is only necessary to modify the original system slightly. In the field of video processing, the trajectory of moving objects is one of the most widely used features for data retrieval. The trajectory of a moving object can be exploited to perform a coarse video search when a video database is very large. Calderara et al. (2009) developed a system that performs video surveillance and multimedia forensics by using the underlying feature trajectory to solve problems. Su et al. (2007) proposed a motion flow-based video retrieval system that links continual motion vectors with similar orientations to form motion flows. They also designed an algorithm to convert a group of motion flows into a single trajectory based on their group tendency. With a slight modification, the technique can be applied in video forensics. For example, the command "search all vehicles driving east" becomes a simple event detection problem if the trajectories of all moving objects are calculated in advance.

Since the 9/11 attack on the United States, counterterrorism strategies have been given a high priority in many countries. Surveillance camcorders are now almost ubiquitous in modern cities. As a result, the amount of recorded data is enormous, and it is extremely difficult and time-consuming to search the digital video content manually. The problem was highlighted by Worring and Cucchiara (2009) who presented a paper titled "Multimedia in Forensics" at the 2009 ACM Multimedia Conference in Beijing. They observed that "traces used to be fingerprints, fibers, documents and the like, but with the proliferation of multimedia data on the Web, surveillance cameras in cities, and mobile phones in everyday life we see an enormous growth in multimedia data that needs to be analyzed by forensic investigators." Clearly, the investigators should be highly trained. However, in a 2010 report, the United States Academy of Sciences observed that "much of forensic science is not rooted in solid science." Meanwhile, Neufeld and Scheck (2010) reported that "unvalidated or improperly used forensic science contributed to approximately half of the 251 convictions in the United States that have been overturned after DNA testing since 1989." In 2005, Saks and Koehler (2005) challenged the conventional forensic science training process. Specifically, they stated that "... in normal science, academically gifted students require four or more years of doctoral training. During the training process, methodological rigor, openness, and cautious interpretation of data form the major components of a training process." However, they found that "96% of the positions in forensic science are held by persons with bachelor's degrees, 3% with master's degrees, and only 1% with PhD degrees." For experts trained to conduct research in the field of video forensics, an in-depth knowledge of image/video processing is essential.

In video forensics, mining for criminal evidence in videos recorded by a heterogeneous collection of surveillance camcorders is a major challenge. This is a new interdisciplinary field, and people working in the field need video processing skills as well as an in-depth knowledge of forensic science; hence, the barrier for entering the field is high. Mining surveillance videos directly for criminal evidence is very different from conventional crime scene investigations. In the latter, detectives need to actually visit the crime scene, check all available details and collect as much physical evidence as possible, for example, blood samples, hair, fingerprints, DNA, and weapons. In contrast, to conduct crime scene investigations directly from surveillance videos, forensic experts need to develop software that facilitates the automatic detection, tracking, and recognition of objects in the videos. Since the videos are captured by heterogeneous camcorders, to perform evidence mining on these videos is more challenging.

In the remainder of this chapter, we consider some video forensics-related scenarios in Section 18.2, describe existing technologies needed by experts in the field of video forensics in Section 18.3, and discuss important technologies that need to be developed in Section 18.4.

18.2 SOME VIDEO FORENSICS-RELATED SCENARIOS

In this section, we consider some scenarios that are closely related to the field of video forensics. From these scenarios, it is possible to identify some image/video processing-related techniques that are needed to facilitate the investigation of crime scenes.

Scenario 1

A group of thieves stole a car with an apple logo and used it in a bank robbery. The license plate number could not be seen in the video of the crime scene because of the viewing angle problem. Therefore, detectives used a rotation, scaling, and translation-invariant object detection algorithm (RST-invariant) to search for the apple logo in videos recorded by the surveillance camcorders mounted in the neighborhood of the crime scene. If the detectives can identify the car in some of the video clips and see the license plate number clearly, then the search time can be reduced significantly.

Scenario 2

A lady was robbed and the surveillance camcorders in the vicinity recorded the whole incident. However, the camcorder mounted at the actual location of the robbery only captured a profile of the thief. Therefore, the police analyzed the gait of the suspect and compared it with the gaits of all pedestrians recorded by the camcorders in the vicinity. Then, the gaits that were closest to the suspect's gait formed a candidate set. Since the search space could be reduced significantly by comparing the subject's gaits, the police only needed to check the height of the subjects, the color of their clothes, and other features to identify the suspect. Of course, a frontal shot of the suspect's face would be the ideal view.

Scenario 3

The general manager of a high-tech company was killed on the fifth floor of the company's building. The suspect took the elevator to the first floor and was picked up by a white van. The surveillance camcorder at the front door of the building captured an image of the side of the van, but did not provide any clue about the license plate number. Detectives checked the surveillance videos captured by neighboring camcorders and found a good shot of the license plate, but the characters were blurred due to the distance. The police contacted a famous image processing laboratory, which used a systematic method to distinguish the license plate number. The police then used the license plate number to identify the owner of the van and traced his/her cellular phone record. Based on the communications between the suspect and the person who gave instructions, the police were able to solve the crime.

Scenario 4

In the early hours of July 3, 2010, an old lady was hit by a speeding car at the intersection of Brown Avenue and Jackson Road. The car then sped away westbound along Jackson Road. The police retrieved the surveillance videos captured by all eastbound buses that were in the neighborhood during that period. They used these videos to analyze potential escape paths that the suspects may take.

From the above scenarios, it is clear that a number of video processing-based technologies are definitely needed. In the following, we discuss some existing forensics-related video processing technologies and others that are under development.

18.3 EXISTING TECHNOLOGIES THAT SUPPORT VIDEO FORENSICS

In this section, we discuss blurred license plate image recognition, invariant object recognition, trajectory analysis of moving objects, and video inpainting. These technologies can assist forensics experts to perform evidence mining on videos.

18.3.1 BLURRED LICENSE PLATE IMAGE RECOGNITION

To control the size of a video, the resolution of a video frame should not be too high. Under the circumstance, a suspect car or a human subject grabbed from a video directly is usually blurred. In Anagnostopoulos et al. (2006), the authors presented a survey of license plate recognition methods. They divided the license plate recognition procedure into three steps: (1) license plate detection, (2) character segmentation, and (3) character recognition. Existing license plate recognition methods process or modify the content of an image or video to some extent. In forensic science, the content of a video clip or an image captured from a crime scene is regarded as evidence. Conventional video/image processing techniques may modify the content of a video or an image so that the data presented as evidence are not exactly the same as the original version. The techniques are not suitable for forensic investigations because, in a court case, the defense lawyer could argue that the original evidence has been modified, and the prosecutor would find it difficult to refute the argument. Therefore, an appropriate image/video processing technique must not

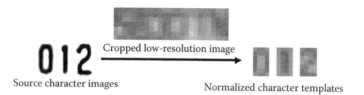

FIGURE 18.1   Normalizing source character images into character template images.

modify the content (i.e., potential evidence) in any way. In other words, the technique should be passive rather than active.

In a previous work (Hsieh et al., 2010), we proposed a filter that satisfies the above design requirement. Our approach can recognize the characters in blurred license plate images. Only one license plate image is required and character segmentation is not necessary. First, we use single-character templates to identify the positions of the characters and estimate the corresponding character list. Second, we train the character set, which includes all digits (from 0 to 9) and all English characters (from A to Z) to obtain 36 single-character templates, as shown in Figure 18.1. The training process is adaptive so it takes the height and average intensity of an extracted license plate image into consideration. Then, we slide the 36 single-character templates over the blurred license plate image and calculate the distance of all intensity difference during the sliding operation. As shown on the left-hand side of Figure 18.2a, we

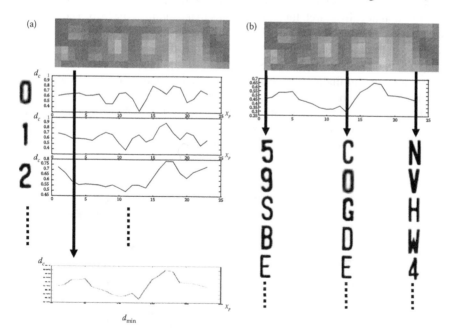

FIGURE 18.2   Candidate character positions and corresponding character estimation: (a) the calculated similarity curves and the minimum distance curve and (b) the estimated candidate positions and part of their corresponding characters.

calculate the minimum distance at each position and form a minimum distance curve (the bottom of Figure 18.2a). The local minima of a minimum distance curve indicate the potential positions of the characters (Figure 18.2b). However, the results obtained by sliding single-character templates are not reliable due to the blurred nature of the image; therefore, we need to expand the size of the sliding window to double-character templates and then multiple-character templates. We also have to consider special symbols, which are very common on license plates. The symbols may vary and may be located in different positions on license plates. We solved the special symbol problem in Hsieh et al. (2010) and refined the recognition results by expanding the single-character templates to multiple-character templates. Since a license plate image may be captured at a slanted angle, we also handle the perspective effect. Figure 18.3 shows the image of a car and its correct license plate number 5E-1340 (8th place out of more than 10 million combinations). This technique could help the police solve the problem described in Scenario 3.

18.3.2 ROTATION-, TRANSLATION-, AND SCALING-INVARIANT OBJECT RECOGNITION

Rotation-, translation-, and scaling-invariant (RST-invariant) object recognition is a very important technique in video forensics. In scenario 1, a group of thieves stole a car with an apple logo and used it to commit a crime and escape. The police retrieved all surveillance videos recorded in the vicinity of the crime scene and used the apple logo to automatically search for all video clips that contained the logo. The most difficult part of this type of problem is that the search involves videos recorded by a heterogeneous collection of camcorders. Therefore, the designed features should be independent of cameras. Lowe (1999) proposed a scale-invariant feature (SIFT) that is very effective in finding the corresponding points of two images. However, the requirement for high-dimensional feature input restricts its flexibility. For example, it is difficult to retrieve objects with similar structures but slightly different patterns. Compared to conventional interest points-based descriptors, shape descriptors are more powerful for general object detection because they extract more semantic

FIGURE 18.3 An example of recognition performed on perspective projection distorted license plate images. The correct license plate number is ranked 8th in the top 10 list.

FIGURE 18.4   A hand-drawn query model.

information. Ferrari et al. (2006) introduced the kAS family of local contour features to demonstrate its power within a slide window-based object detector. Felzenszwalb and Schwartz (2007) proposed a hierarchical representation to capture shape information from multiple levels of resolution, while Zhu et al. (2008) extended the shape context of selected edges to represent contours on multiple scales.

In Su et al. (2010), we proposed an object detection method that does not need a learning mechanism. The method uses a local descriptor-based algorithm to detect objects in a cluttered image. Given a hand-drawn model as a query (as shown in Figure 18.4), we can detect and locate objects that are similar to the query model without learning. The algorithm matches partial shapes based on the local polar histogram of edge features and the histogram of gradient information, as shown in Figure 18.5. In contrast to conventional feature descriptors, which usually use more than one scale to perform matching, the scale of our descriptor is determined based on two random high-curvature points (HCPs) on edges. The two HCPs decide the pose and size of each polar histogram to ensure invariance with respect to rotation, scaling, and translation. Figure 18.5a shows a circular descriptor determined by two HCPs, P1 and P2, and Figure 18.5b shows the edge-based descriptor that partially represents the target shape. Figure 18.5c shows the gradient histogram, which can be used with the edge-based descriptor to describe a target object. Finally, we locate the target objects in real images by a voting process. Figure 18.6 shows some examples of how target objects are detected. This technique can be applied to perform object search in a large database.

18.3.3   Trajectory Analysis on Moving Objects

In this section, we discuss how the trajectory of a moving object in a video can be used for event detection in a large surveillance video database. Since the amount of video content collected from a large number of surveillance camcorders is enormous, it is extremely difficult to examine the videos manually. A large number of motion-based video retrieval systems have been proposed in the last decade (Chang et al., 1998; Dagtas et al., 2000; Fablet et al., 2002; Ma and Zhang, 2002; Manjunath et al., 2002). The existing motion descriptors can be classified into two types: (1) statistics-based and (2) object-based. Fablet et al. (2002) used causal Gibbs models

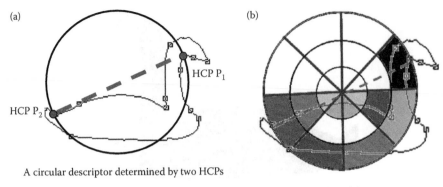

(a) A circular descriptor determined by two HCPs

(b) Edge-based descriptor

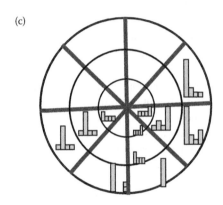

(c) Histogram that describes gradient information

FIGURE 18.5 Our algorithm matches partial shapes based on the local polar histogram of edge features and the histogram of gradient information. (a) A circular descriptor determined by two HCPs, (b) local polar histogram of edge features, and (c) the histogram of gradient information.

to represent the spatio-temporal distribution of the dynamic content in a video shot. In Ma and Zhang (2002), a multidimensional vector is generated by measuring the energy distribution of a motion vector field. In Manjunath et al. (2002), some fundamental bases of motion activity, such as intensity, direction, and spatial-temporal distribution, have been adopted to retrieve content from video databases. Chang et al. (1998) proposed object-based motion descriptors. They grouped regions that are similar in color, texture, shape, and motion together to form a video object. Then, the query-by-sketch mechanism is applied to retrieve video content from a database. The constituents of their database comprise a set of trajectories formed by linking the centroids of video objects across consecutive frames. Dagtas et al. (2000) proposed using a combination of trajectory—and trail-based models to characterize the motion of a video object. They adopted a Fourier transform-based similarity metric and a two-stage comparison strategy to search for similar trajectories.

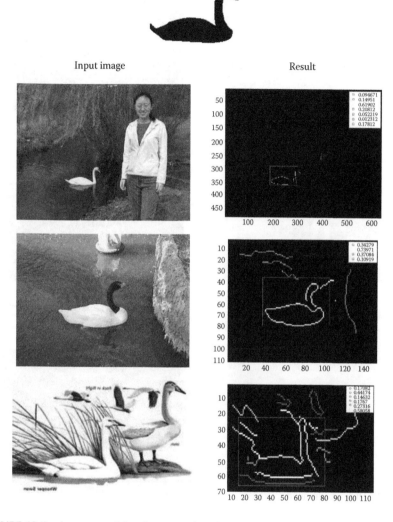

FIGURE 18.6 A query model and some retrieved results.

In Su et al. (2007), we proposed a motion flow-based video retrieval scheme that detects events in a video database based on the trajectories of moving objects. We exploit the motion vectors embedded in MPEG bitstreams to generate "motion flows," which are then used to detect events. By using the motion vectors embedded in videos directly, we do not need to consider the shape of a moving object and its corresponding trajectory. Instead, we simply link the local motion vectors across consecutive video frames to form motion flows, which are recorded and stored in a video database. The event detection process can be executed by comparing the motion flow of a query video with all the motion flows in a video database.

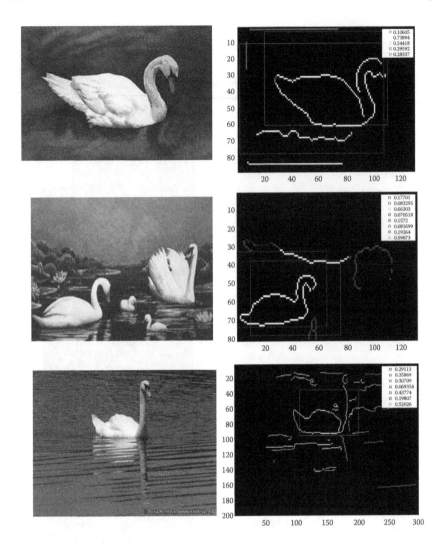

FIGURE 18.6 Continued.

Figure 18.7 shows an event detection process using a trajectory summarized from motion flows. In this example, the user drew a climbing-wall trajectory with the help of a user interface. A total of 108 events were stored in trajectory format in the database. Among the top six retrieved events, only the second event occurred at another location. However, its corresponding trajectory was very close to the query sketch. This example demonstrates that the trajectory of a moving object in a video can be a good descriptor for searching for target events. In video forensics, we often encounter a scenario like "the suspect is heading west in a stolen car." If we can record the trajectories of all moving objects in the surveillance videos, we can simply convert the query to "search all trajectories heading west." In this way, we can save a lot of manpower during the event detection process.

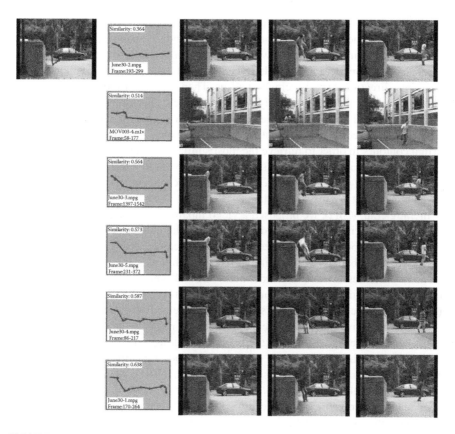

FIGURE 18.7 Event detection using trajectory information. A detected query trajectory (left) and some retrieved database events with similar trajectories (right).

18.3.4 VIDEO INPAINTING

Video inpainting has become a very popular research topic in recent years because of its powerful ability to fix/restore damaged videos and the flexibility it provides for editing home videos. Most inpainting techniques were developed to deal with images (Efros and Leung, 1999; Bertalmio et al., 2000; Criminisi et al., 2004). Conventional image inpainting techniques can be categorized into three groups: texture synthesis-based methods (Efros and Leung, 1999), partial differential equation-based (PDE-based) methods (Bertalmio et al., 2000), and patch-based methods (Criminisi et al., 2004). The concept of texture synthesis is derived from computer graphics. The idea is to insert a specific input texture into a damaged region. In contrast, PDE-based methods try to propagate information from the boundary of a missing region toward the center of that region. The above two methods are suitable for repairing damaged images that have thin missing regions. Patch-based methods, on the other hand, are much better for repairing damaged images because they yield high-quality visual results and maintain the consistency of local structures. Since patch-based methods are quite successful in image inpainting, Wexler et al. (2007) applied a similar concept to video inpainting. However, the issues that need to be dealt with in video inpainting are much more challenging. In Ling et al. (2011), we presented a framework for completing objects in a video. To complete an occluded object, we first sample a 3-D volume of the video into directional spatio-temporal slices. Then, we perform patch-based image inpainting to complete the partially damaged object trajectories in the 2-D slices. The completed slices are combined to obtain a sequence of virtual contours of the damaged object. Next, we apply a posture sequence retrieval technique to the virtual contours in order to retrieve the most similar sequence of object postures in the available nonoccluded postures. In video forensics, this technique can be used to restore moving object in the foreground if some consecutive frames that contain the moving object are missing or damaged.

In Tang et al. (2011), we proposed a video inpainting technique and used it as a video enhancement tool to repair digital videos. The quality of many surveillance videos is very poor due to the resolution of the camcorders used for recording. Therefore, some better video inpainting-related techniques are desperately needed. In recent years, researchers have extended some well-known image inpainting techniques to the repair of videos. An intuitive approach involves applying image inpainting techniques to each video frame so that the completed frames are visually pleasing when viewed individually. However, this approach neglects the issue of continuity across consecutive frames, so the quality of the resulting video is usually unsatisfactory. To resolve the problem, both spatial and temporal continuity must be considered in a video inpainting process. In Tang et al. (2011), we proposed two key techniques, namely, *motion completion* and *frame completion*, to perform better video inpainting job. Since accurate motion information is the key to achieving good video inpainting results, we begin by constructing a motion map to track the motion information in the background layer and then repair the missing motion information in the motion completion step. Motion completion recovers missing motion information in damaged areas to maintain good temporal continuity. Frame completion, on the other hand, repairs damaged frames to produce a visually pleasing video with good spatial continuity and stable luminance.

18.4 SOME REQUIRED TECHNIQUES THAT MAY HELP PROMOTE VIDEO FORENSICS

To deal with video forensics-related problems, forensic experts need to develop more useful software tools to facilitate forensics-related investigations. In this section, we shall mention a number of related techniques of this sort.

18.4.1 PEOPLE COUNTING IN VIDEOS

The number of human subjects in a short video clip provides a clue for conducting a coarse search of a large surveillance video database. In video forensics research, automating the video search process with the assistance of computers is a key objective. Among different types of features that can be used in searches, determining the existence of human subjects or counting the number of human subjects in videos are two very useful screening criteria. Knowing whether or not a video clip contains human subjects can reduce the search space significantly, while knowing the number of human subjects can improve the accuracy of annotating the video content. For example, if three suspects rob a bank and escape in a vehicle traveling east, the police can retrieve the videos captured by all surveillance camcorders along the escape route. Searching the videos manually would be extremely time-consuming. However, if the police use an annotation scheme that can determine the number of people in each video clip, they could easily select the clips that contain at least three people as fine search targets.

We proposed two people counting schemes in previous works (Chen et al., 2009; Su et al., 2009). In Chen et al. (2009), we introduced an online boosted people counting system for electronic advertising machines. The system can count the number of people that watch an advertisement on a large wall TV during a given period. More specifically, the system integrates face detection, face matching, dynamic face database management, and machine learning techniques to count people in real time. In Su et al. (2009), we proposed a vision-based people counting system that utilizes the symmetry feature to detect human subjects in a video clip. Since the symmetry feature is very easy to detect, the system can run in real time and achieve a high success rate. However, the above approaches can only partially assist a general-purpose people counting scheme, which must detect human subjects in a variety of postures, such as sitting, standing, running, walking, or lying down. In addition, the size of human subjects in the field of view may vary significantly. The people counting task may also be affected by the lighting conditions, the viewing angle, and the weather. Therefore, counting the number of people in a video is a challenging task. Recently, we propose a cross-camera people counting scheme which is able to deal with perspective effect as well as occlusion (Lin et al., 2011). Our scheme can adapt itself to a new environment without the need of manual inspection. The proposed counting model is composed of a pair of collaborative Gaussian processes, which are respectively designed to count people by taking the visible and occluded parts into account. The first Gaussian process exploits multiple visual features to result in better accuracy and the second Gaussian process instead investigates the conflicts among these features to recover the underestimate caused by occlusions. Our proposed system achieved promising performances.

18.4.2 RECOGNIZING VIDEO OBJECTS USING FEATURES EXTRACTED FROM A VIDEO SHOT (CLIP)

In video forensics, it is often necessary to retrieve every possible piece of evidence from the video clip of a crime scene. Since all the evidence that may help solve the crime is embedded in the video clip, it is necessary to extract features from the clip directly. Therefore, the problem becomes: "What features would be the most effective for solving the crime?" To efficiently manage video information, including proper indexing, efficient storage, and fast retrieval, it is required to develop better video indexing and fast search algorithms. In the last decade, many crucial algorithms, such as shot change detection (Su et al., 2005), shot representation (Chang et al., 1999), key frame/clip extraction (Lee and Kim, 2003), and video sequence matching (Cheung and Zakhor, 2003), have been developed to enhance video indexing and retrieval.

In Ho et al. (2006), we used motion vector statistics extracted directly from a shot to represent a video shot. The video retrieval scheme we proposed is a coarse-to-fine shot-based approach. The coarse search is to select a reasonably small number of video clips from a database, while avoiding nondetection of correct clips. We check the entropy of the motion vectors from every constituent shot and pick the one with the maximum entropy as the query shot. This shot usually has the most diverse object motions. The first step of the coarse search identifies a set of similar video clips by using shot-level spatio-temporal statistics. Then, in the second step, an adjacent shot to the first query shot is chosen, and the two shots are concatenated to form a two-shot query. A "causality" relation that defines the order of two consecutive shots is introduced to strengthen the discriminating capability. In the coarse search process, we extract object motions and then quantize them into 2-D probability distribution. The feature of this form is the temporal feature. In addition, the color histogram of key frames extracted from the same shot is used as the spatial feature. The joint distance, which sums up the distance of the motion statistics and that of the color histograms, is then used to measure the similarity between the two shots. Following the coarse search, a fine search is performed to enhance the retrieval accuracy. In the fine-search process, we extract color features from a set of selected key frames and use them to refine the ranks of the matched video clips obtained in the coarse search. Finally, we calculate the Bhattacharyya distance and use it to choose the closest shots from the coarse-search outcomes. Using similar strategies in video forensics, it is definitely helpful in a search-space reduction process.

18.5 CONCLUSIONS

Since the 9/11 attack on the United States, counterterrorism strategies have been given a high priority in many countries. Surveillance camcorders are now almost ubiquitous in modern cities. As a result, the amount of recorded data is enormous, and it is extremely difficult and time-consuming to search the digital video content manually. In this chapter, we consider some video forensics-related scenarios, describe existing technologies needed by experts in the field of video forensics, and discuss important technologies that need to be developed. We hope the initiation of video forensics can be a good start and we look forward to seeing the prosperity of its future.

REFERENCES

Anagnostopoulos, C. N., Anagnostopoulos, I., Loumos, V., and Kavafas, E., A license plate-recognition algorithm for intelligent transportation system applications, *IEEE Transactions on Intelligent Transportation Systems*, 7(3), 377–392, 2006.

Bertalmio, M., Sapiro, G., Caselles, V., and Ballester, C., Image inpainting, in *Proceedings of ACM SIGGRAPH*, pp. 417–424, 2000.

Bijhold, J. et al., Forensic audio and visual evidence 2004–2007: A review, in *Proceedings of the 15th INTERPOL Forensic Science Symposium*, Lyon, France, October 2007.

Calderara, S., Prati, A., and Cucchiara, R., Video surveillance and multimedia forensics: An application to trajectory analysis, in *Proceedings of First ACM Workshop on Multimedia in Forensics*, Beijing, China, pp. 13–18, October 2009.

Chang, H. S., Sull, S., and Lee, S. U., Efficient video indexing scheme for content-based retrieval, *IEEE Transactions on Circuits and Systems for Video Technology*, 8(8), 1269–1279, 1999.

Chang, S. F., Chen, W., Meng, H. J., Sundaram, H., and Zhong D., A fully automated content-based video search engine supporting spatiotemporal queries, *IEEE Transactions on Circuits and Systems for Video Technology*, 8(5), 602–615, 1998.

Chen, D. Y., Su, C. W., Zeng, Y. C., Sun, S. W., Lai, W. R., and Liao, H. Y. M, An on-line boosted people counting system for electronic advertising machines, in *Proceedings of IEEE International Conference on Multimedia and Expo*, New York, June 2009.

Cheung, S. and Zakhor, A., Efficient video similarity measurement with video signature, *IEEE Transactions on Circuits and Systems for Video Technology*, 13(1), 59–74, 2003.

Criminisi, A., Perez, P., and Toyama, K, Region filling and object removal by exemplar-based image inpainting, *IEEE Transactions on Image Processing*, 13(9), 1200–1212, 2004.

Dagtas, S., Al-Khatib, W., Ghafoor, A., and Kashyap, R. L., Models for motion-based video indexing and retrieval, *IEEE Transactions on Image Processing*, 9(1), 88–101, 2000.

Efros, A. and Leung, T., Texture synthesis by non-parametric sampling, in *Proceedings of the IEEE Conference on Computer Vision*, vol. 2, pp. 1033–1038, 1999.

Fablet, R., Bouthemy, P., and Perez, P., Nonparametric motion characterization using causal probabilistic models for video indexing and retrieval, *IEEE Transactions on Image Processing*, 11(4), 393–407, 2002.

Felzenszwalb, P. and Schwartz, J., Hierarchical matching of deformable shapes, in *Proceedings of the IEEE Conference on Computer Vision and Pattern Recognition*, pp. 1–8, 2007.

Ferrari, V., Tuytelaars, T., and Van Gool, I. J., Object detection by contour segment networks, in *Proceedings of the European Conference on Computer Vision*, vol. 3, pp. 14–28, 2006.

Franke, K. and Srihari, S. N., Computational forensics: Towards hybrid-intelligent crime investigation, in *Proceedings of the 3rd International Symposium on Information Assurance and Security*, 2007.

Franke, K. and Srihari, S. N., Computational forensics: An overview, in *Proceedings of the International Workshop on Computational Forensics*, Lecture Notes in Computer Science, vol. 5158, pp. 1–10, 2008.

Ho, Y. H., Lin, C. W., Chen, J. F., and Liao, H. Y. M., Fast coarse-to-fine video retrieval using shot-level spatio-temporal statistics, *IEEE Transactions on Circuits and Systems for Video Technology*, 16(5), 642–648, 2006.

Hsieh, P. L., Liang, Y. M., and Liao, H. Y. M., Recognition of blurred license plate images, in *Proceedings of the IEEE International Workshop on Information Forensics and Security*, Seattle, Washington, USA, December, 2010.

Lambert, E., Hogan, N., Nerbonne, T., Barton, S., Watson, P., Buss, J., and Lambert, J., Differences in forensic science views and needs of law enforcement: A survey of Michigan law enforcement agencies, *Police Practice and Research*, 8(5), 415–430, 2007.

Lee, H. C. and Kim, S. D., Iterative key frame selection in the rate-constraint environment, *Signal Processing: Image Communication*, 18, 1–15, 2003.

Lin, T. Y., Lin, Y. Y., Weng, M. F., Wang, Y. C., Hsu, Y. F., and Liao, H. Y. M., Cross camera people counting with perspective estimation and occlusion handling, in *Proceedings of the IEEE International Workshop on Information Forensics and Security*, Brazil, 2011.

Ling, C. H., Lin, C. W., Su, C. W., Liao, H. Y. M., and Chen, Y. S., Virtual contour guided video object inpainting using posture mapping and retrieval, *IEEE Transactions on Multimedia*, 13(2), 292–302, 2011.

Lowe, D. G., Object recognition from local scale-invariant features, in *Proceedings of the International Conference on Computer Vision*, vol. 2, pp. 1150–1157, 1999.

Ma, Y. F. and Zhang, H. J., Motion texture: A new motion-based video representation, in *Proceedings of the 16th International Conference on Pattern Recognition*, vol. 2, pp. 548–551, August 11–15, 2002.

Manjunath, B. S., Salembier, P., and Sikora, T., *Introduction to MPEG-7: Multimedia Content Description Interface*, Wiley, Berlin, June 2002.

Neufeld, P. and Scheck, B., Making forensic science more scientific, *Nature*, 464, 351, 2010.

Saks, M. J. and Koehler, J. J., The coming paradigm shift in forensic identification science, *Science*, 309, 892–895, 2005.

Su, C. W., Liang, Y. M., Tyan, H. R., and Liao, H. Y. M., An RST-tolerant shape descriptor for object detection, in *Proceedings of the International Conference on Pattern Recognition*, Istanbul, Turkey, August 2010.

Su, C. W., Liao, H. Y. M., and Tyan, H. R., A vision-based people counting approach based on the symmetry measure, in *Proceedings of the IEEE International Symposium on Circuits and Systems*, May 2009.

Su, C. W., Liao, H. Y. M., Tyan, H. R., and Chen, L. H., A motion-tolerant dissolve detection algorithm, *IEEE Transactions on Multimedia*, 7(6), 1106–1113, 2005.

Su, C. W., Liao, H. Y. M., Tyan, H. R., Lin, C. W., Chen, D. Y., and Fan, K. C., Motion flow-based video retrieval, *IEEE Transactions on Multimedia*, 9(6), 1193–1201, 2007.

Sun, Z., Bebis, G., and Miller, R., On-road vehicle detection using evolutionary Gabor filter optimization, *IEEE Transactions on Intelligent Transportation Systems*, 6(2), 125–137, 2005.

Sun, Z., Bebis, G., and Miller, R., On-road vehicle detection: A review, *IEEE Transactions on Pattern Analysis and Machine Intelligence*, 28(5), 694–711, 2006.

Tang, N. C., Hsu, C. T., Su, C. W., Shih, T. K., and Liao, H. Y. M., Video inpainting on digitized vintage films via maintaining spatiotemporal continuity, *IEEE Transactions on Multimedia*, 13(4), 603–614, 2011.

Wexler, Y., Shechtman, E., and Irani, M., Space-time completion of video, *IEEE Transactions on Pattern Analysis and Machine Intelligence*, 29(3), 1–14, 2007.

Worring, M. and Cucchiara, R., Multimedia in forensics, in *Proceedings of the ACM Conference on Multimedia*, pp. 1153–1154, Beijing, China, October 19–24, 2009.

Zhu, Q., Wang, L., Wu, Y., and Shi, J., Contour context selection for object detection: A set-to-set contour matching approach, in *Proceedings of the European Conference on Computer Vision*, vol. 2, pp. 774–787, 2008.

19 Using the Self-Synchronizing Method to Improve Security of the Multi-Chaotic Systems-Based Image Encryption

Di Xiao and Frank Y. Shih

CONTENTS

19.1 INTRODUCTION

With rapid development and wide applications in digital technology, the security of digital images has attracted increasing attention. Applying the merits of chaos to protect image security is a promising research field. In order to achieve significant confusion and diffusion, a chaotic image encryption scheme typically includes two stages: the first stage is to permute all the pixels of an image as a whole by using a two-dimensional (2D) chaotic map, and the second stage is to change the entire pixel values of the permutated image sequentially. To the best of our knowledge, this scheme was first proposed by Fridrich (1998). Since then, a series of variants have been developed. For example, Guan et al. (2005) used a 2D Cat map in the substitution stage and a chaotic system in the diffusion stage. Lian et al. (2005) used a chaotic standard map for pixel position permutation and a quantized logistic map for pixel value masking. Chen et al. (2004) employed a three-dimensional Cat map for permuting the pixel position. Compared with the chaotic image encryption scheme, a novel scheme was developed by Huang and Nien (2009). Its novelty lies in the fact that the encrypted image is obtained only by vertical and horizontal pixel shuffling, while the shuffling parameters are generated by iterating four chaotic systems and sorting the chaotic sequences. However, Solak et al. (2010) pointed out a security problem of this scheme and attacked it using the chosen-plaintext/known-plaintext method.

In this chapter, we first analyze the vulnerability of the original multi-chaotic systems-based image encryption scheme, and then propose the corresponding enhancement measures to defeat cryptanalysis (Xiao and Shih, 2010). Theoretical analysis and computer simulation indicate that the modified scheme can overcome the security flaw and still hold the characteristics of the original scheme for realizing image encryption by vertical and horizontal pixel shuffling.

19.2 BRIEF REVIEW ON THE ORIGINAL SCHEME AND ITS SECURITY FLAW

19.2.1 ORIGINAL SCHEME

In the original encryption scheme, four chaotic systems are used: Hénon Map (discrete map), Lorenz (butterfly attractor), Chua (double scroll attractor), and Rössler (spiral attractor) (Huang and Nien, 2009). This scheme consists of two-stage shuffling. In the first stage, designated bits of all the pixels are shuffled in the vertical direction. In the second stage, the bits of each pixel are shuffled among themselves in the horizontal direction. The shuffling parameters are obtained by iterating the above four chaotic systems and sorting the chaotic variable sets. The scheme key includes 12 initial conditions of these chaotic maps. The image size is $m \times n$.

19.2.2 SECURITY FLAW

According to the original scheme, the generation of the shuffling parameters has nothing to do with the pending plain image. Instead, it depends solely on the key, namely the 12 initial conditions of the four chaotic maps. Regardless of different plain images, the same key will always yield the same shuffling parameters, and then the shuffling process for every plain image is the same, which would benefit an attacker. The cryptanalysis proposed in Solak et al. (2010) takes the advantage of this drawback. The key of the original scheme includes 12 initial conditions of the four chaotic maps. Due to the huge key space, it seems impossible for an attacker to reveal these initial conditions. However, those conditions are used to control the generation of the chaotic sequences and the corresponding shuffling parameters, which are used to decide the shuffling process and realize its encryption. If an attacker knows the shuffling parameters, he or she is able to decrypt the encrypted image even though the initial conditions of the chaotic maps are unknown. Thus, in the original scheme, its equivalent key is the shuffling parameters, namely the sorting sequences Fx, Fy, Fz. In Solak et al. (2010), the chosen-plaintext/known-plaintext attacks are used to deduce the equivalent key successfully. Details of these attacks can be found in Solak et al. (2010).

19.3 MODIFIED SCHEME AND ITS PERFORMANCE ANALYSIS

The reason why the chosen-plaintext/known-plaintext attacks can totally or partially break the original scheme is that the generation of the shuffling parameters depends solely on the key and has nothing to do with the pending plain image. In order to defeat these attacks, a close link between the shuffling parameters and the pending plain image must be established. We propose a self-synchronizing method to solve this problem, whose name is derived from the self-synchronizing stream cipher in classical cryptography. In the original self-synchronizing stream cipher, the keystream is generated as a function of both the key and a fixed number of previous ciphertext/plaintext digits (Vanstone et al., 1996). Similarly, in our scheme, we partition an image into two parts, and then utilize the first part to control the encryption of the second part for ensuring that the generation of the shuffling parameters of the second part depends on both the key and the first image part.

19.3.1 MODIFIED SCHEME

19.3.1.1 Partition of the Original Image into Two Parts

An image can be partitioned into two parts in any size and any direction. In other words, there is no limitation on whether the pending image is a square or a rectangle, or the size of image is an odd or an even number. For simplicity, we illustrate the same size and the horizontal direction in this section.

19.3.1.2 Carry Out Self-Synchronizing Image Encryption

The whole encryption process is shown in Figure 19.1a. First, the left part L is used to encrypt the right part R into R'; second, the right part R' is used to encrypt the left part L into L'; third, the left part L' is used to encrypt the right part R' into R''. Finally, the encrypted image is obtained by combining the two parts- L' and R''. The whole decryption process is in an inverse order, as shown in Figure 19.1b. The original plaintext is recovered by combining the two parts—L and R.

Without loss of generality, the detailed process of utilizing image part A to control the encryption of image part B is described below (see Figure 19.2). Note that the red, green, and blue components are processed independently. Image part A and part B represent the left part L and the right part R of the first-round encryption in Figure 19.1a, respectively. Similarly, they represent the right part R' and the left part L of the second-round encryption, as well as the left part L' and the right part R' of the third-round encryption in Figure 19.1a.

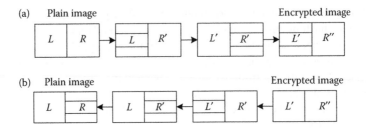

FIGURE 19.1 Whole structure of the modified scheme: (a) encryption process, (b) decryption process.

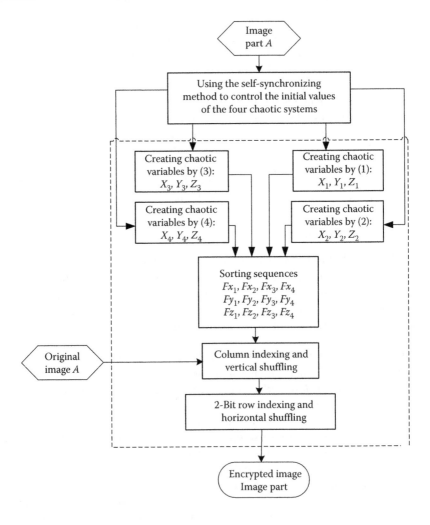

FIGURE 19.2 Utilize image part A to control the encryption of image part B.

1. Using the self-synchronizing method to control the initial values of the four chaotic systems.
   a. In a specific scan order, we map A into a one-dimensional array C by linear transform, in which each element is a number within $[0, 1]$, and the array length s is equal to the number of the pixels in A.
   b. The pixels in A are modulated into the chaotic iteration parameter and orbit using the following method, derived from the chaotic hash functions for text message (Xiao et al., 2005, 2006, 2008; Yi, 2005; Lian et al., 2006).

   We choose the initial value $XX_0 \in [0, 1]$ and the initial parameter $PP_0 \in (0, 1)$ of Piecewise Linear Chaotic Map (PWLCM). Let C_i, $i = 1$, $2, \ldots, s$ denote the linearly transformed pixel value. The iteration process of PWLCM is given as follows:

   1st iteration:
   $$PP_1 = (C_1 + PP_0)/4 \in (0, 0.5), \quad XX_1 = F_{PP_1}(XX_0) \in (0, 1),$$
   2nd–sth iterations:
   $$PP_k = (C_k + XX_{k-1})/4 \in (0, 0.5), \quad XX_k = F_{PP_k}(XX_{k-1}) \in (0, 1),$$
   $(s + 1)$th iteration:
   $$PP_{s+1} = (C_s + XX_s)/4 \in (0, 0.5), \quad XX_{s+1} = F_{PP_{s+1}}(XX_s) \in (0, 1),$$
   $(s + 2)$th–2sth iterations:
   $$PP_k = (C_{(2s-k+1)} + XX_{k-1})/4 \in (0, 0.5), \quad XX_k = F_{PP_k}(XX_{k-1}) \in (0, 1).$$

   Note that the PWLCM in Equation 19.1 is iterated to obtain the next iteration value XX_k under the control of the current initial iteration value XX_{k-1} and the current iteration parameter PP_k. Finally, the chaotic orbit value XX_{2s} is obtained.

   The PWLCM with the current iteration parameter $u \in (0, 0.5)$ is defined as

   $$x(k + 1) = F_u(x(k)) = \begin{cases} x(k)/u, & 0 \leq x(k) < u \\ (x(k) - u)/(0.5 - u), & u \leq x(k) < 0.5 \\ (1 - x(k) - u)/(0.5 - u), & 0.5 \leq x(k) < 1 - u \\ (1 - x(k))/u, & 1 - u \leq x(k) \leq 1 \end{cases}$$

   (19.1)

   c. Extract the 2nd, 3rd, and 4th digits of the current value of XX_{2s} after the decimal point to construct a new decimal integer Y, and then use it to replace the 12th, 13th, and 14th digits of the initial values of the four chaotic systems (Huang and Nien, 2009).

2. By iterating the four chaotic systems based on the obtained initial values and system parameters above, we create chaotic variable sets $X_1 \sim X_4$, $Y_1 \sim Y_4$, and $Z_1 \sim Z_4$, as well as the corresponding chaotic sequences $X_{1_{C,1}} \sim X_{4_{C,1}}, Y_{1_{C,1}} \sim Y_{4_{C,1}}$, and $Z_{1_{C,1}} \sim Z_{4_{C,1}}$, for $C = 1, 2, 3, \ldots, m \times n$.

3. By sorting the obtained chaotic sequences, we generate the indexing sequences $Fx_1 \sim Fx_4$, $Fy_1 \sim Fy_4$, and $Fz_1 \sim Fz_4$. That is

$$Fx_1 = \mathrm{sort}(X_{1_{C,1}}), Fx_2 = \mathrm{sort}(X_{2_{C,1}}), Fx_3 = \mathrm{sort}(X_{3_{C,1}}), Fx_4 = \mathrm{sort}(X_{4_{C,1}})$$

(19.2)

for $C = 1, 2, 3, \ldots, m \times n$, where $sort(*)$ is the sequencing index function. Accordingly, other sorting sequences $Fy_1 \sim Fy_4$ and $Fz_1 \sim Fz_4$ can be produced.

4. Transfer the original R-level, G-level, and B-level matrix ($AR_{m \times n}$, $AG_{m \times n}$, $AB_{m \times n}$) to the corresponding $Arb_{m \times n \times 1}$, $Agb_{m \times n \times 1}$, $Abb_{m \times n \times 1}$. As shown in Equation 19.3, each column of the matrix Arb is vertically shuffled by using one of the sequences Fx_1, Fx_2, Fx_3, Fx_4 to generate the intermediate matrix Aerb as

$$\mathrm{Aerb}(i, j) = \begin{cases} \mathrm{Arb}(Fx_1(i), j), & j = 1, 2, \\ \mathrm{Arb}(Fx_2(i), j), & j = 3, 4, \\ \mathrm{Arb}(Fx_3(i), j), & j = 5, 6, \\ \mathrm{Arb}(Fx_4(i), j), & j = 7, 8. \end{cases}$$

(19.3)

Similarly, the vertical shuffling is applied to transfer Agb and Abb to Aegb and Aebb respectively by the shuffling sequences Fy_1, Fy_2, Fy_3, Fy_4, and Fz_1, Fz_2, Fz_3, Fz_4.

For shuffling the red component of ith pixel horizontally, the sorting sequence of four numbers $Fx_1(i)$, $Fx_2(i)$, $Fx_3(i)$, and $Fx_4(i)$ is used. Similar shuffling of pixel bits is applied to green and blue components with the sorting sequences derived from Fy and Fz, respectively. The obtained three components, Br, Bg, and Bb, are joined into the final encrypted image.

Note that the above steps (2) through (4), namely, chaotic variable creation, chaotic sequence sorting, column vertical shuffling, and row horizontal shuffling, are similar to the original encryption scheme. But, the self-synchronizing method ensures that there is a close link between the shuffling parameters and A during the encryption process of B. Due to high sensitivity of chaos with respect to initial value and parameter, any tiny change in A will lead to totally different shuffling parameters. Consequently, the final encrypted image part B' will definitely be different.

19.3.2 RESISTANCE AGAINST CHOSEN-PLAINTEXT/KNOWN-PLAINTEXT ATTACK

As aforementioned, the reason why the chosen-plaintext or known-plaintext attack can work lies on the fact that the generation of shuffling parameters depends solely on the key and has nothing to do with the pending plain image. As long as the key keeps unchanged, the shuffling parameters generated are always the same regardless of different plain images. Consequently, the flaw may arise from the use of the same key for every plain image. As shown in Figure 19.1, through the alternate encryption of two image parts-L and R in our modified scheme, the whole encryption process establishes very complicated nonlinear connections with both the scheme key and the pending plain image. In other words, the shuffling parameters rely on both of them. Since different plain images correspond to different shuffling parameters, the base of chosen-plaintext/known-plaintext attack has been destroyed.

We conduct the following experiment to quantitatively analyze the dependency of shuffling parameters on plain images. We adopt the same image "Lena $(256 \times 256 \times 3)$" in (Huang and Nien, 2009). Figure 19.3 shows the original image and its red, green, and blue (RGB)-level spectrums. Without lose of generality, we take the left image part L of its red component as a sample.

First, the vertical and horizontal shuffling parameters corresponding to L are generated. Second, after a pixel in L is randomly selected and its least significant bit is changed, the new vertical and horizontal shuffling parameters are generated. Two kinds of parameters are compared, and the average absolute differences of two sequences are calculated by $d = 1/N \sum_{i=1}^{N} |e_i - e_i'|$, where e_i and e_i' denote the ith item of the two sorting sequences with length $N = 256 \times 128 = 2^{15}$. The experiment is repeated 2000 times. The maximum, mean, and minimum of the differences are listed in Table 19.1. The two kinds of differences are very close to the theoretical

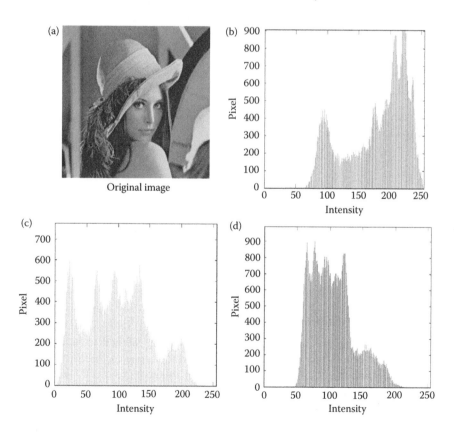

FIGURE 19.3 Lena and its RGB-level spectrums: (a) original image, (b) R-level spectrum, (c) G-level spectrum, and (d) B-level spectrum.

TABLE 19.1

Absolute Differences of Two Shuffling Parameters

		Maximum	Minimum	Mean
Vertical shuffling parameter	Fx_1	10,969.2225	10,913.4902	10,938.8884
	Fx_2	10,968.1901	10,908.5551	10,942.2428
	Fx_3	10,824.0464	10,778.1123	10,802.3767
	Fx_4	10,991.1125	10,879.1596	10,953.8003
Horizontal shuffling parameter		1.2469	1.2524	1.2516

values, 10,922.66 and 1.3333, indicating that even a little change in the pixel value will produce completely shuffling parameters.

The theoretical value 10,922.66 of vertical shuffling parameter difference is calculated as follows. If there are two independent and uniformly distributed random sequences, say e_i or e_i', having equal probability to be an integer of $\{1, 2, 3, \ldots, 2^{15}\}$, then the mean absolute difference is $1/3 \times 2^{15} = 10,922.6667$. Similarly, the theoretical value 1.3333 of horizontal shuffling parameter difference can be obtained when e_i or e_i' has equal probability to be an integer of $\{1, 2, 3, 4\}$.

Lemma

Let $\xi_1 \in (0, 1)$ and $\xi_2 \in (0, 1)$ be independent and uniformly distributed random variables. If $\eta = |\xi_2 - \xi_1|$, then the expected value $E(\eta) = 1/3$.

Proof:

$$E(\eta) = \int_0^1 \int_0^1 |x - y|\, dxdy = \int_0^1 dx \int_0^x (x - y)dy + \int_0^1 dx \int_x^1 (y - x)dy$$

$$= \int_0^1 0.5x^2 dx + \int_0^1 (0.5 - x + 0.5x^2)dx = \frac{1}{3} \tag{19.4}$$

19.3.3 OTHER PERFORMANCE ANALYSIS

In the modified scheme, we use the self-synchronizing method to modulate the image into the chaotic iteration parameter and orbit, so that a close link between the shuffling parameters and the pending plain image can be established. The image encryption processes, including chaotic variable creation, sorting chaotic sequences, column vertical shuffling, and row horizontal shuffling, are essentially similar to the original scheme. In order to compare the performances between the modified and the original schemes, we adopt the given experiments in Huang and Nien (2009).

19.3.3.1 Encryption Result

For the plain image "Lena $(256 \times 256 \times 3)$" in Figure 19.3, its corresponding encrypted image is shown in Figure 19.4. Obviously, not only Lena's outline but also RBG levels are changed dramatically. The encryption results of the modified scheme are similar to those of the original scheme.

19.3.3.2 Key Space

In the modified scheme, the secret key includes 13 initial conditions of the five chaotic maps and the initial parameter $u_0 \in (0, 1)$ of PWLCM. The sensitivity to these initial parameters is shown as follows (Xiao et al., 2008; Huang and Nien, 2009):

Hénon Map: $x_0 = 10^{-15}$, $y_0 = 10^{-14}$, $z_0 = 10^{-14}$
Lorenz: $x_0 = 10^{-15}$, $y_0 = 10^{-15}$, $z_0 = 10^{-14}$
Chua: $x_0 = 10^{-16}$, $y_0 = 10^{-16}$, $z_0 = 10^{-16}$
Rössler: $x_0 = 10^{-15}$, $y_0 = 10^{-15}$, $z_0 = 10^{-15}$
PWLCM: $x_0 = 10^{-16}$, $u_0 = 10^{-16}$.

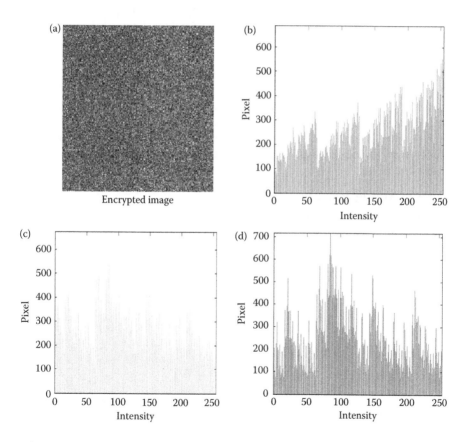

FIGURE 19.4 Encrypted Lena using the modified scheme and its RGB-level spectrums: (a) encrypted image, (b) R-level spectrum, (c) G-level spectrum, and (d) B-level spectrum.

Using the PWLCM in the modified scheme, we extract the 2nd, 3rd, and 4th digits of its final iteration value and replace the 12th, 13th, and 14th digits of the initial values of the other four chaotic systems. Therefore, we can derive the size of the key space, which is approximately 10^{176}. Although this number is smaller than the key space 10^{180} of the original scheme, it is still large enough to resist the brute-force attack.

To speed up, the extra iteration of PWLCM can be implemented in a parallel mode. While the image encryption is realized simply by vertical and horizontal pixel shuffling, the overall performance of the modified scheme is efficient.

19.3.3.3 Correlation Coefficient

Similar to the original scheme, we calculate the coefficient r of two adjacent pixels in the plain and the encrypted images as

$$r = \frac{n\left(\sum_{i=1}^{n} X_i Y_i\right) - \left(\sum_{i=1}^{n} X_i\right)\left(\sum_{i=1}^{n} Y_i\right)}{\sqrt{\left[n\left(\sum_{i=1}^{n} X_i^2\right) - \left(\sum_{i=1}^{n} X_i\right)^2\right]\left[n\left(\sum_{i=1}^{n} Y_i^2\right) - \left(\sum_{i=1}^{n} Y_i\right)^2\right]}} \qquad (19.5)$$

where $n(\sum_{i=1}^{n} X_i Y_i) - (\sum_{i=1}^{n} X_i)(\sum_{i=1}^{n} Y_i)$ is the sample covariance, and $[n(\sum_{i=1}^{n} X_i^2) - (\sum_{i=1}^{n} X_i)^2]$ and $[n(\sum_{i=1}^{n} Y_i^2) - (\sum_{i=1}^{n} Y_i)^2]$ are the standard deviations of X and Y, respectively. The correlation of two horizontally adjacent pixels in the original and the encrypted images is shown in Figure 19.5. The obtained correlation coefficients for horizontal, vertical, and diagonal directions are shown in Table 19.2. These correlation analyses indicate that the proposed modified scheme performs very well in all test directions.

19.3.3.4 FIPS PUB 140-1

Similar to the original scheme, we conduct the monobit test, poker test, runs test, and long-run tests according to the prescription of FIPS PUB 140-1 (1994). The

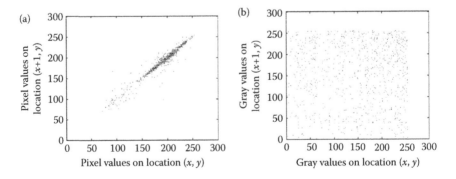

FIGURE 19.5 Correlations of two horizontally adjacent pixels: (a) plain image and (b) encrypted image.

TABLE 19.2

Absolute Differences of Two Vertical Shuffling Indexes

	Plain Image	Original Scheme	Modified Scheme
Horizontal	0.9597	0.1257	0.0631
Vertical	0.9792	0.0581	0.0226
Diagonal	0.9570	0.0504	−0.0192

TABLE 19.3

Test Results of Lena for FIPS PUB 140-1

Monnbit test			OK	x: 9654 < X = 10,098 < 10,346					
			OK	y: 9654 < Y = 9915 < 10,346					
			OK	z: 9654 < Z = 10,038 < 10,346					
Poker test			OK	x: 1.03 < X = 13.0483 < 57.4					
			OK	y: 1.03 < Y = 14.2469 < 57.4					
			OK	z: 1.03 < Z = 13.8736 < 57.4					
Runs test		Length of run	1	2	3	4	5	6+	
		Required interval	2267– 2733	1079– 1421	502– 748	223– 402	90–223	90–223	
	1	OK	x	2477	1291	691	311	170	121
		OK	y	2441	1229	636	317	168	155
		OK	z	2449	1261	643	346	153	148
	0	OK	x	2526	1280	667	306	164	118
		OK	y	2471	1215	633	299	165	163
		OK	z	2492	1266	644	302	151	145
Long run test	1	OK	x: 0, y: 0, z: 0						
	0	OK	x: 0, y: 0, z: 0						

results are listed in Table 19.3, demonstrating that the modified scheme can pass these tests.

19.3.3.5 Number of Pixel Change Rate and Unified Average Changing Intensity

In order to evaluate the variations between the original and encrypted images, we use Number of Pixel Change Rate (NPCR) and Unified Average Changing Intensity (UACI) as measurement:

$$\text{NPCR} = \frac{\sum_{i=0}^{m-1} \sum_{j=0}^{n-1} D(i,j)}{m \times n} \times 100\%$$

TABLE 19.4

Results of NPCR and UACI Tests of Lena

Test Scheme	NPCR (%)			UACI (%)		
Image	R	G	B	R	G	B
Original scheme	99.42	99.60	99.54	27.78	27.66	24.94
Modified scheme	99.54	99.55	99.46	26.27	27.18	23.91

for

$$D(i, j) = \begin{cases} 0, & A(i, j) = AH(i, j) \\ 1, & A(i, j) \neq AH(i, j) \end{cases} \tag{19.6}$$

$$\text{UACI} = \frac{1}{m \times n} \sum_{i=0}^{m-1} \sum_{j=0}^{n-1} \frac{|A(i, j) - AH(i, j)|}{255} \times 100\% \tag{19.7}$$

where A and AH denote the RGB-level matrices of the original and encrypted images, respectively. Table 19.4 shows that the modified scheme has similar performance as the original scheme.

19.4   CONCLUSIONS

In this chapter, the weakness of vulnerability of the original multi-chaotic systems-based image encryption is analyzed. The described technique of enhancement measures is proposed to thwart the reported cryptanalysis. Theoretical analysis and computer simulation indicate that the modified scheme is more secure than the original one. At the same time, it still holds the other merits of the original scheme.

REFERENCES

Chen, G., Mao, Y., and Chui, C., A symmetric image encryption scheme based on 3D chaotic Cat maps, *Chaos Solitons Fractals*, 21(7), 749–761, 2004.

FIPS PUB 140-1, *Security Requirement for Cryptographic Modules*, Federal Information Processing Standards Publication, Gaithersburg, MD, 1994.

Fridrich, J., Symmetric ciphers based on two-dimensional chaotic maps, *Int. J. Bifurcation Chaos*, 8(6), 1259–1284, 1998.

Guan, Z., Huang, F., and Guan, W., Chaos-based image encryption algorithm, *Phys. Lett. A*, 346(1–3), 153–157, 2005.

Huang, C. K. and Nien, H. H., Multi chaotic systems based pixel shuffle for image encryption, *Opt. Commun.*, 282(11), 2123–2127, 2009.

Lian, S., Sun, J., and Wang, Z., A block cipher based on a suitable use of the chaotic standard map, *Chaos Solitons Fractals*, 26, 117–129, 2005.

Lian, S., Sun, J., and Wang, Z., Secure hash function based on neural network, *Neurocomputing*, 69, 2346–2350, 2006.

Solak, E., Rhouma, R., and Belghith, S., Cryptanalysis of a multi-chaotic systems based image cryptosystem, *Opt. Commun.*, 283, 232–236, 2010.

Vanstone, S. A., Menezes, A. J., and Oorschot, P. C., *Handbook of Applied Cryptography*, CRC Press, Boca Raton, FL, 1996.

Xiao, D., Liao, X., and Deng, S., One-way hash function construction based on the chaotic map with changeable-parameter, *Chaos Solitons Fractals*, 24, 65–71, 2005.

Xiao, D., Liao, X., and Deng, S., Parallel keyed hash function construction based on chaotic maps, *Phys. Lett. A*, 372(26), 4682–4688, 2008.

Xiao, D., Liao, X., and Wong, K., Improving the security of a dynamic look-up table based chaotic cryptosystem, *IEEE Trans. Circuits Syst. II*, 53(6), 502–506, 2006.

Xiao, D. and Shih, F. Y., Using the self-synchronizing method to improve security of the multi chaotic systems-based image encryption, *Opt. Commun.*, 283(15), 3030–3036, 2010.

Yi, X., Hash function based on chaotic tent maps, *IEEE Trans. Circuits Syst. II*, 52(6), 354–357, 2005.

20 Behavior Modeling of Human Objects in Multimedia Content

Yafeng Yin and Hong Man

CONTENTS

20.1   INTRODUCTION

Multimedia content analysis is essential in a wide range of applications, including information retrieval, surveillance, robotics, automation, and so on. A general objective of such operation is to extract semantic meanings from multimedia content. Recent developments in image and video processing, audio signal processing, computer vision, and machine learning have all contributed to the advances in this field (Gong and Xiang, 2011).

As shown in Figure 20.1, the analysis of multimedia content is usually built on top of feature extraction, object/event detection, and recognition. A variety of visual features have been introduced and studied in the past, which include color, histogram, edge, shape, Haar-like feature, SIFT (scale-invariant feature transform), and HOG (histogram of oriented gradient). With the extracted features, object detection and recognition based on statistic modeling and machine learning methods are performed as middle-level analysis operations. Examples include target detection, human detection, face recognition, background modeling, and so on. The results of these operations will then be used to support high-level tasks, such as semantic abstraction, semantic description, and semantic query.

Since human objects are of most importance in a majority of multimedia content analysis applications, in the chapter, we will mainly focus on human behavior modeling. In particular, we will introduce two new methods for modeling human motion changes and small human group motions. Human motion modeling and motion change detection are important tasks in intelligent surveillance systems, and they have attracted significant

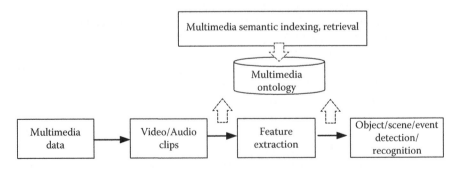

FIGURE 20.1   Multimedia content analysis process.

amount of attentions in recent years. The major challenges in these tasks include complex scenes with a large number of targets and confusors, and complex motion behaviors of different human objects. In real-world applications, variant environmental conditions, such as illumination, reflection, and multiple human objects with occlusion, present further difficulties to human motion modeling. From the learning perspective, existing motion detection algorithms can be divided into two categories: model-based detection and signature-based detection. In model-based motion detection methods, human motions are frequently described by statistical methods (Stauffer and Grimson, 2000; Zhong et al., 2004), while in signature-based motion detection methods, human motions are specified by predefined patterns (Ivanov and Bobick, 1999).

In addition to single person motion recognition, small human groups contain much richer interperson interactions among group members. Compared to dense crowd analysis, in which each person can be regarded as a point in a flow, small groups contain much detailed information about each individual in the group. The major challenges of small group activity analysis include mutual occlusions between different people, varying group size, and interactions within or between groups. Therefore, small group activity recognition demands a structural feature to bridge the local description of single human objects and global description for crowd analysis, as well as addressing both spatial dynamics (varying group size) and temporal dynamics (varying clip length). Unlike single person or dense crowd analysis, small group action recognition requires detection and tracking of each group member rather than extracting feature from the entire scene, as there may be several small groups with different actions in an individual scene.

The rest of this chapter is organized as follows. In Section 20.2, a set of most commonly used visual features for describing human objects will be reviewed, which include Haar-like feature, SIFT features, and HOG features. In Section 20.3, a hierarchical Gaussian process latent variable model (HGPLVM) is introduced for human motion change detection. In Section 20.4, a Gaussian process dynamical model (GPDM) based on a unique social network feature set is introduced for modeling small human group behavior. Selected experimental results are included in these discussions. Conclusions and discussions on future steps are provided in Section 20.5.

20.2   FEATURE CATEGORIZATION

Feature selection plays a very crucial role in multimedia content analysis. Various features and feature extraction methods have been proposed for image and video processing. On one hand, the original high-dimensional data are usually mapped to a low-dimensional feature space during feature extraction, which greatly reduces the computational cost in the later stages. On the other hand, a distinctive feature set can capture the characteristic of visual objects in both spatial domain and temporal domain, which is essential for multimedia semantic analysis. Several commonly used features are introduced in this section.

Besides the basic features, such as image pixel value, edges, and color histogram, more and more distinctive features have been proposed in large-scale machine learning algorithms. Among those features, SIFT (Lowe, 1999), Haar-like (Viola and Jones, 2001), and HOG (Dalal and Triggs 2005) are the most widely used features for human object detection and recognition in recent years. Here, we focus on the features for the behavior modeling with a small number of actors in a controlled environment, such as SIFT, HOG, optical flow, 3D-SIFT, and space–time interest points (STIP) features.

SIFT features were proposed by Lowe in 1999 (Lowe, 1999). Due to their translation, scale, and rotation-invariant properties, SIFT features have been widely applied for object detection and recognition. To characterize these invariant properties in SIFT features, the key points are first identified through different scales and locations of the image. For each key point, an array is sampled around the key point's location and the array is transformed to the direction of the key point. The entire sampled array is then divided into grids of a certain size. In each grid, the gradient of each pixel is calculated and accumulated to different directions in a histogram. The normalized bin of the histogram will then form the final SIFT descriptor. For a 16×16 array with 4×4 grids, and 8 bins for each histogram, the dimension of the SIFT descriptor is $4 \times 4 \times 8 = 128$.

HOG is derived from SIFT features, and it was introduced by Dalal and Triggs (2005). By implementing HOG features with support vector machine (SVM) classifier, it has shown impressive results on human detection. To extract HOG features, the whole detection window is first divided into grids of small cells. Similar to the cells for computing SIFT features, the gradient in each cell is also accumulated at different directions in a histogram and forms a normalized vector description. Usually, 2×2 cells are combined into a block. Different from SIFT calculation, blocks are allowed to overlap with each other in a sliding window fashion in HOG feature calculation. This approach connects all the feature descriptions in local regions. Zhu et al. (2006) showed that the combination of the cascade of rejecters approach and the HOG features led to a fast and accurate human detection system. They used AdaBoost for selecting the best blocks for human detection. They claimed that their system could achieve close to real-time performance.

While most aforementioned features are local image features in the spatial domain, which is naturally extended from images to frames in the video, to exploit the characteristic in the temporal domain as well as the spatial domain, many new features have been proposed for video analysis, such as optical flow, 3D-SIFT, and STIP features.

Optical flow is the most used feature for motion recognition (Barron et al., 1994). Optical flow describes the local motion difference between two or more consecutive frames. Based on the partial derivatives with respect to the spatial and temporal coordinates, optical flow shows the motion magnitude and direction at each pixel. By assuming the motion to be small and with an image constraint: $I(x,y,t) = I(x + \delta x, + \delta y, t + \delta t)$, optical flow is computed by

$$I_x V_x + I_y V_y = -I_x \tag{20.1}$$

where I_x and I_y are the derivatives at location (x, y) of the frame at time t, and V_x and V_y are the velocities. As there are two unknowns in Equation 20.1, other constraints are needed to solve this equation. Many approaches have been proposed, such as Lucas–Kanade method (Lucas and Kanade, 1981), Horn–Schunck method (Horn and Schunck, 1981), and so on.

3D-SIFT extends the original SIFT features from 2D image domain to 3D video data cube (Cupillard et al., 2002). The gradient magnitudes in each small cube at different directions are accumulated to form a histogram. 3D-SIFT-based action recognition methods have been shown to outperform many other feature-based recognition methods.

Beyond the aforementioned features, recently, many middle-level features have been proposed for human group behavior analysis, as group behavior involves interactions among different group members, and requires the feature set that can capture local detailed information as well as global structure description. Middle-level features characterize the global properties of low-level features rather than local description. Ni et al. (2009) proposed a middle-level feature set for group structure information above the general low-level feature, which has been developed for small group human action recognition. Yuan et al. (2010) also proposed a middle-level representation for human activity recognition.

20.3 HUMAN MOTION CHANGE DETECTION

Motion change detection has been frequently used in human action recognition and suspicious behavior analysis (Rao and Sastry, 2003). Many existing works use segmentation and tracking for human motion detection (Haritaoglu et al., 2000; Medioni et al., 2001). Boiman and Irani (2005) proposed a probabilistic graphical model for detecting irregularities in the video. Kiryati et al. (2008) extracted the motion features from videos, and used a pretrained motion model to classify different human motions. The abnormal human motion patterns relied on the trained motion trajectories. Sherrah and Gong (2000) presented a platform (VIGOUR) for tracking and recognizing the activities of multiple people. The system can track up to three people and recognize their gestures. However, most of these methods cannot handle human motion change detection with multiple people, due to the increasing computational complexity in tracking multiple targets.

In the following sections, we will introduce a new human motion change detection method as an example for human behavior modeling in surveillance video.

20.3.1 HIERARCHICAL GAUSSIAN PROCESS DYNAMICAL MODEL

A hierarchical Gaussian process dynamical model (HGPDM) integrated with particle filter tracker is introduced for human motion change detection (Yin et al., 2010). First, the high-dimensional human motion trajectory training data are projected to a low-dimensional latent space with a two-layered hierarchy. The latent space of the leaf nodes at the bottom layer represents a typical human motion trajectory, while the root node at the upper layer controls the interaction and switching among the leaf nodes.

The HGPDM is inspired by the HGPLVM (Lawrence and Moore, 2007). Lawrence applied the HGPLVM for modeling the human interaction. The HGPDM approach is different from Lawrence's work in two ways: first, it extends the GPLVM to GPDM in the leaf nodes, which gives a compact representation for the joint distribution of observed temporal data in the latent space. In addition, the latent space of the root node is optimized after the optimizations of leaf nodes, instead of a joint optimiza-tion of all nodes simultaneously (Lawrence and Moore, 2007). In other words, the root node in this framework is a classifier controlling all the switching between the leaf nodes.

The trained HGPDM will then be used to classify test object trajectories which are captured by a particle filter tracker. If the motion trajectory is different from the motion in the previous frame, the root node will transfer the motion trajectory to the corresponding leaf node. In addition, HGPDM can be used to predict the next motion state, and provide Gaussian process dynamical samples for the particle filter framework, and this framework can accurately track and detect the human motion changes despite the complex motion and occlusion. Also, the sampling in the hier-archical latent space has greatly improved the efficiency of the particle filter frame-work. The assumption of this approach is that the normal human is prone to having similar motion trajectory in a specific location, while sudden motion change usually implies suspicious behaviors. If we can learn these similar trajectories in advance, they can be used for motion trajectory classification and motion change detection.

In contrast to the GPDM used in Urtasun (2006), Wang et al. (2008a), Raskin et al. (2008), and Wang et al. (2008b), HGPDM focuses on the global human motion rather than the local motion of human body parts. Therefore, instead of using 3D motion database, it selects 2D motion data of multiple human objects to construct a high-dimensional training dataset, and trains HGPDM. In addition, this framework can address the switching between different motion patterns, which is very critical for change detection of human with complex motion.

20.3.2 HUMAN MOTION CHANGE DETECTION BASED ON HGPDM

The HGPDM method is aimed to learn a general human motion trajectory model for multiple human motion change detection. The pretrained model can robustly detect different human motion changes, and reduce the computational complexity as well as improve the robustness of particle-tracking framework. The flowchart of this motion change detection framework is shown in Figure 20.2.

The basic procedure of the particle filter with the HGPDM is described as follows.

20.3.2.1 Creating HGPDM

The leaf node of HGPDM is created on the basis of the trajectory training datasets, that is, coordinate difference values, while the top node is created on the basis of depen-dence of leaf nodes. The learning model parameters include $\Gamma = \{Y^T, X^T, \bar{\alpha}, \bar{\beta}, W\}$, where Y^T is the training observation dataset, X^T is the corresponding latent variable set, $\bar{\alpha}$ and $\bar{\beta}$ are hyperparameters, and W is a scale parameter.

20.3.2.2 Jointly Initializing Model Parameters

The three nodes of latent variable sets and parameters $X^T, \bar{\alpha}, \bar{\beta}$ are obtained by minimizing the negative log-posterior function $-\ln p(X^T, \bar{\alpha}, \bar{\beta}, W|Y^T)$ of the unknown parameters $\{X^T, \bar{\alpha}, \bar{\beta}, W\}$ with scaled conjugate gradient (SCG) on the training datasets.

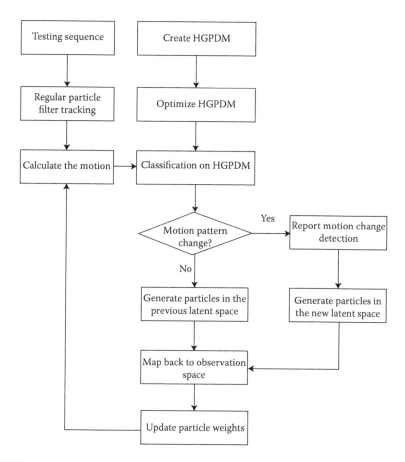

FIGURE 20.2 Human motion change detection and tracking framework.

20.3.2.3   Initializing the Particle Filter Framework

The prior probability is derived from the basis of the created model. In this step, target templates are obtained from the previous frames as reference images for similarity calculation in the later stage.

After initializing the target's position, each target will be tracked by the regular particle filter in the first five frames. Then, the test observation motion pattern data are calculated and projected to the latent coordinate system on the top node of HGPDM by using probabilistic principal component analysis (PPCA).

20.3.2.4   Latent Space Motion Classification and Change Detection

After projecting the current motion data into the latent space, the top node will determine which motion pattern in the leaf node the current motion belongs to. If the motion pattern is not consistent with the previous one, the motion change is reported and the latent space will switch to the new leaf node.

20.3.2.5 Leaf Node Latent Space Particle Sampling and Predicting

Particles are generated in the latent space of leaf node GPDM to infer the likely coordinate change value $(\Delta x_i, \Delta y_i)$.

20.3.2.6 Determining Probabilistic Mapping from the Latent Space to Observation Space

The log-posterior probability of the coordinate difference values of the test data is maximized to find the best mapping in the training datasets of the observation space. In addition, the most likely coordinate change value $(\Delta x_i, \Delta y_i)$ is used for predicting the next motion.

20.3.2.7 Updating the Weights

In the next frame, the similarity between the template's corresponding appearance model and the cropped region centered on the particle is calculated to determine the weights w_i and the most likely location $(\hat{x}_{t+1}, \hat{y}_{t+1})$ of the corresponding target, as well as to decide whether resampling is necessary or not.

The detailed training and test procedures are introduced in the following sections.

20.3.3 HUMAN MOTION TRAINING IN HGPDM

The reason for utilizing the HGPDM for change detection is that the root node can control all the interactions and switchings between each leaf node. Therefore, it can model more complex motion patterns. Its structure is described in Figure 20.3.

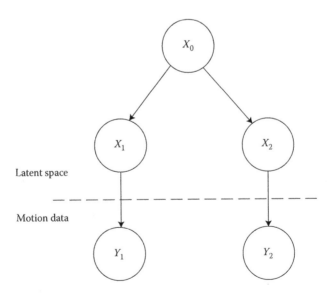

FIGURE 20.3 A hierarchical Gaussian process dynamical model. The root node X_0 controls all the interactions between the latent space of leaf nodes in $X_1, \ldots, X_n$. $Y_1, \ldots, Y_n$ denotes the observation data associated with $X_1, \ldots, X_n$.

20.3.3.1 Extracting Human Motion Trajectory Data

During the training phase, the motion data are manually extracted from the KTH motion database (KTH dataset, 2004). At each frame of a walking cycle, a point at the central location of the human body is selected. Then, coordinate difference between two consecutive frames forms a motion vector. Supposing that the total frame number is 30, the dimension of one person's motion data is 30 by 2. In the training set, 20 different people are chosen and each person walks in a different direction. The total training data are then 30 by 40 dimensional. To demonstrate the performance of the HGPDM framework, two typical motion trajectories, walking and running, are selected from the KTH motion database. In the walking and running categories, different human motions at different directions are extracted as the training data.

20.3.3.2 Establishing Trajectory Learning Model on the Leaf Node

GPDM is applied to learn the specific trajectories of the moving human. The probability density function of latent variable X and the observation variable Y are defined by

$$P(Y_k|X_k) = \frac{|W|^N}{\sqrt{(2\pi^{ND}|K_Y|^D)}} \exp\left(-\frac{1}{2}tr(K_Y^{-1}YW^2Y^T)\right) \tag{20.2}$$

where W is the hyperparameter, N is the number of Y sequences, D is the data dimension of Y, and K_Y is the kernel function.

In this study, the radial basis function (RBF) kernel given by the following equation is used for the HGPDM model:

$$k_Y(x,x') = \exp\left(-\frac{\gamma}{2}\|x-x'\|^2\right) + \beta^{-1}\delta_{x,x'} \tag{20.3}$$

where x and x' are any latent variables in the latent space, γ controls the width of the kernel, and β^{-1} is the variance of the noise.

Given a specific surveillance environment, certain patterns may be observed and are worth exploring for future inferences. To initialize the latent coordinate, the d (dimensionality of the latent space) principal directions of the latent coordinates is determined by deploying PPCA on the mean-subtracted training dataset Y^T, that is, $Y^T - \overline{Y^T}$.

Given Y^T, the learning parameters are estimated by minimizing the negative-log-posterior using SCG (Riedmiller and Braun, 1992).

20.3.3.3 Optimizing the HGPDM

As the structure described in Figure 20.2, Y_1 and Y_2 denote the high-dimensional multiple human walking and running data and X_1 and X_2 represent the corresponding latent space in the leaf nodes.

The joint probability distribution of Y_1 and Y_2 is given by

$$P(Y_1, Y_2) = \int p(Y_1|X_1) \times \int p(Y_2|X_2) \times \int p(X_0|X_1, X_2) dX_3 dX_2 dX_1 \qquad (20.4)$$

where each conditional distribution is given by the Gaussian process. The major advantage of the GPDM is that each training datum is associated with likelihood in the latent space. Then, the maximum A-posteriori probability (MAP) method is applied to find all the values of latent variables. For this simple model, we are trying to optimize the parameters by maximizing

$$\mathrm{logp}(X_0, X_1 X_2|Y_1, Y_2) = \mathrm{logp}(Y_1|X_1) \times \mathrm{logp}(Y_2|X_2) \times \mathrm{logp}(X_0|X_1, X_2) \quad (20.5)$$

The training process for optimizing the HGPDM is described as follows:

1. Initialize each leaf node: Project the walking and running training data to (X_1, X_2) through PPCA.
2. Initialize root node: Initialize the root's latent variable through PPCA and its dependence of (X_1, X_2).
3. Jointly optimize the parameters of each GPDM.

Once the latent space has been optimized, supervised clustering is applied at the top latent space to group the training data to the leaf node. Supposing we have N leaf nodes, and $Y = \{y_i, i = 1, 2, \ldots, M\}$. Each leaf node is associated with mean μ_j and variance σ_j, and each training sample y_i is assigned to the leaf node by

$$\hat{X}_{y_i} = \arg\max_{j=1,2,\ldots,N} p(x'|\mu_j, \sigma_j), \quad i = 1, 2, \ldots, M \qquad (20.6)$$

One example of the learned running latent space in the leaf node is shown in Figure 20.4.

20.3.4   HGPDM with Particle Filter in Testing Phase

After jointly optimizing the HGPDM, the trained HGPDM can be used to identify the human motion trajectories captured by particle filter tracker. In the meantime, the classified motion pattern provides the most similar motion trajectory for efficient particle sampling. The process of the HGPDM framework is described as follows.

20.3.4.1   Initializing the Particle Filter Framework

A particle filter is a Monte Carlo method for nonlinear, non-Gaussian models, which approximates the continuous probability density function by using a large number of samples. In the HGPDM framework, a histogram was used as appearance modeling for its simplicity and efficiency. The red, green, blue (RGB) histogram of the tem-

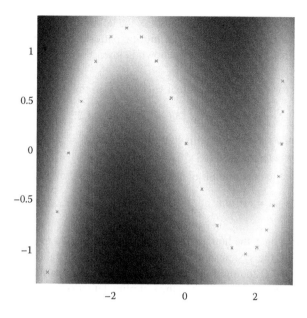

FIGURE 20.4   Trained running latent space.

plate and the image region under consideration are obtained, respectively. The likelihood $P(Z_t | k_t, \hat{Y}_t)$ is defined to be proportional to the similarity between the histogram of the template and the candidate, and is measured by the Bhattacharya distance.

At this step, target templates are obtained by using background subtraction. The obtained target templates will be used for similarity calculation in the testing stage. After initializing the target's position for the first frame, each target will be tracked by using regular particle filter in the first five frames. This is based on the assumption that the human motion does not change at the very beginning. Then, the test observation motion data of the first five frames are calculated and projected to the latent coordinate system on the top node of HGPDM.

20.3.4.2   Latent Space Motion Classification and Change Detection

Since HGPDM was constructed in the latent space, at the beginning of the test process, the target observation data of the first five frames have to be projected to the same two-dimensional latent space in order to be compared to the upper level of trained HGPDM. The purpose of projecting the test data from the observation space to the latent space is to initialize the testing data in the latent space and obtain a compact representation of the similar motion patterns in the training dataset. This projection is achieved by using PPCA, same as the first stage in HGPDM learning. The feature vector of each frame contains the coordinate change values for every target being tracked in that frame. For n targets, the feature vector will contain $n \times 2$ pairs of coordinate change values. The PPCA projection will reduce this $n \times 2$-dimensional feature vector to a 1×2 latent space vector. After projecting test motion data from observation to latent space, the upper level of hierarchical GPDM will be used to identify the most similar motion patterns in the leaf node. The classification algorithm in the HGPDM is shown below:

1. *Select* the top K most likely latent variables x_i in the root node by

$$x_i = \arg\max_{x \in I} K(p(y'|x)), \quad i = 1,2,...,K \tag{20.7}$$

2. *Compute* the relative normalized probability as $\pi_i = p(y'|x_i)/\sum_{i=1}^{K} p(y'|x_i)$
3. Establish the latent variable x' as $x' = \sum_{i=1}^{n} \pi_i x_i$
4. Determine the corresponding latent space in the leaf node $\hat{X}$ as in

$$\hat{X} = \arg\max_{j=1,2,...,N} p(x'|\mu_j, \sigma_j), \quad i = 1,2,...,K \tag{20.8}$$

At each frame, once the human motion pattern is classified to a different category by Equation 20.8, the system does not immediately transfer the latent space to the corresponding one. If the motion pattern continues to be identified as another category for five consecutive frames, then the sampling latent space is switched from one leaf node to the other, and the human motion change is reported to the system. Otherwise, the temporary human motion change is considered as noise and the particle filter framework will still perform sampling in the latent space which is determined in the last frame. The next possible position is predicted by determining the most similar trajectory pattern in the leaf node and using the corresponding position change value plus noise.

20.3.4.3 Particle Filter Tracking and Update

After determining the general categories of the human motion pattern in the root node of HGPDM, the particles are propagated in the latent space of the leaf node. The next possible position is predicted by determining the most similar trajectory pattern in the training database and using the corresponding position change value plus noise. The number of particles is reduced from over 100 to about 20 by deriving the posterior distribution over latent functions. Each point on this 2D latent space in Figure 20.4 is a projection of a feature vector representing 20 training targets, that is, 20 pairs of coordinate change values. The gray-scale intensity represents the precision of mapping from the observation space to the latent space, and the lighter the pixel appears, the higher the precision of mapping is.

Thereafter, the latent variables are mapped in a probabilistic way to the location difference data in the observation space. The estimation maximization (EM) approach is employed to determine the most likely observation coordinates in the observation space after the distribution is derived.

The nondecreasing log-posterior probability of the test data is given by $P(Y_k | X_k)$. K_Y is a kernel matrix defined by the RBF kernel function. The log-posterior probability is maximized to search for the most probable correspondence on the training datasets. The corresponding trajectory pattern is then selected for predicting the following motion.

The weights of the particles are updated in terms of the likelihood estimation based on the appearance model. The importance weight equation is given by

$$P(\hat{Y}_t | Z_t, k_t) = \frac{P(Z_t | k_t, \hat{Y}_t) P(k_t, \hat{Y}_t)}{P(Z_t)}, \quad w_t \propto P(Z_t | k_t, \hat{Y}_t) P(k_t, \hat{Y}_t) \quad (20.9)$$

where $\hat{Y}_t$ denotes the estimation data, Z_t denotes the observation data, k_t represents the identity of the target, and w_t represnts the weight of a particle.

20.3.5   Experimental Results on Human Motion Change Detection

The HGPDM particle filter framework was implemented using MATLAB®. Lawrence's Gaussian process software provides the related GPDM functions for conducting simulations (Lawrence, 2005).

During the training phase, the human motion data are extracted from the KTH action dataset (KTH dataset, 2004), which includes running and walking videos of different people. The walking and running motion trajectories are extracted manually and projected to the leaf nodes of hierarchical latent space GPDM. After joint optimization, the root node represents the interaction and switching of leaf nodes, while the leaf nodes denote different types of trajectory motion. The learned motion model are then used to classify test object trajectories, predict the next motion state, and provide Gaussian process dynamical samples for the particle filter framework.

The motion change detection framework is evaluated on the video dataset (Kiryati et al., 2008) and the IDIAP dataset (Smith et al., 2005). The first testing sequence extracted from the IDIAP dataset contains two targets and one of them runs after the other from left to right. The motion data are captured by the particle filter and classified in the HGPDM. Human motion change is reported at the 58th frame. Figure 20.5 shows the corresponding frames sampled from the tracking results and Figure 20.6 shows the posterior probability of left target motion belonging to the running GPDM latent space. According to Figure 20.6, the posterior probability is higher than 0.5 before the 58th frame, while lower than 0.5 after the 58th frame. This means that the motion trajectory is switched from running to walking in the HGPDM near the 58th frame. The sampling frames in Figure 20.5 verify that the human motion change is correctly detected. It needs to be noticed that there are several sparks in the posterior probability curve due to the noisy motion data captured by the particle filter.

The second sequence contains four people with occlusion and complex motion patterns. The person with blue bounding box first walked to the road center, and then circled around the bike ramp in the middle of the road. Both the direction and the velocity were varied at this circle motion. Motion changes are reported at the 42th and 129th frame. Figure 20.7 shows the posterior probability of motion change. The posterior probability is below 0.5 near the 42th and 129th frame, as the person made

FIGURE 20.5   Sampling frames of 3, 19, 58, and 98 in the first testing sequence.

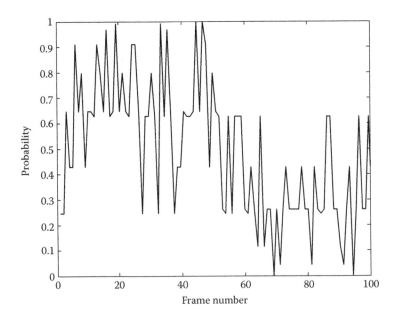

FIGURE 20.6   Posterior probability of running motion change.

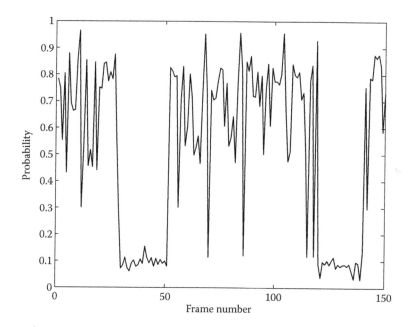

FIGURE 20.7   Posterior probability of circle motion.

a turn near these two frames. The sampling frames in Figure 20.8 indicate that the motion change is correctly detected.

The change detection framework is also evaluated on the dataset (Kiryati et al., 2008), and the human motion change detection results with the abnormal motion recognition are compared. The experimental results indicate that the HGPDM framework can detect all the motion change correctly, which are corresponding to the defined abnormal motions (Kiryati et al., 2008). One of the testing sequences shown in Figure 20.9 was defined as the jumping episode (Kiryati et al., 2008). The motion change detection of the left target is reported at the 18th and 60th frame. According to Figure 20.8, the woman stopped from running near frame 9 and began to walk after jumping toward the man near frame 60. The change detection system can capture the motion change successfully through the switching between different latent space nodes.

As all the abnormal motion patterns in Kiryati et al. (2008) were predefined, their system can only detect all the trained trajectories, while the HGPDM-based change detection framework can be used for general motion change detection of tracked targets over time.

These experimental results demonstrate the efficiency and accuracy of this motion change detection framework. They indicated that this method can correctly detect each motion change and robustly track multiple targets with complex motion at the same time. As the HGPDM is a general motion trajectory model, this framework can not only be used for suspicious behavior analysis and irregular motion detection, but can also be used for key frame detection and scene change detection in multimedia content analysis.

FIGURE 20.8   Sampling frames of 1, 42, 65, 87, 129, and 150 in the second testing sequence.

20.4   HUMAN GROUP BEHAVIOR ANALYSIS

Compared with single human action recognition and human crowd analysis, action recognition of small human group with <10 people has more practical applications in surveillance systems. Most public safety scenarios consist of small group activities. However, relatively few research works have been reported on this topic, due to the difficulties of describing varying number of participants and mutual occlusions among group members. Ni et al. (2009) introduced three types of localized causalities for human group activities with different numbers of people, and their experiment results showed that dynamic human interactions could be used to classify group actions. They provided feature vectors of different sizes to describe different group activities, which needed to train specific classifiers using different

FIGURE 20.9 Sample frames of 1, 18, 25, 33, 60, and 85 in the third testing sequence.

input samples with different lengths. Chang et al. (2010) proposed a bottom-up method to form a group and calculated the similarity of different groups. Ge et al. (2009) also developed a hierarchical clustering algorithm for small group detection in a crowded scene. Guimera et al. (2005) proposed a collaboration network structure to determine the team performance, and the experiment result indicated that team assembly mechanism could be used for predicting and describing the group dynamics.

Most human activity recognition methods fall into two large categories, that is, non-target-oriented and target-oriented approaches. Non-target-oriented methods extract features on the entire image region regardless of each target's position and the number of targets. Non-target-oriented methods are widely used in single human action recognition, as the background is often fixed and the individual's actions dominate the entire scene. However, target-oriented methods require tracking of each target in the scene and then modeling targets' behavior according to the motion flow. As the extracted features are much more accurate in the scene, object-oriented methods can be more effective for multiple human activities recognition and human group behavior analysis under a complex environment.

To exploit the property of human action, Hospedales et al. (2009) implemented the dynamic Bayesian network (DBN) or other models to drive the potential characteristic behind the simple features, which sought to connect the extracted features by probabilistic distribution. However, in the large-scale human action recognition, people always clustered to groups, which means there are intergroup interaction as well as intragroup interaction. Treating the entire scene as a whole may not be a good choice for such a recognition task. Therefore, target-oriented approaches are becoming more popular in the large-scale video analysis.

Tracking plays an essential role in the target-oriented methods for human action recognition, as it can provide the central location information for each target in the video sequence. With the development of multiple-camera system, robust and accurate tracking of multiple human is not difficult. Two of the most widely used tracking algorithms are mean-shift- and particle filter-based tracking methods.

After targets tracking, many machine learning algorithms have been introduced for high-level human action understanding. Ryoo and Aggarwal (2008) proposed a hierarchical recognition algorithm for the recognition of high-level group activities, and each group activity is based on activities of individual members and group activity is represented with language-like description, while each representation need to be encoded by human experts. To handle recognition and understanding on video clips with variable length, a probabilistic model (GPDM) is built for each human behavior during the training phase, and the posterior probability is computed for each test sample.

Vaswani et al. (2003) utilized the statistical shape theory to detect abnormal group activity. Each person in a frame is modeled as a point and the shape is formed based on the connections of a fixed number of landmarks. The normal shape sequence was modeled as a stationary Gaussian Markov model, and abnormal activity was detected by comparing the tracking results with pretrained models. Cupillard et al. (2002) proposed a high-level group behavior recognition algorithm based on the hierarchy of operators; each operation was associated with a method to recognize human behaviors. Each group behavior was defined by different combination of operators. This provided the flexibility for different users. Also, the author fused the detected motion information from multiple cameras to form a global graph to improve long-term tracking for each target. Intille and Bobick (1999) introduced the belief networks for probabilistically representing and recognizing multiple-agent action based on temporal structure description. The experiment results had demonstrated their approach could find the logic connection of each agent during football play.

To recognize group human behaviors, the aforementioned approaches have exploited both the feature description and semantic understanding from different perspectives. However, due to the difficulty of understanding spatial dynamics as well as temporal dynamics, group human behavior recognition still remains a challenging problem in the image-processing field.

In the following sections, we will introduce a new group human behavior modeling method based upon social network features and probabilistic learning as an example for multimedia content analysis.

20.4.1 Group Human Behavior Recognition Framework

We will demonstrate a multimedia content analysis system focusing on the group human action recognition, which includes feature extraction, human detection and tracking, and group human behavior classification. This system introduces a novel structural feature set to represent group activities as well as a probabilistic framework for small group activity learning and recognition.

As shown in Figure 20.10, the whole learning procedure of our system consists of four stages. First, a robust mean-shift-based tracker is applied to track each individual in a small group sequentially. Then, the output coordinates of each tracker will be clustered and allocated to different small human groups. Based on social network feature description, the structural features are extracted from each video clip in the third stage. The feature vectors from each frame will form a feature matrix for each video clip. To understand the group human behavior, a corresponding GPDM is trained for each group behavior. The group activity feature matrix for each video clip will be projected to a low-dimensional latent space and the posterior conditional probability is computed with each trained model to identify different human group behaviors.

The detailed procedure of social network features-based probabilistic framework is introduced below.

20.4.1.1 Adaptive Mean-Shift Tracking

One of the important factors for small group human activities analysis is the accuracy and robustness of tracking each individual in the human group. As multiple camera systems develop, the accurate tracking of each individual can be well addressed. In the social network features-based probabilistic framework, the adaptive mean-shift tracking (Collins et al., 2005) on two different datasets is applied.

Compared to the general mean-shift tracking, online feature selection is applied during the adaptive mean-shift tracking. The feature set (Collins et al., 2005; Yin and Man, 2009) consisted of a linear combination of pixel valves at R, G, B channels: $F \equiv \omega_1 R + \omega_2 G + \omega_3 B$, where $\omega_i \in [-2, -1, 0, 1, 2]$, $i = 1, \ldots, 3$. By pruning all redundant

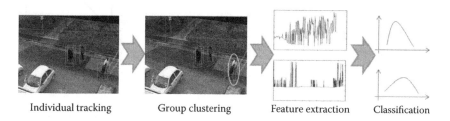

| Individual tracking | Group clustering | Feature extraction | Classification |

FIGURE 20.10 Overview of group activity recognition framework.

coefficients of ω_i, the feature set was cut down to 49. Linear discriminative analysis (LDA) was then used to determine the most descriptive feature for target tracking.

In order to reduce the computational complexity during tracking, the feature set is updated every 50 frames instead of being updated at each frame. In addition, the single mean-shift tracking algorithm is extended for multiple targets tracking. As the cameras were fixed in these two datasets, a simple motion detector is applied to detect each new person coming into the scene. Once a person comes in the scene, a new tracker will be allocated and track that person over time. Due to the difficulties in the reliable multiple targets tracking, each target will be reinitialized manually if the tracking algorithm fails for some reason.

20.4.1.2 Small Group Clustering

After obtaining all the positions of each target, a group clustering algorithm (Chang et al., 2010) will be applied to locate small groups. The closeness of each person is calculated and the minimum span tree (MST) clustering is used to obtain the distribution of each group. After that, the hierarchical clustering method (Chang et al., 2010) is applied to locate the mass center of each small group.

20.4.1.3 Social Network-Based Feature Extraction

Inspired by the social network analysis (SNA) (Wasserman and Faust, 1994), several structural features are extracted to capture the dynamic properties of a small group structure. It is believed that the dynamic structure and its theoretical framework can help us to model the group scenario in the real world. To our best knowledge, this is the first time that SNA-based feature is used to model group behavior in the surveillance videos. Similar to the original definitions of *betweenness*, *closeness*, and *centrality* (Krebs, 2000; Blunsden and Fisher, 2010) in SNA, several group structure features which are derived from SNA with modification in the group activity recognition are defined, including motion histograms, closeness histogram, and centrality histogram.

According to the previous section, suppose we have a group activity clip of m frames and the size of the feature matrix is $26 \times m$. Figure 20.11 shows the

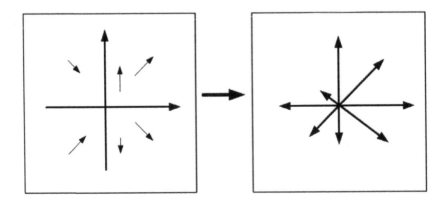

FIGURE 20.11 Motion histogram illustrations.

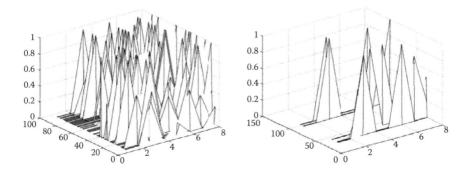

FIGURE 20.12   The overlapped central histogram of GroupFighting (left) and InGroup (right) from two video clips.

centrality feature of two different group activities. The assumption of this framework is that in the normal situation, the motion distribution of a group is prone to have a Gaussian distribution. If we treat the centrality feature in Figure 20.12 as a Gaussian process, then the centrality histogram at each frame is a sampling of this process. Different group activities can be seen as a set of Gaussian processes with different means and covariance matrices. Therefore, the Gaussian process can be used to model the dynamics in the temporal dimension. However, the size of covariance matrix will increase as the number of samples increases. In addition, since the Gaussian process just captures the general properties of the social network-based structural feature, a more specific characteristic of human motion needs to be addressed.

20.4.1.4   Human Group Behavior Modeling

In order to describe the dynamic property of the group behavior, the GPDM (Wang et al., 2008a) is adopted to represent different group activities.

As shown in Figure 20.10, the small group activity recognition framework consists of four stages: adaptive mean-shift tracking, small group clustering, group feature extraction, and group activities recognition.

20.4.2   SMALL GROUP ACTIVITY RECOGNITION

The small group activity recognition can be divided into two phases: group activity training and group activity classification. In the training stage, for each small group activity $\{A_i, i = 1, \ldots, n\}$, a GPDM $\{\Lambda_i, i = 1, \ldots, n\}$ will be trained. Suppose we have k samples of a group activity A_i, the length of each sample is m, then we have k feature matrices of size $26 \times m$. To learn a specific GPDM for A_i, the mean value $\bar{Z}$ of k feature matrices will first be computed, and will be used for activity model training.

GPDM is applied to learn the specific trajectories of a group activity. The probability density function of the latent variable X and the observation variable $\bar{Z}$ are defined by the following equations. The basic procedure of the GPDM training is described below.

20.4.2.1 Creating GPDM

GPDM $\Lambda = \{\bar{Z}^T, X^T, \bar{\alpha}, \bar{\beta}, \Omega\}$ is created on the basis of the trajectory training data-sets, that is, extracted structural feature, where $\bar{Z}^T$ is the training observation data, X^T is the corresponding latent variable sets, and $\bar{\alpha}$ and $\bar{\beta}$ are hyperparameters.

20.4.2.2 Jointly Initializing Model Parameters

The latent variable sets and parameters $\{X^T, \bar{\alpha}, \bar{\beta}\}$ are obtained by minimizing the negative log-posterior function $-\ln p(X^T, \bar{\alpha}, \bar{\beta}, \Omega | \bar{Z}^T)$ of the unknown parameters $\{X^T, \bar{\alpha}, \bar{\beta}, \Omega\}$ with SCG on the training datasets.

20.4.2.3 Training GPDM for Each Group Activity

For each group activity $\{A_i, i = 1, \ldots, n\}$, repeat procedure 1 and 2, and create a corresponding GPDM: $\{\Lambda_i, i = 1, \ldots, n\}$. After training, we have a set of GPDMs: $\{\Lambda_i, i = 1, \ldots, n\}$ for the human group activities. When a new human group activity Z^* comes in, the conditional probability will be computed with respect to each trained GPDM, and the one with the highest conditional probability will be selected.

20.4.2.4 Calculating the Conditional Probability with Each Trained GPDM

For each trained GPDM $\{\Lambda_i\}$, compute $X_i^{\mathring{a}}$ by using the learned parameters: $\{\bar{\alpha}_i, \bar{\beta}_i\}$. This can be obtained by minimizing the negative log-posterior function $-\ln p(X^T, \bar{\alpha}_i, \bar{\beta}_i, \Omega | Z^{\mathring{a}})$ with SCG on the training datasets. After that, the conditional probability $P\left(Z_i^{(*)}, X_i^{(*)} | \Lambda_i\right)$ is calculated.

20.4.2.5 Selecting the GPDM with the Highest Conditional Probability

The new group activity can be determined by

$$\operatorname{argmax}_{i=1,\ldots,n} P\left(Z_i^{(*)}, X_i^{(*)} | \Lambda_i\right) \tag{20.10}$$

As discussed in the previous section, the length of the new observation can be different from the size of the training data, which means that the number of frames in test clips can be different with training clips. Therefore, the trained model can address the dynamics in the temporal dimension. As the duration of an activity may change under different situations, it is important that the classifier can handle the testing sequences with varying lengths.

20.4.3 Experimental Results on Human Group Behavior Recognition

The social network-based group behavior modeling method was evaluated on two group activity datasets. The first one is the recently released BEHAVE dataset (Blunsden and Fisher, 2010), which contains the ground truth for each group activity. The second dataset is IDIAP dataset (Smith et al., 2005), which was originally captured for multiple target tracking.

20.4.3.1   Results on BEHVAE Dataset

The BEHVAE dataset consists of 76,800 frames in total. This video dataset is recorded at 26 frames per second and has a resolution of 640×480. Different activities include InGroup, Approach, WalkTogether, Split, Ignore, Following, Chase, Fight, RunTogether, and Meet. There are 174 samples of different group activities in this dataset.

To focus on the small group activity analysis, 118 samples from all the group activities dataset are selected, excluding those samples with less than three people in the scene. The selected group activities include InGroup (IG), WalkingTogether (WT), Split (S), and Fight (F) as the group activities for classification. For each activity, the samples are divided into 10 categories, nine for training and one for testing. The classification result is shown in Table 20.1. Two of the learned GPDMs are shown in Figure 20.13. Each point in the latent space corresponds to a feature vector in a single frame. The distribution of InGroup activity is prone to have some local clusters in the latent space, while the distribution of GroupFight activity is similar to a random distribution.

Some selected frames for InGroup and GroupFighting are shown in Figure 20.14. Although the number of group size is varying, the framework can still recognize the group activity correctly.

The result of social network features-based probabilistic method is also compared with the best recognition results (Blunsden and Fisher, 2010). The training and testing data are divided 50/50; the comparison results in Table 20.2 indicate the competi-

---

TABLE 20.1

Classification Results of the Social Network Features-Based Probabilistic Method

Small Group Action Type	IG	WT	F	S
The social network features-based probabilistic approach	94.3%	92.1%	95.1%	93.1%

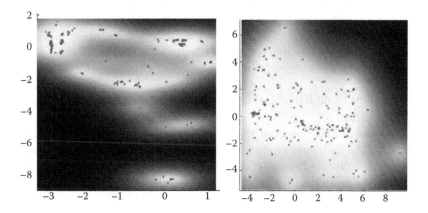

FIGURE 20.13   Visualization of trained GPDMs, the left one is the InGroup and the right one is GroupFight.

FIGURE 20.14 Sampling frames of InGroup and GroupFight, the left column is the InGroup, the middle column is GroupFight, and the right column is WalkTogether.

TABLE 20.2
Comparison of Classification Results

HMM-Based Method (Blunsden and Fisher, 2010)	Social Network Features-Based Probabilistic Approach
93.67%	93.12%

tive performance of the social network features-based probabilistic framework. It should be noted that the recognition rate is the average rate for all activities. For the hidden Markov model (HMM)-based method (Blunsden and Fisher, 2010), the time window size is 100, which means that their method required at least 200 frames to recognize an action type, while the social network features-based probabilistic framework can handle small group action recognition regardless of time durations through the probabilistic recognition approach.

20.4.3.2 Results on IDIAP Dataset

The IDIAP dataset is first used for multiple targets tracking (Smith et al., 2005). The dataset contains 37,182 frames in a total of 46 clips that are manually selected with different lengths for human group activity recognition. As there is no Fight activity in the IDIAP dataset, only three other activities, InGroup, WalkTogether, and Split, are evaluated. To validate the robustness of the social network features-based probabilistic framework, the trained GPDMs from the BEHVAE dataset for activity recognition are directly applied on the IDIAP dataset for testing, and the overall average classification rate is 90.3%. The experiment results indicate that the social network features-based probabilistic framework is robust enough to identify human group

FIGURE 20.15 Sampling frames of InGroup, WalkTogether, and Split.

activities under different scenarios. Some of the sample frames from IDIAP dataset are shown in Figure 20.15.

20.5 DISCUSSIONS AND FUTURE DIRECTIONS

Human behavior modeling is very challenging in intelligent surveillance systems. In this chapter, we introduced two new modeling methods for human motion change detection and small human group behavior modeling.

The HGPDM (Yin et al., 2010) extends the GPDM (Lawrence, 2005; Lawrence and Moore 2007) and incorporates particle filtering in general modeling of human motion patterns. The hierarchical structure in HGPDM can address complex human motions and model the changes of such motion patterns. On one hand, the particle filter is capable of tracking nonlinear and non-Gaussian human motions. This enables HGPDM to utilize the captured human motion data in classification and change detection over time. On the other hand, the hierarchical structure of GPDM ensures the finding of the most similar motion pattern in the leaf nodes. Therefore, the sampling in the prelearned HGPDM model improves the efficiency of the particle filter framework.

We also introduced a conditional GPDM based on a set of unique structural features for small human group activity recognition. The social network-based structural feature set is very effective in describing the interactions among multiple group members. The feature set can characterize both the global distribution of a group as well as local motion of each individual. In addition, this feature set can keep a fixed length while handling varying group size and group location, which is very important for recognition. Multiple GPDMs are pretrained for known group activities. The conditional probabilities on these GPDMs are computed for each test sequence, and the one with the highest probability is selected as the group activity type.

Experiment results have demonstrated the efficiency and accuracy of these models in human action change detection and group behavior modeling. Overall, these are examples of recent progresses toward semantic analysis of multimedia content. There are still many interesting research topics to be explored, including new features, modeling tools, and learning schemes. For example, social networking has become a popular network service with billions of users generating media-rich content every day. It becomes possible and desirable to automatically analyze these contents and recognize as well as understand the social behaviors of the users in each of their social groups. Video-based human action and group behavior modeling can be exploited for such purpose. In addition, multimodality modeling methods, which can effectively fuse video-based features and decisions with outputs from other modalities, such as geo-locations, social connections, wireless environment, and so on, will be able to provide a much more comprehensive understanding of the object of interest. Furthermore, semantic representations of visual concepts need to be further extended from attribute-oriented ontologies to rich behavioral descriptions.

REFERENCES

Barron, J. L., Fleet, D. J., and Beauchemin, S. S., Performance of optical flow techniques, *International Journal of Computer Vision*, 12, 43–77, 1994.

Blunsden, S. and Fisher, R. B., The BEHAVE video dataset: Ground truthed video for multi-person, *Annals of the British Machine Vision Association*, 2010(4), 1–11, 2010.

Boiman, O. and Irani, M., Detecting irregularities in images and in video, *Proceedings of the IEEE International Conference on Computer Vision*, pp. 462–469. Beijing, China, October 2005.

Chang, M. C., Krahnstoever, N., Lim, S., and Yu, T., Group level activity recognition in crowded environments across multiple cameras, In *Workshop on Activity Monitoring by Multi-Camera Surveillance Systems*, Boston, USA, September 2010.

Collins, R. T., Liu, Y., and Leordeanu, M., Online selection of discriminative tracking features, *IEEE Transactions on Pattern Analysis and Machine Intelligence*, 27(10), 1631–1643, 2005.

Cupillard, F., Bremond, F., and Thonnat, M., Group behavior recognition with multiple cameras, *Proceedings of the IEEE Workshop on Applications of Computer Vision*, pp. 177–183, Orlando, FL, USA, December 2002.

Dalal, N. and Triggs, B., Histograms of oriented gradients for human detection, *Proceedings of the IEEE Conference on Computer Vision and Pattern Recognition*, vol. 2, pp. 886–893, San Diego, CA, USA, June 2005.

Ge, W. Collins, R. T., and Ruback, B., Automatically detecting the small group structure of a crowd, In *Workshop on Applications of Computer Vision*, pp. 1–8, Snowbird, UT, USA, 2009.

Gong, S. and Xiang, T., *Visual Analysis of Behaviour: From Pixels to Semantics*, 376 pages, Springer, London, 2011.

Guimera, R., Uzzi, B., Spiro, J., and Amaral, L., Team assembly mechanisms determine collaboration network structure and team performance, *Science*, 308(5772), 697–702, 2005.

Haritaoglu, I., Harwood, D., and David, L. S., Real-time surveillance of people and their activities, *IEEE Transactions on Pattern Analysis and Machine Intelligence*, 8(22), 809–830, 2000.

Horn, B. K. P. and Schunck, B. G., Determining optical flow, *Artificial Intelligence*, 17, 185–203, 1981.

Hospedales, T., Gong, S., G., and Xiang, T., A Markov clustering topic model for mining behavior in videos, *Proceedings of the International Conference on Computer Vision*, pp. 1165–1172, Kyoto, Japan, September 2009.

Intille, S. S. and Bobick, A. F., A framework for recognizing multi-agent action from visual evidence, *Proceedings of the Conference of the Association for the Advancement of Artificial Intelligence*, pp. 518–525, Orlando, Florida, July 1999.

Ivanov, Y. and Bobick, A., Recognition of multi-agent interaction in video surveillance, *Proceedings of the International Conference on Computer Vision*, pp. 169–176, Kerkyra, Greece, September 1999.

Kiryati, N., Raviv, T., Ivanchenko, Y., and Rochel, S., Real time abnormal motion detection in surveillance video, *Proceedings of the International Conference on Pattern Recognition*, pp. 1–4, Tampa, FL, USA, December 2008.

KTH action database, http://www.nada.kth.se/cvap/actions, 2004.

Krebs, V., The social life of routers, *Internet Protocol Journal*, 3, 14–25, 2000.

Lawrence N. D.'s Gaussian process software, http://www.cs.man.ac.uk/neill/software.html, 2005.

Lawrence, N. D. and Moore, A. J., Hierarchical Gaussian process latent variable models, *Proceedings of the International Conference on Machine Learning*, pp. 481–488, Corvallis, OR, USA, June 2007.

Lowe, D. G., Object recognition from local scale-invariant features, *Proceedings of the International Conference on Computer Vision*, vol. 2, pp. 1150–115, Kerkyra, Greece, September 1999.

Lucas, B. D. and Kanade, T., An iterative image registration technique with an application to stereo vision, *Proceedings of the Imaging Understanding Workshop*, pp. 121–130, Washington, USA, April 1981.

Medioni, G., Nevatia, R., and Cohen, I., Event detection and analysis from video streams, *IEEE Transactions on Pattern Analysis and Machine Intelligence*, 23, 873–889, 2001.

Ni, B. B., Yan, S. C., and Kassim, A. A., Recognizing human group activities with localized causalities, *Proceedings of the IEEE Conference on Computer Vision and Pattern Recognition*, pp. 1470–1477, Miami, FL, USA, 2009.

Rao, S. and Sastry, P., Abnormal activity detection in video sequences using learnt probability densities, *IEEE Convergent Technologies for the Asia-Pacific Region*, vol. 1, pp. 369–372, Bangalore, October 2003.

Raskin, L., Rivlin, E., and Rudzsky, M., Using Gaussian process annealing particle filter for 3D human tracking, *EURASIP Journal on Advances in Signal Processing*, 2008, 1–14, 2008.

Riedmiller, M. and Braun, H., RPROP—A fast adaptive learning algorithm, *Procedings of the International Symposium on Computer and Information Sciences (ISCIS VII)*, Antalya, Turkey, November 1992.

Ryoo, M. S. and Aggarwal, J. K., Recognition of high-level group activities based on activities of individual members, *Proceedings of the IEEE Workshop on Motion and Video Computing*, pp. 1–8, Copper Mountain, CO, USA, January 2008.

Sherrah, J. and Gong, S., A system for tracking and recognition of multiple people and their activities, *Proceedings of the International Conference on Pattern Recognition*, vol. I, pp. 179–182, Barcelona, Spain, September, 2000.

Smith, K., Perez, D. G., and Odobez, J., Using particles to track varying numbers of interacting people, *Proceedings of the IEEE Conference on Computer Vision and Pattern Recognition*, pp. 962–969, San Diego, CA, USA, June, 2005.

Stauffer, C. and Grimson, W., Learning patterns of activity using real-time tracking, *IEEE Transactions on Pattern Analysis and Machine Intelligence*, 22(8), 747–757, 2000.

Urtasun, R., 3D people tracking with Gaussian process dynamical models, *Proceedings of the IEEE Conference on Computer Vision and Pattern Recognition*, pp. 238–245, New York, NY, USA, June 2006.

Vaswani, N., Chowdhury, A. R., and Chellappa, R., Activity recognition using the dynamics of the configuration of interacting objects, *Proceedings of the IEEE Conference on Computer Vision and Pattern Recognition*, vol. II, pp. 633–640, Madison, WI, USA, June, 2003.

Viola, P. A., and Jones, M. J., Rapid object detection using a boosted cascade of simple features, *Proceedings of the IEEE Conference on Computer Vision and Pattern Recognition*, vol. 1, pp. I: 511–518, Kauai, HI, USA, September 2001.

Wasserman, S. and Faust, K., *Social Networks Analysis: Methods and Applications*, Cambridge University Press, 1994.

Wang, J. M., Fleet, D., and Hertzmann, A., Gaussian process dynamical models for human motion, *IEEE Transactions on Pattern Analysis and Machine Intelligence*, 30(2), 283–298, 2008a.

Wang, J., Yin, Y., and Man, H., Multiple human tracking using particle filter with Gaussian process dynamic model, *EURASIP Journal on Image and Video Processing*, 2008, 1–11, 2008b.

Yin, Y. and Man, H., Adaptive mean shift for target-tracking in FLIR imagery, *Proceedings of the International Conference on Wireless and Optical Communications Conference*, Newark, NJ, USA, April 2009.

Yin, Y., Man, H., and Wang, J., Human motion change detection by hierarchical Gaussian process dynamical model with particle filter, *Proceedings of the International Conference on Advanced Video and Signal-Based Surveillance*, pp. 304–317, Boston, USA, September 2010.

Yuan, F., Prinet, V., and Yuan, J., Middle-level representation for human activities recognition: The role of spatio-temporal relationships, In *ECCV Workshop on Human Motion: Understanding, Modeling, Capture and Animation*, Crete, Greece, September 2010.

Zhong, H., Shi, J., and Visontai, M., Detecting unusual activity in video, *Proceedings of the IEEE Conference on Computer Vision and Pattern Recognition*, vol. 2, pp. 819–826, Los Alamitos, CA, USA, July 2004.

Zhu, Q. A., Yeh, M. C., Cheng, K. T., and Avidan, S., Fast human detection using a cascade of histograms of oriented gradients, *Proceedings of the IEEE Conference on Computer Vision and Pattern Recognition*, pp. II: 1491–1498, New York, NY, USA, June 2006.

Index

Printed and bound by CPI Group (UK) Ltd, Croydon, CR0 4YY

18/10/2024

01776257-0014